The Evolution of Complex Hunter-Gatherers

INTERDISCIPLINARY CONTRIBUTIONS TO ARCHAEOLOGY

THE ARCHAEOLOGIST'S LABORATORY
The Analysis of Archaeological Data
E. B. Banning

AURIGNACIAN LITHIC ECONOMY
Ecological Perspectives from Southwestern France
Brooke S. Blades

DARWINIAN ARCHAEOLOGIES
Edited by Herbert Donald Graham Maschner

EARLIEST ITALY
An Overview of the Italian Paleolithic and Mesolithic
Margherita Mussi

EMPIRE AND DOMESTIC ECONOMY
Terence N. D'Altroy and Christine A. Hastorf

EUROPEAN PREHISTORY: A SURVEY
Edited by Saurunas Miliasuskas

THE EVOLUTION OF COMPLEX HUNTER-GATHERERS
Archaeological Evidence from the North Pacific
Ben Fitzhugh

FAUNAL EXTINCTION IN AN ISLAND SOCIETY
Pygmy Hippotamus Hunters of Cyprus
Alan H. Simmons

A HUNTER-GATHERER LANDSCAPE
Southwest Germany in the Late Paleolithic and Neolithic
Michael A. Jochim

HUNTERS BETWEEN EAST AND WEST
The Paleolithic of Moravia
Jiri Svoboda, Vojen Lozek, and Emanuel Vlcek

MISSISSIPPIAN COMMUNITY ORGANIZATION
The Powers Phase in Southeastern Missouri
Michael J. O'Brien

The Evolution of Complex Hunter-Gatherers

Archaeological Evidence from the North Pacific

Ben Fitzhugh

University of Washington
Seattle, Washington

Kluwer Academic / Plenum Publishers
New York, Boston, Dordrecht, London, Moscow

Library of Congress Cataloging-in-Publication Data

Fitzhugh, Ben.
The evolution of complex hunter-gatherers : archaeological evidence from the North Pacific / by Ben Fitzhugh.
p. cm. — (Interdisciplinary contributions to archaeology)
Includes bibliographical references and index.
ISBN 0-306-47753-X (hardbound); ISBN 0-306-47853-6 (paperback)
1. Eskimos—Alaska—Kodiak Region—Hunting. 2. Eskimos—Alaska—Kodiak Region—Antiquities. 3. Hunting and gathering societies—Alaska—Kodiak Region 4. Paleoecology—Alaska—Kodiak Region. 5. Excavations (Archaeology)—Alaska—Kodiak Region. 6. Social evolution—Alaska—Kodiak Region. 7. Kodiak Region (Alaska)—Antiquities. I. Title. II. Series.
E99.E7F544 2003
979.8′4—dc21
2003051330

ISBN 0-306-47753-X (hardbound)
0-306-47853-6 (paperback)

233 Spring Street, New York, New York 10013

http://www.wkap.nl

10 9 8 7 6 5 4 3 2 1

Printed in the United States of America.

Preface

In 1987, I had the good fortune to join in the excavation of a phenomenal archaeological site on the western coast of Kodiak Island, in Alaska. The New Karluk site (a.k.a., "Karluk One") was perched on the edge of the small village of Karluk at the mouth of the river of the same name, once one of the most productive salmon rivers in the North Pacific. I had just completed my sophomore year of college, and under the direction of Richard Jordan, I enthusiastically joined several other students in the Kodiak Archaeology Project's New Karluk excavation. I had participated in my father's archaeological research in Eastern Canada since early childhood, but the Karluk dig was unlike any archaeology I had experienced before. For three months, we peeled back layers of grass, wood, and earth floors separated by remnants of ancient sod roofs. Due to the unusual preservation and richness of the site, at every turn we uncovered perishable items such as bent-wood bowls, masks, dolls, puffin-beak rattles, grass baskets, fragments of fiber netting, locks of hair, and food waste. Preservation was so excellent, in fact, that we often exposed grass blades still green after hundreds of years, which once exposed to air would turn brown before our eyes. Floor upon floor were stacked over three meters high above the cobble beach that supported them, and every day brought awesome discoveries, not the least of which was the ability to observe household space laid out almost as if the house had been just abandoned.

The archaeological record of New Karluk reinforced and qualified 18th and 19th century accounts of traditional "Koniag" lifeways—and their change over 800 years of site occupation. In the abundant faunal remains and preserved subsistence artifacts, we found evidence of a sophisticated maritime hunter-fisher-gatherer technology and economy. The size of the site, the consistent reconstruction of houses in the footprint of previous foundations, and the density of material remains convinced us that we were digging through what had been an unusually sedentary hunter-gatherer village. The size of the houses and the elaboration of material culture further suggested that large groups—probably extended families and slaves—lived in these houses and that they lived well. Dance-masks, dolls, and gambling paraphernalia (so designated by direct analogue to ethnographic practices) indicated the importance of competitive and ritualized social engagements. Plate armor and shields confirmed the importance of warfare.

Surveys of the river and nearby bays indicated that regional population densities had been high at the time New Karluk was occupied. Also, there were hints of fairly complex settlement patterns, based in analyses of Karluk River/Lagoon and Uyak Bay site distributions. Working upriver early in the summer of 1987, Jordan identified "Olmec-style" mounds at one site. Whatever the true significance of this (still unassessed) observation, the Kodiak Archaeology Project was instrumental in redirecting Kodiak archaeology towards questions of social and political development on the archipelago, developments that anthropologists assumed unlikely for hunter-gatherer groups only a decade earlier.

The Kodiak Archaeological Project was ground breaking in an even more important respect—its heavy commitment to and involvement with Kodiak Alutiiq communities and its explicit goal of generating discoveries that could help address Alutiiq questions in addition to the more typical academic concerns. This project developed at a time when Native leaders were actively seeking to strengthen the social, cultural, and educational core of Alutiiq communities. The Kodiak Area Native Association, under the leadership of Gordon Pullar, promoted Kodiak archaeology as a means of enriching and supplementing other aspects of Alutiiq heritage. This was part of a cultural revival that arose in reaction to Russo-American and Anglo-American political, cultural, and educational hegemony in the two centuries after Russian conquest. The insights—and objects—derived from the Karluk excavation helped to stimulate interest in pre-conquest lifestyles, history, and art. The establishment of the Alutiiq Culture Center and later its successor, the Alutiiq Museum and Archaeological Repository, emerged directly from this enthusiasm.

I remained fascinated by the Karluk experience during my early graduate studies at the University of Michigan. I was particularly interested in the evolutionary and historical processes that led to the relatively complex hunter-gatherer lifestyles inferred from the New Karluk site and complementary ethnohistoric documents. Ultimately this led me to explore the archaeological history of Kodiak for a dissertation project. This book is built upon that effort. The initial argument and analyses were presented as a Ph.D. dissertation at the University of Michigan (1996). This book significantly develops and revises that work.

In the pages that follow, I have sought to justify, contextualize, model, and evaluate an argument for Kodiak social evolution from colonization more than 7500 years ago to "complex hunter-gatherers" of the last few hundred years. I do this from a behavioral ecological frame of reference that includes social and cultural factors as components of dynamic ecosystems. My perspective draws insight as well from mutually inclusive aspects of cultural ecology, neo-Marxism, and practice theory, although these are ultimately rationalized in behavioral ecological terms.

This book is the culmination of close to fifteen years of intellectual development, research, discussion, and thinking about Kodiak and North Pacific archaeology from the perspective of anthropological and ecological theory. This journey spans from my first trip to Kodiak as a student in 1987, into my recent years on the faculty at the University of Washington. Inevitably, I owe much for what is put forward here to my family, teachers, colleagues, students, and friends. I could not possibly thank them all individually—though my gratitude to all is great. Nevertheless, I want to recognize some of those who have been particularly supportive in this endeavor.

First and foremost, I thank the people of Old Harbor, Alaska, who embraced me as family and supported my research with interest and enthusiasm. Among, them, I could not have done any of the research reported here without the overwhelming support of Mary and Sven Haakanson, Sr. and their son, fellow archaeologist and confidant, Sven Jr. Mary and Sven Sr. welcomed my family and crews into their home and shared generously of their knowledge and hospitality. Sven Sr. was a devoted community leader and a generous and loving person. He told great stories—and jokes—and always made me feel at home. His recent passing is a sad loss to all who knew him. Over the years, Sven Jr., now the director of the Alutiiq Museum and Archaeological Repository, has given me valuable advice, introduced me to good people, and helped me to understand his community better. Indeed the entire Haakanson family has always been wonderful to me. Like Mary and Sven, Phyllis and Glen Clough, Wanda and Gary Price, and Allen Christiansen shared their detailed knowledge of the land and waterways, lent critical equipment, helped us get our boat in and out of the water each year, gave us space to store our gear, fed us lots, built us *banyas* (sweat baths), and let us stay in their houses at various times. My deepest gratitude extends as well to other Old Harbor residents, including Freddie and Glenna Christiansen, Emil Christiansen, Sr., Carl Christiansen, Harold Christiansen, Lisa-Ann Christiansen and Phil Reed, Jeff Peterson, Ralph Christiansen, Jr., Rick and Wilma Berns, Patrick Andrewvitch, Tony Azuyak, Sr., George Inga, Sr., Wayne Arndt, Ray Krumrey, Carla Lam, Steve Rounsaville, Merv Finstad, Phil Johnson, Charlie Powers, and Mike and Judy Dobbins.

In Kodiak, the staff of the Alutiiq Museum and Archaeological Repository (and previously the Alutiiq Culture Center) have been generous in intellectual, logistical, and material support. This includes Sven Haakanson, Jr., Rick Knecht, Philomena Haussler-Knecht, Amy Steffian, Libby Pontii-Euphemio, and Patrick Saltonstall. These friends and others in Alaska, including Pete Cummiskey, Steve Hall, and Debbie and Tom Corbett, have provided shelter, stimulating discussion, and good company on numerous occasions.

The Sitkalidak Archaeological Survey crew between 1993 and 1995 included Lauren Bigelow, Laada Bilaniuk, Jennifer Bither, Donald Clark, William and Lynne Fitzhugh, Keri Kennerly, Nilcole Misarti, Megan Partlow, Libby Pontii-Euphemio, Deborah Robinson, Brent and Catherine Shaffer, and Laura Staro. Old Harbor 7th through 12th grade students, teachers, and other community members provided the inspiration and labor for the field research at the Old Harbor Lighthouse site during short excavations in 1995 and 1996. Subsequent investigations at the Tanginak Spring site, reported briefly in this volume, were ably assited by Kris Bovy, Michael Cannon, Aksel Casson, Chin Yung Chao, Jennie Deo, Michael Etnier, Catherine Foster, and Robert Kopperl. Gary Carver also consulted during a brief field visit. A total of 44 undergraduates and 6 Old Harbor teens did the bulk of the excavating at this site, and many students assisted in lab analyses at the University of Michigan and the University of Washington.

Advisors at various stages in my dissertation research and in this revised study, both officially and unofficially, include Thomas Amorosi, Donald Clark, Terry Hunt, Daniel Fisher, William Fitzhugh, Richard Ford, Joan Gero, Steve Harrell, Richard Jordan, Stephen Loring, Allen McCartney, Preston Miracle, Madonna Moss, Elizabeth Sobel, John Speth, Amy Steffian, Robert Whallon, and Bruce Winterhalder. Of these, William Fitzhugh (my lifelong mentor), John Speth (my committee Chair), Robert Whallon (committee member), and Bruce Winterhalder (undergraduate advisor and subsequent consultant) have been particularly influential in the formation and refinement of my interests, theoretical and substantive approaches, and archaeological methods.

This research was funded by generous grants from the Old Harbor Native Corporation, the National Science Foundation (OPP-9311676), the Wenner-Gren Foundation for Anthropological Research (Small Grant #5774), the University of Michigan's Museum of Anthropology Griffin Fund and Rackham School of Graduate Studies, and the University of Washington's Department of Anthropology and Education Outreach Program. Proprietary images have been made available by Koniag, Inc., the University of Alaska Rasmusen Library, the Alaska State Library, the University of Washington, the National Archives, and the National Anthropological Archives (Smithsonian). The Smithsonian Institution Department of Anthropology loaned an 18 foot research vessel for the duration of the research.

This volume benefits from the thorough editorial assistance of Laada Bilaniuk, Jane Brem, Lynne Fitzhugh, William Fitzhugh, Larkin Hood, and Larissa Lumba. James Taylor scanned and edited many of the photographs, arranged for access to archival images, and shot all of the artifact photographs. Laada Bilaniuk gets lion's praise for her help and patience. She joined in the fieldwork at various stages, read multiple drafts, dragged me off to Eastern Europe when I was getting too focused—or not focused enough, and helped in countless other ways. Finally, our daughter Laska provided necessary balance, diversion, and perspective.

Contents

Chapter 8

Chapter 9

Chapter 10

The Evolution of Complex Hunter-Gatherers

Chapter **1**

The Evolution of Complex Hunter-Gatherers

1.1. INTRODUCTION

When Russians made contact with the natives of Kodiak (Kodiak Alutiiq or Koniag)[1] in the second half of the 18th century, they encountered one of the most densely populated and militarily impenetrable societies in the North Pacific (Burch, 1988b). For two decades, unsuccessful trading ventures and military resistance forced Russian fur traders to bypass the Kodiak Archipelago in search of more profitable interactions elsewhere (Black, 1988). Kodiak's Alutiiq and their neighbors, the Eastern Aleutian (Fox) islanders as well as several Northwest Coast "tribes," such as the Tlingit, shared characteristics such as intertribal warfare, prestige economies including long distance trade in prestige valuables, institutionalized social inequality (social ranking), and slavery. Like these other neighboring groups, Kodiak's inhabitants subsisted on a mix of fish, sea mammals, shellfish, birds, and plant products. In the traditional terminology of anthropologists, they were hunter-gatherers, but not typical of hunter-gatherers as they have been best known to ethnographers of the 20th century. These were relatively complex hunter-gatherers.

This book is an exploration of the evolution of complex hunter-gatherers in the North Pacific, based on an archaeological study from the southeast region of the Kodiak Archipelago in the central Gulf of Alaska. Hunter-gatherers are groups mainly living by the capture and collection of wild foods. Complex hunter-gatherers are hunter-gatherers who are organized in relatively more complex ways than others. Sometimes this complexity is manifested in technological, economic, social, or political dimensions; but there is no necessary reason why complexity in one dimension will necessarily correspond to complexity in another dimension (Oswalt, 1987). Here the main focus is on the development of social and political complexities, which have been documented in the ethnohistoric and ethnographic literature of North Pacific societies. Specifically, this book examines the processes of change that led small-scale, mobile, egalitarian bands to become dense sedentary societies with political rank and hierarchy, endemic warfare, slavery, and competitive political economies while retaining a hunting and gathering mode of subsistence. To achieve this we will consider the development of increasingly complex social organization as well as the emergence of social inequality from egalitarianism. In considering these social and political dimensions, we will also track a number of other features such as technological, economic, and demographic change.

1.1.1. Social Complexity—A Definition

Complexity is defined here as a condition in which a system is composed of greater internal differentiation (of component parts) than another system to which it is being compared (Cohen, 1985; Price and Brown, 1985). In the case of social evolution, complexity can refer to greater structural differentiation in either or both horizontal and/or vertical dimensions (e.g., Johnson 1982). Horizontal differentiation involves the segmentation of society into structurally integrated subgroupings (e.g., clans, moieties, domestic units, families, households, age grades, secret societies, etc.), while vertical differentiation refers to rank and hierarchical organization. One of the greatest challenges facing anthropologists is explaining cross-cultural variation in social complexity. Anthropological archaeologists have considered this challenge one of their key missions for more than fifty years (Childe, 1951, 1957).

1.1.2. Complex Hunter-Gatherers—A Definition

It is recently recognized that hunter-gatherers occasionally live in ranked and stratified societies with high population densities, sedentism, warfare, and slavery. This recognition qualifies long-standing arguments about the importance of food production economies in social evolution (compare Childe, 1951 with Arnold, 1996a, 1996b, 1996c; Feinman, 1995; Price and Brown, 1985; Price and Feinman, 1995; Ruyle, 1973). More investigators now think that social differentiation and organizational complexity among food-producing societies have often (although not necessarily: see Arnold 1996c:84) evolved out of non-egalitarian, complex hunter-gatherer social formations (Price and Feinman, 1995). If this is the case, understanding the processes that promote and obstruct emergent complexity among hunter-gatherers should help illuminate the dynamics that underlie the evolution of even more complex societies, such as states and empires.

But what are "complex hunter-gatherers?" As defined in the last section, complexity is a relative measure of structural differentiation. To be useful, this definition requires us to consider any particular organization in relationship to a series of other organizations. Put simply, a hunter-gatherer society is complex if it is demonstrably more socially differentiated (horizontally and/or vertically) than other societies under comparison (including former or subsequent conditions of the same society). Attempts to create a universally applicable "complex hunter-gatherer" type become problematic, especially when the goal is explaining evolutionary processes of social change (as opposed to some ranking of historically unrelated cases). Nevertheless, there is broad agreement that many hunter-gatherer societies share certain attributes as a result of their contextual circumstances and mode of production. Characteristics of the majority of hunter-gathering societies documented in the past century include: a relatively large degree of residential mobility (within a landscape of intimate familiarity), small living units, a high degree of resource and information sharing, a pervasive egalitarian ethos, and a tendency to evade rather than initiate conflict and aggression (Berndt and Berndt, 1964; Boas, 1964; Boehm, 1993; Brody, 2001; Cashdan, 1980; Gould, 1982; Hill and Hurtado, 1996; Leacock, 1978; Leacock and Lee, 1982; Lee, 1972; Lee and DeVore, 1968). In contrast, there are some hunter-gatherer societies with the following characteristics: a relatively high degree of residential permanence, higher population densities, multi-seasonal food storage, competition over the rights to productive resource locations and accumulated surplus, status asymmetry, and organized warfare (Ames, 1981, 1985, 1991, 1995, 2001; Arnold, 1992, 1996c, 2000; Donald, 1984; Drucker, 1983; Hayden, 1994; Maschner, 1991, 1992; Price and Brown, 1985; Ruyle, 1973; Townsend, 1980, 1983; Widmer, 1988).

For the purposes of this book and within its chosen comparative parameters, the following definition can be offered:

> Complex hunter-gatherer societies are social groups primarily engaged in a foraging mode of production with institutionalized inequality (rank or stratification) and an organizational structure integrating multiple family units into larger political formations.

This definition is similar to one advocated by Arnold (1996c). She suggests that the term "complex hunter-gatherers" refers to an organizational characteristic of those hunting and gathering societies that feature "social and labor relationships in which leaders have sustained or on-demand control over non-kin labor and social differentiation is hereditary" (Arnold, 1996c:78). Differences in our definitions may legitimately relate to differences in the degree of social differentiation we seek to explain. By maintaining semantic flexibility and calling attention to the concept of relative differentiation, I seek to avoid the hazard (common to typological definitions) of assuming that societal evolution necessarily passes through a series of stepwise transformations (see Dunnell, 1980; Feinman and Neitzel, 1984; O'Shea and Barker, 1996).

Complex hunter-gatherers fitting my definition have been documented along the Northwest Coast of North America (e.g., Ames, 1981, 1994, 1995; Ames and, 1999; Drucker, 1939, 1951, 1983; Hayden, 1994, 1995; Matson and Coupland,

1995; Mitchell and Donald, 1988; Suttles, 1968), in southern California (e.g., Arnold, 1992; Lambert and Walker, 1991), and in northern Florida (Goggin and Sturtevant, 1964; Marquardt, 1988; Widmer, 1988). Groups such as the Tlingit, Coast Tsimshian, Kwakwaka'wakw, Chumash, and Calusa had developed large villages with concentrated populations and centralized political control. Despite considerable variation from society to society, these groups shared inherited status differentiation (rank) and maintained slaves (war captives) who could be sacrificed or traded for material wealth and prestige (Donald 1984, 1997; Mitchell 1984). In several of these cases, as many as three distinct social "classes" were maintained by restrictive systems of inheritance that supplemented rank structuring (Ames, 1995:161–163; Ruyle, 1973; see also Halpin and Seguin, 1990:275). The multiple dynamic implied by simultaneous rank and stratification in a single society casts doubt on Fried's (1967) more simplistic unilineal model of political evolution (Feinman, 1995:263–264) and challenges us to look for more sophisticated models.

1.2. THE NORTH PACIFIC RIM

The North Pacific Rim is a region of considerable importance in the study of hunter-gatherer groups characterized by relatively high degrees of social and political complexity. This is because a large proportion of these ethnographically documented hunter-gatherers are found along the coasts of western North America, southern Alaska, and Japan. These groups form a nearly continuous circuit of ethnographically documented populations characterized by high densities, relatively complex social structures, and elaborate trade-based political economies. Chiefs and their close relations often claimed title to a wide range of productive resource locations. Chiefly power was manifest in the ability to control the labor of kin and non-kin in the production of surpluses. Surpluses and trade valuables formed the basis of social competition between rival elites and challengers, as demonstrated in the elaborate competitive potlatch feasts well documented in ethnographies of the late 19th and early 20th centuries (e.g., de Laguna, 1988). Prior to European contact, competitive warfare was endemic and the spoils of war included stored surplus subsistence products, valuables, and slaves (often women and children). Slaves often spent their lives in the service of their captors. Slaves also could be exchanged for high prices with elite trading partners or killed at will.

In the history of anthropological attempts to characterize and understand hunting and gathering societies, complex hunter-gathers are something of an anomaly. Most hunter-gatherers in the past 100 years lived in ecologically marginal regions where agriculture potential was low. These groups tended to maintain population densities an order of magnitude or more lower than agricultural populations. They lived in small groups, tended towards residential mobility and inter-group fluidity, shared goods freely, went to great lengths to avoid appearances of superiority, and worked actively to maintain egalitarian relations. In societies such as the !Kung, Ju/'huasi, Mbuti, Hadza, Ache, and Inuit, property ownership and competitiveness were considered socially destructive and were actively avoided and thwarted (Boehm, 1993; Lee, 1972, 1990). These more or less egalitarian hunter-gatherer groups have become the archetypes of small-scale hunter-gatherer societies, and are often viewed as exemplary of the typical mode of hunter-gatherer life before the invention and spread of farming societies that transformed many parts of the world (Lee and Devore, 1968).

The North Pacific Rim groups defy the stereotypic image of hunter-gatherers that emerged in the late 1960s (from studies of the more egalitarian hunter-gatherers just mentioned). Their lifestyles raise the following questions: How could societies of the North Pacific Rim develop such political and social complexity? How do their circumstances compare with the egalitarian societies of Africa, Australia, South America, and the Arctic? Why would people sacrifice the benefits of small-scale egalitarian life-styles, especially those who stood to lose the most autonomy in the process? These are some of the questions this book seeks to answer. The range of complexity observed in hunter-gatherer groups begs for explanation. This book joins a small number of other case studies that attempt to describe and explain long-term trajectories of change in hunter-gatherer societies (e.g., Ames, 1994; Ames and Maschner, 1999; Arnold, 1992, 1993, 1995, 1996c, 1997, 2000; Arnold, et al. 1997; Bettinger, 1999; Fagan and Maschner, 1991; Fitzhugh and Habu, 2002; Gibson, 2001; Kelly, 1991, 1995; Maschner, 1992, 1997; Widmer, 1988; see also Matson and Coupland, 1995).

1.2.1. Causality

A host of facilitating factors contribute to the evolution of increasing complexity and social inequality among North Pacific hunter-gatherers. In attempts to understand emergent complexity, anthropologists and anthropological archaeologists have argued for several of these: population pressure, resource abundance, storage, sedentism, labor control, and warfare. These factors are often found to be consistent with available evidence, but are individually insufficient as mono-causal explanations.

It is perhaps no accident that the most socio-politically complex foraging societies frequently developed on highly productive coastal margins that were often associated with major near-shore upwelling systems (Yesner, 1980). The natural ecological productivity and resource concentrations of these areas are unparalleled. A simple—possibly too simple—explanation attributes the emergence of relatively high social complexity among hunter-gatherers in these areas to the uniquely high environmental productivity of these regions. Where food is abundant, population densities can increase without ecological limits and labor can be diverted to craft specialization, social competition, and warfare. But why should people be drawn to these forms of competitive behavior in the absence of stress? Was it boredom? Probably not, given the evidence that egalitarian hunter-gatherers often had plenty of "leisure" time in which to get bored, create crafts, and engage in games or competitions (Lee, 1972; Sahlins, 1968; but see Bird-David, 1992; Smith, 1988; Winterhalder, 1993).

Perhaps seasonal variability was a factor; that is, the abundance of resources available in summer compared with the less ample resources of winter may have prompted stresses. Individuals might have been willing to cede power and privilege to others with talent in managing the accumulation of resources for redistribution in times of hardship. Possibly the stresses were more social than economic. For example, the organizational demands of social interaction under crowded conditions could have prompted social changes leading to complexity, or the difficulty of coordinating economic activities in concentrated populations may have been a factor (e.g., Ames, 1985; Johnson, 1982).

Hayden (1994, 1995) posits a genetic mechanism for social differentiation in high density populations. He believes that resource abundance and ensuing population growth allow for the more predictable occurrence of unusually competitive personality types (accumulators). These accumulators can manipulate less competitive individuals for personal gains in prestige, power and wealth. These competitive personalities were somehow suppressed in smaller populations with more limited resources. Boehm (1993) has provided a compelling argument for how individuals in small-scale societies might maintain egalitarian relations through the suppression of such competitive tendencies. Why this suppression would cease under higher population densities remains a mystery.

Maybe the causes are, in fact, more universal. Possibly human societies possess the imperative of historical transformation towards increasing complexity through the resolution and rebirth of social struggles (e.g., Bender, 1978, 1985, 1990; Friedman and Rowlands, 1978; Meillassoux, 1978, 1981). Woodburn (1980) and Testart (1982) both see large-scale commitment to storage as the key transformative leap that spurs transformation, a suggestion with considerable merit in empirical cases such as those discussed in this book, but, perhaps for different reasons.

A prevalent trend in current social science sees socio-cultural change arising from the iterative relationship between individual behaviors and the systems of cultural meaning that both emerge from and condition behavior, a process called "structuration" by Giddens (1984). Such a perspective has stimulated at least one compelling model for the evolution of inequality (Clark and Blake, 1994) based on the notion that situations of subsistence abundance allow some individuals (aggrandizers) unprecedented opportunities to gain power, prestige and wealth through the monopolization of exchange networks, social debts, and corporate productivity. This view is similar in some ways to Hayden's arguments outlined above. At the most proximate (immediate or conscious) levels, as some scholars have emphasized, historical accidents (contingencies) and cultural traditions play important roles in the specific cases of social evolution (Aldenderfer, 2002; A. Cannon, 2002). But neither factor is sufficient to explain cross-cultural similarities in processes of change, or even to provide compelling accounts of social change based on archaeological data, where much cultural detail is missing.

Some will argue that the question of the social evolution of the North Pacific societies is simply wrong-headed. They will see any study of emergent complexity to presume the discredited view that social evolution is progressive and inevitably marches from simple to complex. These critics may even feel that any attempt to explain emergent complexity supports the idea that more complex societies today (and the individuals within them) are somehow "more evolved" than contemporary simpler societies. Such misunderstandings need to be dispelled in order to focus on legitimate issues underlying the evolution of social and political variation. All humans, in whatever society, ethnic group, or race, have experienced the same amount of evolution and history. There is no evidence to suggest, nor any reason to believe, that intelligence, creativity, compassion, or moral capacity differ based on membership in a particular social group or kind of group. Likewise, there is nothing inherently better about complex societies relative to simple ones. Indeed, as Brody has recently argued (2001), there is something very attractive about life in organizationally simple hunter-gatherer societies as compared to more complex ones. Complexity, as it is used here, refers only to a structural principle. It implies nothing about inherent quality of life, civility, or morality. This runs contrary to popular myths embracing notions of the greater good of "civilization" as opposed to "savagery." The persistence of relatively simple social organizations, at and beyond the "margins" of agriculture today, is testimony to the evolutionary durability of these kinds of organizations in certain environments. Their loss, if in fact inevitable, will result from a globally consuming socioeconomic and political environment rather than an expression of their inherent worth.

At the risk of using "social evolution" (a term engendering many misconceptions), there remain compelling reasons to pursue social evolutionary questions and the causes underlying evolutionary processes. Evolutionary history has tended to result in more complex structures, not only in human societies, but also in biological and ecological systems. While individual trajectories commonly follow unpredictable trajectories (and these trajectories cannot be assumed to be progressive and thus resulting in better standards of living), it is clear that complexity has been an emergent or developmental condition of evolution, not an initial condition. This fact alone makes the emergence of organizational complexity a compelling research topic. Despite attention directed towards this topic in the past, the question of social evolution within the hunter-gatherer economic mode has had little examination. In most social evolutionary models, hunter-gatherer societies are presumed to lie at the simple end of an evolutionary continuum that proceeds towards increasing complexity only when some form of agriculture is established.

In the last two decades of the 20th century, anthropologists began to pay attention to the great variability exhibited in the social and political life of small-scale societies past and present (e.g., Ames, 1991; Arnold, 1996c; Braun and Plog, 1982; Habu, n.d.; Hayden, 1995; Kelly, 1991; Price and Brown, 1985). This focus opened a new frontier for the evolutionary study of hunter-gatherer societies. With the expectation of variability, we can now look to the archaeological and historical record of hunter-gatherers with a greater sensitivity to systematic or processual changes in small-scale societies of the past. This perspective provides a better position to understand broader questions of social evolution. Dispelling the facile myth that an agricultural revolution necessarily underlies all cases of increasing complexity, we are better able to consider what actually stimulates and inhibits the increases (or decreases) of complexity in its various dimensions.

1.3. THEORETICAL ORIENTATION

Knowledge is advanced through the interplay of ideas and observations (theory and data) in a recursive and continuous cycle. Contrary to popular understanding, "facts" do not speak for themselves. Explanations of states and processes require the development of logically interconnected concepts embedded in networks of mutually reinforcing bodies of knowledge. The conceptual realm of theory (the web of logically consistent and empirically constrained ideas) is the glue that holds observations together in a way that makes sense of the world. When observations fail to make sense, we lack appropriate theory to understand them. And when theories fail to make sense of observations, we know that there is something amiss in our theoretical understanding of the phenomena in question. As a method of understanding the world, science brings theory and data together in a conscious effort to improve and expand knowledge.

In this book, I use a perspective drawn from a synthesis of evolutionary ecology and processual archaeology to explain the emergence of relatively complex hunter-gatherer societies around the North Pacific Rim. Unlike many previous models, my approach assumes that evolutionary change at the system level is generated by individual behaviors that are partially responsive to natural selection. I also assume that individuals make context-sensitive decisions in efforts to improve their conditions by whatever appropriate means. Social evolution is seen as the consequence of a dynamic interplay between individual adaptive behaviors and the socio-ecological contexts they help to shape.

In my view this process is an ultimate (i.e. evolutionary scale; *sensu* Tinbergen, 1963; see Alcock and Sherman, 1994) form of "structuration" (Giddens, 1984). The behaviors of individuals are conditioned by cognitive biological inheritance (instinctive decision rules retained by natural selection) to seek optimal fitness consequences through strategic and often creative actions, actions that often change the adaptive environment in the process. In this sense, human agency is at the heart of social change influenced by a web of dynamically interwoven relationships. These dynamics are made comprehensible by an evolutionary theory of behavior (O'Connell, 1995) that provides both a broadly and cross-culturally applicable set of motivations (fitness maximization, sometimes glossed as "adaptation" or "adaptive behavior") and a focus on adaptive flexibility to socio-ecological context. In this context, cultural values emerge from social interaction. Social interactions and cultural values then supply the immediate motivations and symbolically rich contexts within which individuals pursue fitness goals (see Boyd and Richerson, 1985). An evolutionary ecological approach allows us to examine dynamic interrelationships between culture and biology and to avoid the stark dichotomies of nature and culture that often polarize anthropological and popular thought.

The similarity between this view and the more proximate (cultural, ideological) models such as those of practice theory (Brumfiel, 1994; Clark, 2000; Ortner, 1994) is hardly coincidental. There is a growing, and I believe appropriate, tendency in contemporary science (both social and physical) to view macro-scale patterns and long-term trends as the consequence of the accumulation of incremental actions made by numerous individuals. Evolutionary ecology, practice/praxis theory, chaos theory (a.k.a. nonlinear systems theory and complex adaptive systems theory, agent based modeling, etc.) and other approaches to physical, biological, and social organization and change have been enveloped in this broader philosophical trend.

I embrace a form of processual archaeology in the structure of the argument developed throughout this book. In saying this, I am explicitly NOT embracing any form of cultural evolutionism that fundamentally defines culture as "extrasomatic means of adaptation" (Binford, 1962). Instead I view the essential elements of a processual approach as focussing on social processes at archaeologically observable scales of time and space. In this study I examine a 7500 year sequence of social evolution in the North Pacific. I disavow any commitment to a rigid cultural typology, unilineal or progressive evolution, or group selection. In so doing I am able to reconceptualize social evolution according to mechanisms more appropriate to evolutionary ecological theory. In the end, many of the proposed processes of change mirror previous models and provide more compelling mechanisms for their engagement. In other cases, the mechanisms and predicted processes are decidedly different.

The study of complex hunter gatherers in recent decades has sought both to describe and explain variation in hunter-gatherer social formations as well as to contribute to a broader understanding of socio-cultural change (see Fitzhugh, 2002b). These overlapping goals are commonly pursued through a combination of cross-cultural comparison and model building followed by archaeological evaluation of model prediction in particular pre/historic sequences. This book follows a similar approach. Specifically, Chapters 2 and 3 establish the environmental and culture-historical framework for the study. Chapters 4 and 5 define the boundary conditions for the archaeological study by establishing the differences in adaptation and social organization of Kodiak hunter-gatherers in ethnographic times compared to 7500 years ago. In Chapter 6, a narrative model develops expectations for the processes that led Kodiak occupants from their initial (and relatively simple) form of organization to the complexity documented ethnographically. Chapters 7 through 9 examine the results of an intensive archaeological survey project in Southeast Kodiak used to evaluate model predictions. Chapter 10 considers the successes and limitations of the ecological model and generalizes the Kodiak findings. These findings contribute insight into the archaeological history of Kodiak and the North Pacific and the emergence of hunter-gatherer complexity more generally.

Chapter 2

The Kodiak Environment

The focus of this book is the Kodiak Archipelago of Alaska and, in particular, the southeast region around Sitkalidak Island. At the time of contact with Russians, people on Kodiak were among the most densely packed as well as socially, militarily, and politically complex groups known in the Gulf of Alaska region. Before examining the archaeology and ethnography of this region, it is necessary to understand some basic aspects of the environment where the Kodiak Alutiiq made their home. This chapter outlines the climatic, geological, and ecological character of this region. As noted in Chapter 1, Kodiak is representative of the North Pacific Rim in general while being unique in several ways.

2.1. THE PHYSICAL ENVIRONMENT

The Kodiak Archipelago is a mountainous mass of land jutting out of the central Gulf of Alaska, southeast of the Alaska Peninsula (Figures 2.1, 2.2, 2.3). Once nearly covered by glacial ice (Karlstrom, 1969), Kodiak is now characterized by deep glacially carved valleys and fjords, sharp ridges and peaks, and a reticulated coastline. The easterly flowing Alaskan Stream (derived from the warm Japanese Current and its westerly flowing coastal extension the Alaska Current), moderates Kodiak's maritime climate and leaves most bays and inlets ice-free throughout the winter while damping temperature fluctuations year round.

In opposition to the westerly flow of the Alaska Current, dominant weather patterns track along the Aleutian Island chain and over Kodiak from the west. The Aleutians are a well known source of low pressure cyclones; these storm systems typically follow the island arc from southwest to northeast, and then across the

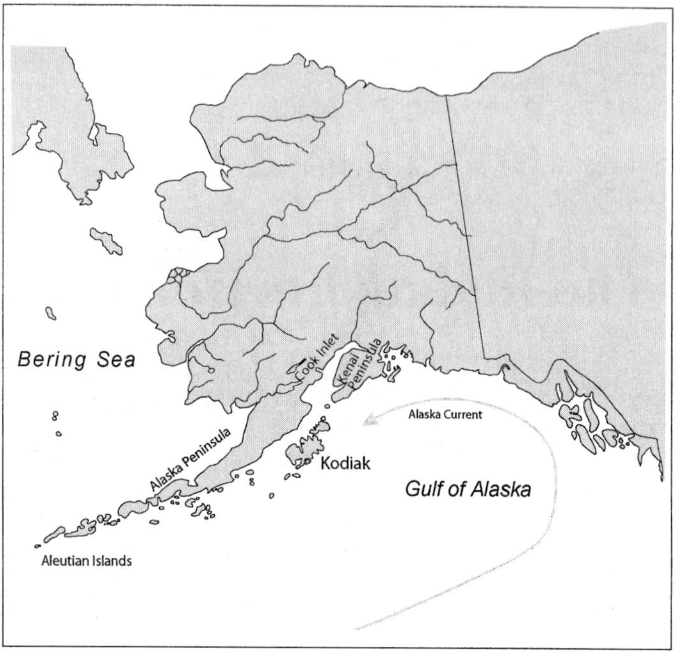

Figure 2.1. Map of the Gulf of Alaska showing the location of the Kodiak Archipelago and the main Gulf of Alaska ocean currents.

northern Gulf of Alaska. These low pressure systems drop an average of 200 cm of precipitation annually on parts of Kodiak and the Alaska Peninsula (Wilson and Overland, 1986:figure 2–19). In today's winters these storms move into the region roughly every four or five days and can generate surface winds in excess of 90 km/hr (25 m/s) (Wilson and Overland, 1986:39), making boat transportation hazardous even to large vessels. Kodiak is situated only two degrees of longitude south of the Arctic Circle and experiences only six and a half hours of daylight at the winter solstice. Summers are milder and drier than winters. Increased daylight (up to 18 hours around the summer solstice) warms the climate; in turn, these changing meteorological conditions lead to fewer intense storms, weaker winds, and greater overall visibility (Wilson and Overland, 1986).

2.2. GEOLOGY

The Kodiak Archipelago is part of a ridge system running northeast to southwest extending from the Chugach Range on the Kenai Peninsula through the Kodiak Archipelago and Chirikof Island. This ridge is formed by folding and uplift caused by the subduction of the Pacific plate below the North American

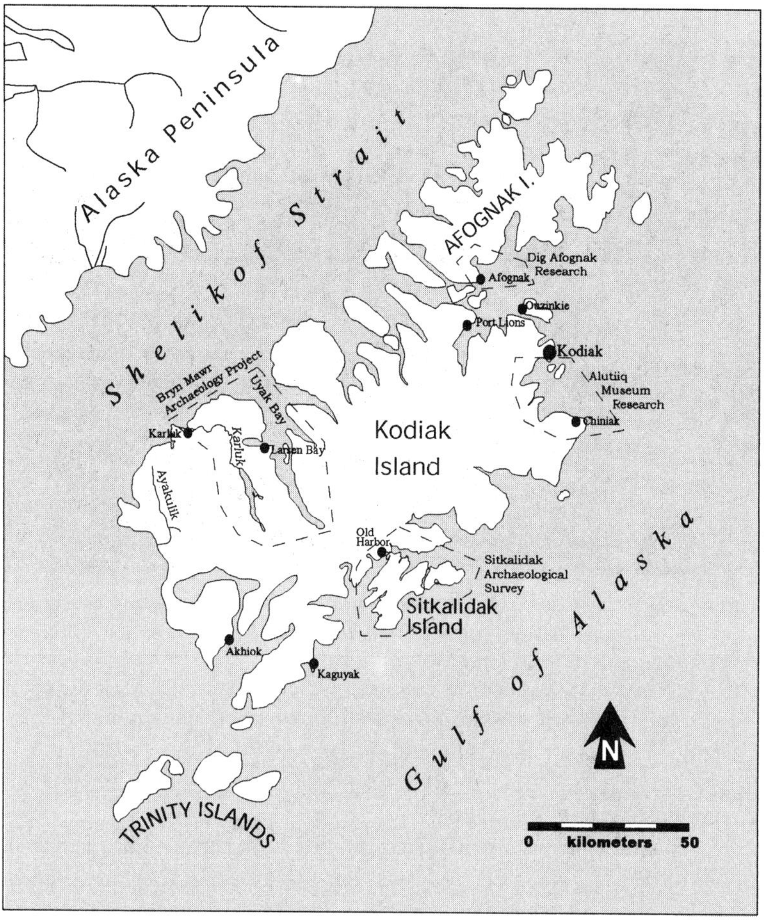

Figure 2.2. Map of the Kodiak Archipelago showing the Sitkalidak Island research area and other areas of sustained archaeological research.

plate and by the accretion of chunks of the Pacific plate (technically called terranes) to the margin of the North American plate at this juncture (Gilpin, 1995; Jacob, 1986).

Geological processes have exposed three distinct geological zones ranging in age from the Mesozoic to the Tertiary and oriented northeast to southwest along the main axis of the Kodiak and Kenai ridge system (Karlstrom, 1969:26–27). The youngest band lies along the Pacific Coast and includes Sitkalidak Island (Figures 2.4 and 2.5). This band is dominated by sandstone, shale, and conglomerate. The central band consists of graywacke-argillite rocks dating to the Late Cretaceous and includes slate of variable metamorphic grade. The northwestern

Figure 2.3. View of Kodiak Island mountains. A fresh water lake at the center of Sitkalidak Island is seen in the foreground. Photo by author.

band, including the northwestern coast of the archipelago, is separated from the central band by a major fault line and is considerably older (Mesozoic or pre-Mesozoic). Rocks in this group include ellipsoidal lava, limestone, phyllite, greenstone, and black, green, and red chert.

Situated on the front lines of a violent subduction zone, Kodiak is tormented by a series of related geological processes that continue to impact the islands and their inhabitants. The most dramatic effects are created by tectonic movement along the subduction zone; this movement generates earthquakes and tsunami waves that devastate littoral and coastal habitats as well as human populations. "Great earthquakes" (magnitudes > 7.8 Mw; Jacob, 1986:154) are rare events (1 or 2 per millennium) with devastating effects. The most recent of these shook the northern Gulf of Alaska on Good Friday, March 27, 1964, measuring 9.2 on the Richter scale (Hood, 1986:6). The "Good Friday Earthquake" generated a series of tsunami waves that severely damaged villages, towns, cities, and the Kodiak-based commercial fishing fleet.

In addition to immediate destructive effects, large seismic events alter the coastline and lead to the submergence of some near-shore environments while raising others (Gilpin, 1995). Littoral habitats are most dramatically affected. For example, shellfish beds are raised or lowered beyond their limits of ecological tolerance (Schink et al., 1983). Occupied settlements and archaeological sites are

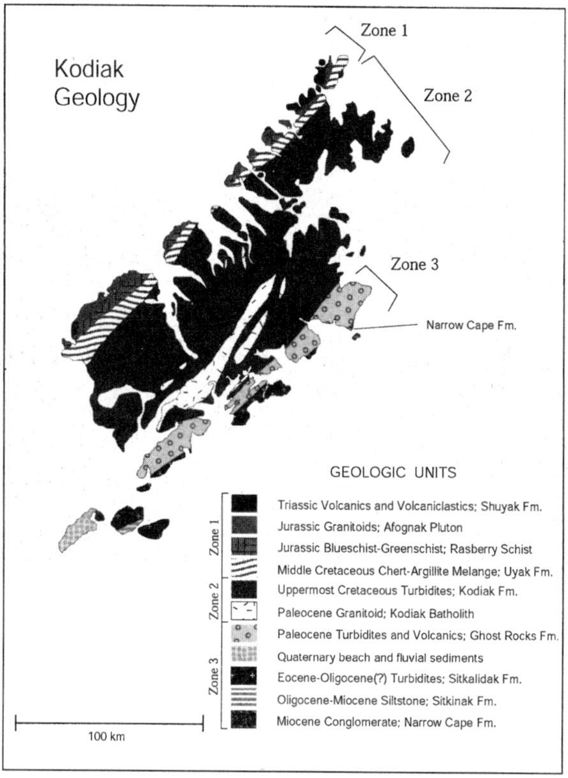

Figure 2.4. Kodiak geology is a product of a complex process of plate margin compression, faulting, folding and uplift at the Aleutian Trench Subduction Zone, combined with various sources of igneous intrusion, sedimentation, and metamorphosis (after Gilpin, 1995).

also endangered. Coastal submergence following earthquakes in AD 1788 and 1792 were partially responsible for the relocation of the Russian colony from Three Saints Bay to Saint Paul's Harbor in AD 1793 (Crowell, 1997:66–67). This move was necessary, in part, because submergence of the spit of land around the harbor at Three Saints Bay opened the anchorage to southerly winds and ocean swells. In addition, countless archaeological sites around Kodiak have been partially or completely lost to erosion following coastal submergence and the resulting exposure to pounding surf (Clark, 1994:138).

Adding to the perils of unpredictable earthquakes and tsunamis, volcanic eruptions on the Alaska Peninsula and in the Aleutian Chain periodically spread ash across parts of the archipelago. In 1912 the giant eruption of Novarupta in the Katmai group dropped approximately 17 meters of ash on parts of the Alaska Peninsula, and between 30 and 60 cm of fine ash dust on some parts of Kodiak

Figure 2.5. Geologist Gary Carver points to chert nodules in a conglomerate formation exposed at Tanginak. Photo by author.

over 100 miles away. This eruption produced over 10 times the volume of ejected material (tuff and clastic debris) and 1000 times more ash than the Mt. St. Helens eruption did in 1980 (Jacob, 1986:174). Studies of the Katmai eruption's effects on the vegetation of Kodiak reported residual silting of ponds, bogs, and salt marshes for several years (Griggs, 1918).

Despite the immediate danger of respiratory distress, hazardous drinking water, and potential damage to plants, fish, and shellfish habitat, longer-term consequences of ash-fall events are thought to be ecologically beneficial (Dumond, 1979). Volcanic ash is often rich in minerals that nourish the soil and coastal waters. Enhanced productivity of some species can be expected within two to three years following an ash-fall (Griggs, 1918). As long as viable strategies were in place to escape localized ash-falls, humans might also have benefited in the long run from these events.

2.3. ECOLOGY[1]

2.3.1. Terrestrial Ecozone

I divide Kodiak ecology into terrestrial, littoral, marine, and riverine ecozones. The terrestrial zone is characterized by a seasonally lush sub-arctic tundra broken by patches of poplar, alder, willow, and birch.[2] Today a thick spruce-hemlock forest covers low elevation sections of the northern half of the archipelago, from Shuyak Island in the north to Ugak Bay on the east of Kodiak and Uganik Bay on the west (Russell, 1991). This coniferous cover is thought to be relatively recent, probably colonizing the northern archipelago less than 1000 years ago. Some scholars believe that this forest is advancing slowly down the archipelago, with occasional assistance from volcanic ash-falls that deepen and enrich local soils (Griggs, 1934; Knecht, 1995:76). Others argue that the tree line on Kodiak is stable and at equilibrium with the climatic conditions in the area (Hu et al., 1998; Linda Brubaker, personal communication 1999).

Southern Kodiak, including the study area in the vicinity of Sitkalidak Island, is best described as a mosaic tundra. Balsam and black poplar provide patchy tree cover in localized groves at low elevations and in well-watered stream drainages. Pockets of shrub willow, alder, and birch are more widespread in valleys and on the sides of hills, but they seldom exceed a few meters in height. Poorly drained lowland valleys and terraces support wet tundra bogs and marshes. Better drained coastal and riverine land-forms alternate between patches of high bushes and

Table 2.1. Land mammals of the Kodiak Archipelago[a]

Scientific name	Common name	Indigenous or introduced?
Microtus oeconomus	Northern vole	indigenous
Citellus parryi	Ground squirrel	indigenous or early prehistoric introduction
Vulpes vulpes	Red fox	indigenous
Myotus lucifagus	Brown bat	indigenous
Mustela erminea	Ermine	indigenous
Lutra canadensis	Land or River otter	indigenous
Ursos arctos	Kodiak brown bear	indigenous
Canis familiaris	Domesticated dog	prehistoric introduction
Odocoileus hemionus	Sitka black-tail deer	introduced–Russian phase
Castor canadensis	Beaver	introduced–Russian phase
Oreamnos americanus	Mountain goat	introduced–American phase
Rangifer tarandus	Caribou	introduced–American phase
Cervus canadensis	Elk	introduced–American phase
Ondatra zibethicus	Muskrat	introduced–American phase
Lepus americanus	Snowshoe hare	introduced–American phase
Tamiasciurus hudsonicus	Red squirrel	introduced–American phase

[a] *Sources:* Rausch 1953, 1969; Haggarty et al. 1991:63.

expanses of tall herb-tundra. Higher windswept elevations support alpine tundra communities throughout the archipelago (Haggarty et al., 1991:60; Viereck and Little, 1991:22–23). Among the various plant communities, some species have been economically important for traditional life in the area (Russell, 1991). These include seasonally lush herbs, grasses, leaves, berries, wood, and roots and bulbs. Products from these plants continue to be used in the local diet and as raw materials for traditional crafts and as fuel for smokehouses and bath houses (*banyas*).

Although the terrestrial ecosystem in the Kodiak Archipelago is seasonally lush in plant life, until recently it was poorly endowed with indigenous fauna. The principle reason for this scarcity is the biogeographical isolation of the island system. According to most estimates, only eight land mammals (besides humans) successfully colonized the region from the period of deglaciation 13,000 years ago until Russian contact (Haggarty et al., 1991:62, table 4; Rausch, 1969). These included the northern vole, ground squirrel, red fox, brown bat, ermine, land otter, and Kodiak brown bear (Figure 2.6). In addition, the domesticated dog was undoubtedly introduced to the island by the native inhabitants as it appears in some prehistoric archaeological faunal assemblages (Clark, 1974a; Kopperl, 2002). Other than the dog, these species would have colonized the archipelago from the mainland either by walking across rafts of ice shortly after deglaciation, flying, swimming, or "hitchhiking" on sea-born flotsam. Most paleontologists discount

Figure 2.6. Two of Kodiak Islands sparse terrestrial fauna, the long-tailed weasel and fox. Photo by author.

the possibility that some of these species (most notably the brown bear) survived the last ice age in the small refugia at the southwestern tip of Kodiak Island (Karlstrom, 1969). Nevertheless, it is likely that some of the arctic tundra plants did survive and that they would have been the first to recolonize the barren terrain left in the wake of deglaciation.

Since AD 1784 several significant additions were made to the list of Kodiak land mammals. Russian colonists introduced Sitka black-tail deer and beaver. Americans later imported mountain goat, caribou, elk, muskrat, snowshoe hare, and red squirrel (Rausch, 1969; Haggarty et al., 1991:63). Prior to these introductions, human subsistence was drawn largely from the marine, littoral, and riverine ecozones. Clark (1974a:30, table III) has reported that terrestrial fauna account for 8%–37% of the mammal remains excavated from several prehistoric sites around the Kodiak Archipelago. These percentages would be considerably lower if whale (noted in some cases as "very common") or non-mammalian remains (e.g., fish, shellfish, birds) were included in Clark's statistics. Of the terrestrial fauna recovered from archaeological contexts, red fox is by far the dominant taxon, followed by dog. Neither the fox nor dog are thought important to subsistence when compared to marine resources.

2.3.2. Littoral Ecozone

The littoral ecozone is one of great importance to Kodiak human ecology. Though limited in area compared to the terrestrial and marine systems it divides, the coastal margin supports a large percent of the biomass available for human use; in addition this ecozone has been the most accessible for coastal-dwelling, marine-oriented people. The littoral zone is defined as the near-shore environment extending from the shoreward edge of terrestrial vegetation down to the base of the intertidal zone and into the upper subtidal zone. This zone grades from sharp craggy cliffs, stacks, and rock piles in exposed locations, to mixed cobble, pebble, and sand beaches in intermediate locations, to fine mud flats at the heads of bays, coves, and lagoons. These microenvironments support varying communities of epifaunal and infaunal mollusks, gastropods, crustaceans, sea stars, urchins, sea anemone, and seaweed species. Rocky, exposed intertidal reef zones are dominated by blue mussels and barnacles in the mid-intertidal zones. Chitons and limpets cling to the lower intertidal substrate. Deep tide pools and shallow subtidal zones in these areas support various species of whelk, periwinkle, and chiton as well as sea urchins, a taxon prized for its year-round production of calorie-rich roe. Smaller species of fish and crab as well as juveniles of larger deeper-water species habituate the shallow subtidal and intertidal zones; in these areas, predators are fewer and water temperatures are often warmer.

The more sheltered low-energy shores (with unconsolidated sedimentary substrates) support mixed communities of butter clams, Pacific littleneck clams, cockles, geoducks, horse clams, and soft shell clams. Mussels often blanket the upper

Figure 2.7. Laska Fitzhugh and Laada Bilaniuk rescue octopi from a low energy tide flat at extreme low tide. Photo by author.

intertidal beaches on cobble or pebble substrates in mid-energy environments. Octopi sometimes reside in dens beneath rocks in the lower intertidal zone (Figure 2.7). On the rare high-energy beaches, razor clams and Alaska surf clams are easily found buried in the lower intertidal zone. Sea grasses, sea lettuce, and angiosperms supply vitamins and other rare nutrients to foragers and support their own extensive communities of small gastropods and other microfauna.

In addition to the resident littoral communities, the near-shore environment is also visited periodically by both terrestrial and marine species. In late summer, beaches make easier highways than the dense, overgrown terrestrial zone, and the chance of finding a beached whale, seal, or sea lion is an added attraction for terrestrial carnivores. Red foxes frequent the beaches at low tide in search of shellfish, an unwary octopus, or beached carrion. Bears visit the beaches when not otherwise occupied on salmon streams, foraging the high country or denning in the hills. Sitka deer are now commonly observed on beaches, where they are easily hunted from boats for local subsistence purposes. Coming from the marine sphere, spawning herring visit the shallow subtidal and intertidal zones in the spring, and their milky white spawn clouds the water. This high-caloric spawn often fixes itself to seaweed and washes onto beaches where it is collected for food by people around the Gulf of Alaska and Northwest Coast (Emmons, 1991:118). In spring and summer bottom fish migrate from the deeper continental shelf into shallower waters. In summer and fall salmon congregate in near-shore areas as they wait for the right conditions to begin their terminal migrations up their natal spawning streams. In appropriate seasons (mostly late spring through early fall), fishing from shore with nets or hook and line can yield salmon, herring, lingcod, flounder, juvenile halibut, sculpin, and a host of similar species.

On some of the more rocky coastal sections (out of human's easy reach and away from heavy boat traffic), harbor seals and sea lions establish rookeries and haul-outs. These animals prefer locations that are separated from the terrestrial ecozone by cliffs or open water (for defense against terrestrial predators such as brown bears), and protected by rocks, seaweed, and surf against free-ranging marine predators such as toothed whales and humans. Harbor seals, sea otters, and sea lions are year-round residents of the archipelago. Northern fur seals migrate past the islands to and from summer breeding grounds in the Pribiloff Islands in the southern Bering Sea (Clark, 1986; Etnier, 2002).

A host of marine birds reside on rocky islets and cliffs. They include puffin, scoter, cormorant, murre, murrellet, guillemot, tern, kittiwake, gull, oystercatcher, and eagle (Table 2.2). Late spring is nesting season for these birds and their aggressive, defensive behavior creates risky, uncomfortable circumstances for any creature approaching their colonies. Nevertheless, Alutiiq people sometimes harvest eggs that provide an important caloric supplement after a long winter. Local residents of the village of Old Harbor report that small bird islets were sometimes burned to increase nesting potential for gulls and to facilitate collection of gull eggs in spring (Mary and Sven Haakanson, Sr., personal communication 1993). Sea birds were hunted on occasion for food or materials. Bird parkas were manufactured from puffin, murre, cormorant, and guillemot skins, while puffin beaks were prized for dance rattles and clothing decoration (Davydov, 1976:3).

2.3.3. Marine Ecozone

Compared to the littoral zone, the marine ecosystem has been of equal, if not greater, importance for Kodiak Islanders. In contrast to the limited terrestrial biota, the maritime environment is seasonally one of the most productive in the

Table 2.2. **Waterfowl and marine birds and of the Kodiak Archipelago**[a]

Scientific name	Common name	Habitat	Season on Kodiak
Cygnus spp.	Whistling swan	marsh wetlands	year-round
Anas crecca	Green-winged teal	marsh wetlands	year-round
Anas acuta	Northern pintail	marsh wetlands	year-round
Anas platyrhynchos	Mallard	marsh wetlands	year-round
Bucephala spp.	Goldeneye	coast, estuary, stream	winter/yr-round
Bucephala albeola	Bufflehead	coast, estuary, stream	fall to spring
Mergus spp.	Merganser	coast, estuary, stream	year-round
Histrianicus histrianicua	Harlequin	coast, estuary, stream	year-round
Melanitta spp.	Scoter	coast, estuary, stream	winter/yr-round
Somateria spp. Polysticta spp.	Eider	rocky coasts, ocean; summer- island, tundra	year-round
Chen canagica	Emperor goose	rocky shore, mudflats, seaweeds	fall to spring
Branta canadensis	Canadian goose	wetlands, bays, fields	spring to fall
Lagopus lagopus	Willow ptarmigan	shrub tundra	year-round
Lagopus mutus	Rock ptarmigan	alpine tundra	year-round
Dendragapus canadensis	Spruce grouse	Spruce forrest	year-round
Phalacrocorax auritus	Double-crested cormorant	cliff	year-round
Phalacrocorax pelagicus	Pelagic cormorant	cliff	year-round
Phalacrocorax urile	Red-faced cormorant	cliff	year-round
Rissa tridactyla	Black-legged kittiwake	cliff	year-round
Larus glaucescens	Glaucous-winged gull	flat ground	year-round
Larus canus	Mew gull	flat ground; trees	year-round
Sterna paradisaea	Arctic tern	flat ground	year-round
Sterna aleutica	Aleutian tern	flat ground	year-round
Uria aalge	Common murre	cliff ledge	year-round
Uria lomvia	Thick-billed murre	cliff ledge	year-round
Haematopus bachmani	Black oystercatcher	boulder rubble	year-round
Cepphus grylle	Black guillemot	boulder rubble	year-round
Cepphus columba	Pigon guillemot	boulder rubble	year-round
Brachyramphuis marmoratus	Marbled murrelet	burrow; talus	year-round
Brachyramphuis brevirostris	Kittlitz's murrelet	burrow; talus	year-round

Table 2.2. (*Continued*)

Scientific name	Common name	Habitat	Season on Kodiak
Cyclorrhynchus spp.	Auklet	burrow; talus	year-round
Fratercula corniculata	Horned puffin	rock crevice	year-round
Fratercula cirrhata	Tufted puffin	burrow	year-round
Haliaeetus albicilla	Bald eagle[b]	clifftops, trees	year-round

[a] *Sources:* Barsch 1985; Partlow, written communication 1991; Peterson 1990; Sowles et al. 1978.

[b] Bald eagles are ubiquitous and opportunistic predator/scavengers. They are not isolated to coastal habitats but can be found in the littoral, riverine and terrestrial ecozones. Mew gulls and Arctic terns habituate both coastal and inland ecozones. All others listed are exclusively coastal species.

world. The Gulf of Alaska is a major upwelling center, drawing nutrients from the terminus of a density-driven deep water current fed by the Atlantic, Antarctic and Indian Oceans and then moving north through the Pacific (Reeburgh and Kipphut, 1986:79). Upon reaching the near-surface photosynthetic zone in the Gulf of Alaska, these nutrients are quickly incorporated into blooms of primary production that feed a diverse chain of sea life. Weak coastal upwelling (controlled by winds, currents, salinity, and temperature) supplements the Gulf of Alaska upwelling system and increases the rate and density of primary production in near coastal regions (Sambrotto and Lorenzen, 1986; Hood, 1986). This process augments the increased release of nutrients into Gulf waters by high annual freshwater run-off, earthquake-induced mass-wasting, and volcanic ash-falls. This diverse coastline, surrounded by nutrient enriched waters, gives Kodiak its extremely productive and diverse marine and littoral ecozones. This abundance, however, is tempered by significant risks. Paralytic shellfish poisoning, generated by certain common diatoms, poses serious hazards for littoral foragers. The deadly, unpredictable toxin contaminates shellfish beds within a day and leaves some bivalves lethal for up to three years, thus dramatically increasing the hazards associated with intertidal foraging (Fitzhugh 1995).

Seals, sea lions, and sea otters are often found individually or in small groups in bays and coves, near stream mouths and points, and in open water (Table 2.3; Crowell 1994; Henning et al. 1978). Several species of baleen and toothed whales migrate into the Gulf of Alaska in the spring and remain through the summer, while others (especially gray whales) migrate past Kodiak in spring and fall on their way to and from summer feeding grounds in the Bering Sea. Blue, fin, sei, sperm, giant bottlenose and right whales remain far offshore, while porpoises, gray, humpback, minke, and orca whales visit the near-shore waters of the archipelago. These whales are often seen feeding and playing in the bays and straits where local upwelling and plankton productivity is high. A small beluga population lives in the outer Cook Inlet region (Henning et al., 1978:105–107).

Figure 2.8. A sports fisherman prepares to butcher a modest sized halibut on the beach at Larsen Bay. Photo by author.

Halibut, sometimes exceeding 135 kilograms, live in deep waters up to five kilometers from shore (Figure 2.8). In spring and summer they move into shallower water to spawn and are often found in deep "holes" in bays and straits. Juvenile halibut remain in shallower waters throughout the year. Pollock and black cod also live at the bottom of deep bays and straits throughout the year. Black bass are found near the bottom, at intermediate depth, in rocky environments with kelp beds; today these fish are more often used for halibut bait than food (Jeff Peterson, personal communication 1994).

During the past century, the near-shore marine ecozone has been most heavily utilized as a fishing ground for salmon, halibut, herring, cod, pollock, and various species of crabs such as Alaskan king, dungeness, and tanner (Campbell, 1992). Aided by sophisticated nets, traps, trawler rigs, and boats, 20th century fishermen intercept these species in high volume.

Traditional technologies limited the harvesting of many of these marine species, while others, such as crabs, were avoided altogether for unknown reasons. Deep-water fish such as halibut and cod were caught by hook and line from kayaks and open boats and are heavily represented in some archaeological faunal assemblages (Figure 2.9; Amorosi, 1986, 1987, 1988; Kopperl, 2002; Partlow, 2000). Meanwhile, herring and salmon would have been taken from shore (at stream mouths and in streams) by nets and spears. Stone weirs were often used

Table 2.3. Sea mammals of the Kodiak Archipelago[a]

Scientific name	Common name	Habitat	Season near Kodiak	Harvested prehistorically?
Phoca vitulina	Harbor seal	nearshore, rocky coast	year-round	yes
Callorhinus ursinus	Northern fur seal	offshore, shelf waters	fall & spring	yes
Eumetopias jubatus	Stellar sea lion	nearshore, rocky coast	year-round	yes
Enhydra lutris	Sea otter	nearshore, rocky coast	year-round	yes
Balaenoptera musculus	Blue whale	offshore, shelf waters	July & August	unknown
Balaenoptera physalus	Fin whale	offshore & bays		unknown
Eubalaena glacialis	Right whale	offshore, shelf waters	summer	unknown
Physeter catodon	Sperm whale	offshore, deep waters	summer (males only)[b]	unknown
Balaenoptera borealis	Sei whale	offshore, deep waters	summer	unknown
Eschrichtius robustus	Gray whale	near & offshore	May–June Oct–Nov	yes
Megaptera novaeangliae	Humpback whale	near & offshore	spring, summer, fall	yes
Berardius bairdi	Giant bottlenose whale	offshore, deep waters	unknown	probably not
Mesoplodon stejnegeri	Bering Sea beaked whale	offshore, deep waters	unknown	probably not
Balaenoptera acutorostrata	Minke whale	near & offshore	summer	probably
Orcinus orca	Killer whale	near & offshore	year-round	probably not
Delphinapterus leucas	Beluga whale	near-shore shallow	year-round Cook I-Afognak	unknown
Phocoenoides dalli	Porpoise	near & offshore	year-round	yes

[a] *Sources:* Henning et al. 1978; Crowell 1994a.
[b] Female sperm whales are seldom seen in waters north of 50 degrees N latitude (Henning et al. 1978).

Table 2.4. **Important marine fish found around the Kodiak Archipelago**[a]

Scientific name	Common name	Habitat	Availability
Hippoglossus stenolepis	Pacific Halibut	Continental shelf and on floors of deep bays	late spring to fall (juveniles = year-round)
Gadus macrocephalus	Pacific cod	Continental shelf and on floors of deep bays	year-round
Microgaddus proximus	Pacific Tomcod	Floors of shallow and deep bays	year-round
Theragra chalcogramma	Pacific Pollock	Continental shelf and on floors of deep bays	year-round
Sebastes spp.	Rockfish	Inshore and coastal distributions	year-round
Sebastes alutus	Pacific Ocean Perch	Shallow waters near mouths of rivers and streams	year-round
Clupea harengus	Pacific Herring	Schools nearshore during spawning	year-round but spring spawning
Platichthys stellatus	Starry flounder	Abundant in shallow waters near mouths of rivers and streams	year-round
Limanda aspera	Yellowfin sole	Common in waters less than 50 fathoms	year-round
Lepidopsetta bilineata	Rock sole	Not recorded in source	year-round
Ophiodon elongatus	Lingcod	Steep rocky submarine areas. inshore and coastal distribution	year-round
various	Sculpin	Shallow water	year-round
Pleurogrammus monopterygius	Atka mackerel	Kelp beds	early spring and summer
Hexagrammos spp.	Greenlings	Inshore and coastal distributions	year-round

[a] *Source:* Kessler 1985.

along larger rivers to channel fish for capture. Sea mammals were hunted in open water and at rookeries by kayakers using harpoon-tipped darts and lances. The Kodiak Alutiiq used a cooperative surround-and-exhaust strategy, extremely effective in sea otter hunting. For this reason, the Russian-America Company transported Alutiiq and Aleut hunters throughout the Russian-American colony (and as far as the Kuril Islands, Kamchatka, and the California Channel Islands) to hunt for the prized otter pelts on their behalf (Clark, 1984a; Knecht, 1995; Ogden, 1941:48–53, 1991:39–40; Shubin, 1994).

Figure 2.9. Fox Aleut halibut fishing as illustrated by Henry Elliot. Courtesy of the National Anthropological Archives (SI-NAA 7119–7; neg. 73–10872).

Whale hunting on Kodiak, unlike better known traditions in northern Alaska, was accomplished by several independent hunters or pairs of hunters in kayaks who attempted to spear a whale under a flipper or a fluke with a poison-tipped lance. The poison used was probably aconite, rendered from the monkshood plant (*Aconitum* spp.; Crowell, 1994:223). If properly struck, the whale flipper would soon become paralyzed from the poison, and the whale would sink and drown. With luck, the dead whale would float to the surface within a few days and drift ashore near the hunters' village (Crowell, 1994).

2.3.4. Riverine Ecozone

The riverine ecozone is unique because it provides a medium for the relatively rich marine biomass to penetrate into the terrestrial zone where it becomes concentrated and is more easily harvested by humans and other terrestrial predators.

Table 2.5. Anadromous and fresh water fish species around the Kodiak Archipelago[a]

Scientific name	Common name	Characteristic	Availability	Spawning habitat
Oncorhynchus nerka	sockeye, red salmon	schoolers	summer	fresh lakes
O. tshawytscha	king, chinook	loners	late spring	fresh lakes
O. gorbuscha	pink, humpback	schoolers	summer, fall	brackish-fresh strm/lake
O. kisutch	silver, coho	schoolers	fall	larger streams
O. keta	dog, chum	schoolers	summer	fresh stream
Salmo gairdnerii	steelhead	anadromous & fresh	can be year round residents in larger streams	fresh streams & lakes
Salvelinus malma	Dolly varden	anadromous & fresh	see *Salmo g.*	fresh streams & lakes

[a] *Sources:* Barsch 1985; Kessler 1985; Ropell 1982.

Anadromous fish (most notably the five species of Pacific salmon, as well as dolly varden, steelhead, and sturgeon) seasonally occupy the larger streams and lakes around Kodiak (Table 2.5). In some of the larger rivers, such as the Karluk and Ayukulik, fish can be found almost year round (Knecht, 1995); significant populations, however, are only present during the summer and fall spawning runs. Fish runs in the numerous smaller streams and creeks are more limited in diversity and timing. Both pink and chum salmon spawn in river gravel, and their fry migrate into the shallow marine waters within days or weeks of hatching. This enables these species to utilize many of the smaller streams. Pink salmon will spawn in just about any drainage regardless of water volume or substrate. As a result, this species is the most populous and available throughout Kodiak waters. The more typical short, steeply graded, smaller streams tend to have runs of shorter duration and lower volume, limiting their potential for human subsistence. King, red, and silver salmon are more particular than pink salmon, as they require larger rivers and streams with still-water lakes for spawning and maturation of juveniles (Barsch, 1985). As a result, their spawning distributions are more selective but also more productive and longer in duration. In addition to increased productivity of spawning adults throughout the summer and fall, these larger stream systems host maturing juveniles from two to four years after hatching, making them potential fishing streams year-round (Barsch, 1985). Growing fat on salmon, bears also visit streams and rivers in high densities during the summer and fall.

Waterfowl are both seasonal and year-round residents of the marshy wet tundra and lake habitats (Table 2.2). In the Karluk River drainage whistling swans, dabbling ducks, teals, pintails, and mallards are year-round residents of the marshlands, around 150–200 meters above sea level (Barsch, 1985). Emperor geese and a variety of diving ducks spend the winter in the lower reaches of the river valley and brackish lagoon where they feed on small fish, mollusks, crustaceans, other small invertebrates, and seaweeds. These waterfowl include ducks such as goldeneye, bufflehead, merganser, harlequin, and scoter. Several of these species move out onto the coast or migrate north in other seasons. Freshwater and brackish lakes, ponds, rivers, and wetlands also attract migrant ducks and geese in the fall and spring.

Of land-based fowl, ptarmigan are found throughout much of the shrub tundra, alpine tundra, and coastal forest zones. Bald eagles are ubiquitous, nesting in trees and bushes, atop cliffs, and on sea stacks during the summer. Of the nonhuman predators, bald eagles are the most universal, utilizing land, stream, shore, sea, and sky. They will prey on anything they can catch and scavenge anything else.

2.4. TEMPORAL DIMENSIONS OF ENVIRONMENTAL VARIABILITY

Scales of temporal resource variation constitute an important aspect of the Kodiak environment. Temporal variation relevant to human occupation and subsistence can be broken into a range of scales. This study considers four variations: seasonal, subseasonal, inter-annual, and long-term. Seasonal variation provides the basic structure of predictable resource variability. The other three scales constitute variation in temporal predictability of seasonal patterns, and a source of environmental risk at various scales.

2.4.1. Seasonal Scale Variability

The Kodiak year can be broken into four seasons based on the availability and behavior of biological resources triggered by changes in daylight and meteorological conditions. The ethnohistorically reported Alutiiq subsistence cycle is illustrated in Figure 2.10.

Winter lasts from late November until April and is the least productive period. Wild food is scarce, days are short, and weather is predictably poor. During this period, shellfish are one of few dependable and convenient subsistence resources other than stored goods. (The availability of over-wintering fish stocks at the Karluk and Ayakulik are an exception.) Small bottom fish, such as flounder, can sometimes be caught close to shore. In calm weather, cod, juvenile halibut, and other fish can be caught in some of the moderately protected bays. Over-wintering ducks and geese as well as several species of annually residing sea birds can be

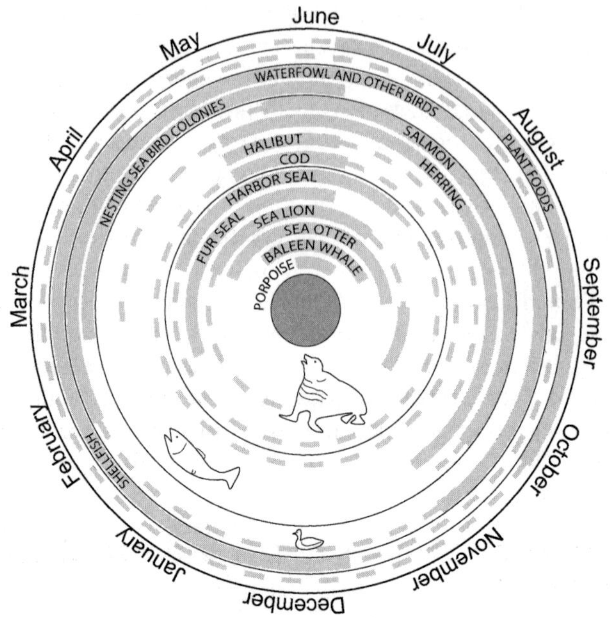

Figure 2.10. The Alutiiq seasonal subsistence cycle as recorded in 17th and 18th century accounts. (adapted from Haggarty et al. 1991:table 6).

hunted between fall and spring. Seals can sometimes be found during the winter. Writing of his 1790 visit to Kodiak, Merck (1980:105) suggested that seal hunting was one of the few traditional economic activities engaged between February and April at the southern end of the archipelago. Only exceptionally good weather would have drawn boaters out to the exposed coast in search of sea mammals during this period, and traveling would have been significantly curtailed by the combined danger of gale force winds, icing, and fog (Steffian, n.d.; see also Wilson and Overland, 1986).

Spring extends from late April until June. During this period the days grow longer, storm systems gradually weaken, and vegetation begins to grow. Birds mate and nest in cliffs, rocks, or between patches of new vegetation. Deep-water fish move into shallower waters and herring school in the bays and straits waiting for appropriately warm water temperatures to spawn. Hungry bears emerge from their dens in search of food; they usually remain high in the hills foraging on vegetation until the salmon start to run in late June and early July. Seals generally form rookeries in the late spring, and both seals and sea lions deliver pups in this season.

Traditional Alutiiq spring pursuits included sea mammal hunting (seal, sea lion, sea otter, and whale), shellfish and egg collecting, halibut and cod fishing, capture of spawn-rich herring and the collection of the herring spawn itself. Spring was the principal season for raiding marine bird colonies for eggs and bird skins.

Guillemots hunted in this season were prized for the manufacture of feather parkas (Davydov, 1976:4).

The onset of summer is symbolized by the return of the salmon. The solitary king salmon arrive in late May and June, followed by large schools of pink and sockeye in July. Silver salmon and late-run sockeye arrive between September and October. In general the coastal waters and streams are full of salmon from July into late August or September. Kodiak streams, however, are highly variable in the species they support and in the size and duration of their runs (Alaska Department of Fish and Game, 1985). To fatten up for winter, bears move onto salmon streams during the summer runs.

Summer months are characterized by greater primary productivity in near-shore waters. This productivity feeds plankton and brings baleen whales close to shore and into the bays. Seals and sea lions can sometimes be found in protected waters chasing schools of salmon or herring. In turn, orcas, porpoises, and other toothed whales follow the larger fish and pinnipeds farther into protected straits and bays during this season.

On land, vegetation growth produces succulent herbs such as nettle and cow parsnip, which are commonly collected and eaten by the Alutiiq (Russell, 1991). From late July through August, salmonberries, blueberries, cranberries, crow berries, and a host of others ripen, attracting birds, bears, foxes, and humans. Kodiak Alutiiq traditionally mixed these plant products with sea mammal oil (whale, seal, or sea lion) to preserve them for the winter (Davydov, 1976:16).

Summertime brings the calmest water and best weather for boating. Alutiiq hunters and fishers ventured into open bays and to outer-coast areas in search of whales, sea mammals, marine birds, and halibut. Summer was the easiest time for long-distance voyaging for trade or war (Townsend, 1980).

Fall extends from September through November. It is the time for silver salmon to run in the mid-sized and larger streams and late-run sockeye salmon to run in the largest streams. Spawned-out pink salmon are available in some larger streams, and they remain a delicacy among the Alutiiq elders even today (Mary Haakanson, personal communication 1993). Herring and mature halibut are found close to shore into October. Migratory and year-round waterfowl are found in the interior wetlands and lagoon estuaries and would sometimes be hunted with nets along low flyways. In the old days, Alutiiq people sometimes broke the ice forming on shallow fresh-water lakes to attract migratory ducks and geese to land within hunting or netting range (Sven Haakanson Sr., personal communication 1993).

For the protohistoric Alutiiq, late summer to fall was a critical time for the production of winter stores. Preparing for the six-month winter required much labor for catching and processing fish, sea mammals (including small whales), and plant products before they spoiled. As winter set in, the Alutiiq traditionally opened their potlatch season and exhausted most of their stores by January. For the remainder of the winter, they relied on shellfish, wintering waterfowl, nearshore fish, and the occasional sea mammal.

2.4.2. Subseasonal Variation

From the perspective of human ecology in the Kodiak Archipelago, the most relevant environmental variation that occurs at the subseasonal scale is day-to-day variations caused by changes in weather and in the timing and duration of biological phenomena. In winter, storms characteristically cross the archipelago at least once per week (Wilson and Overland, 1986). Storm intensity and storm path are products of the point-of-origin of the storm and the position of upper-level steering winds. These factors affect the strength and direction of surface winds as the storm-cell passes the archipelago (Wilson and Overland, 1986:37). In turn, wind direction and contrasting air temperatures between coastal and ocean waters influence cloud cover, rain, fog, and icing. Occasionally North Pacific high pressure forms a blocking ridge in mid-winter or mid-summer. These blocking ridges deflect low pressure cyclones north into the Bering Sea or south into the Pacific, leaving the Gulf of Alaska with extended periods of unusually cloud-free weather (Wilson and Overland, 1986:37–38). These conditions change rapidly and affect boating as well as all of the subsistence and social activities that depend on water travel.

Biological phenomena fluctuate markedly by season but frequently they also vary in timing and intensity within a season. For example, salmon sometimes delay their entrance into spawning streams for weeks because stream levels decline or the high tides needed to get into streams are too low (Figure 2.11). Tides also determine the accessibility of the intertidal zone. People forage for lower-intertidal or shallow-subtidal shellfish only during extreme low tides. The highest tidal ranges recur once every two weeks and are most pronounced in the spring and fall near the equinox. Because the largest high tides occur when the earth, moon, and sun are in line (full moon or new moon), the associated lowest low tides during these periods pass in the early morning and late evening. Tidal variation and seasonal variation in daylight and darkness thus limit the most productive shellfish collection to a few days every few months; foraging in darkness or extended twilight is sometimes feasible. Tidal patterns also affect boat mobility and the movement and feeding patterns of subsistence resources on a sub-daily time scale.

2.4.3. Inter-annual Scale Variation

Environmental variation on a broader scale than seasonal/annual patterns includes both predictable cycles of biological productivity and unpredictable fluctuations in climatic, biological, or geological conditions. Multi-year climatic cycles affect mean-annual surface temperatures and precipitation (Wilson and Overland, 1986). These cycles relate to changes in the frequency, intensity, and tracking patterns of low pressure storm cells and changes in the prevalence of high pressure. Considering the effects of cloud cover on insulation and temperature on

Figure 2.11. Salmon caught in shallow intertidal zone of the Ocean Bay river, waiting for a high enough tide to enter the stream there. Photo by author.

evaporation as well as the joint contribution of precipitation and warmer temperatures on freshwater discharge, it is not surprising that temperature, precipitation, and run-off data indicate a close relationship between these three phenomena (Wilson and Overland, 1986:figures 2.16, 2.20, 2.22). The data summarized by Wilson and Overland suggest a three to eight year periodicity in these fluctuations for the Gulf of Alaska region.[3]

Variations in cumulative weather patterns and their products affect biological productivity and forager mobility. For example, low precipitation and weak freshwater discharge can make it impossible for anadromous fish to recolonize the streams where they spawn and are most easily harvested; in turn, poor spawning years influence the productivity of future generations of fish, and several years of low rainfall can have devastating consequences for the sustainability of stream runs. In addition, variation in freshwater run-off directly alters estuarian and coastal salinity and indirectly tampers with the reproduction, feeding patterns, and survival of littoral and near-coastal species (e.g., algae, seaweeds, bottom fish, and shellfish). Mobile species (such as birds, sea mammals, and some fish) can move to other areas during poor years, leaving human foragers with few local subsistence opportunities.

Two other cycles of inter-annual climatic variability affecting Kodiak include El Niño/La Niña Southern Oscillation (ENSO) and the North Pacific Inter-decadal Oscillation (PDO/NPO) (Finney et al., 2000; Niebauer, 1998). These oceanographic cycles can significantly alter upwelling patterns and primary production on decadal scales. Recently, Finney and colleagues (2002; Mann et al., 1998) identified fluctuations in salmon productivity on the Karluk River/Lake system in western Kodiak and elsewhere that include strong climatic signatures at the decadal to millenial scale.

The scale of variation in geophysical events tends to be longer than the patterns discussed so far. The most significant of these derive from tectonic activities. At least eight earthquakes greater than 7.4 Mw have been recorded on or near Kodiak since AD 1788, based on historic and instrumental data (Jacob, 1986:figure 6.10). These data indicate that, on average, during this period at least one significant earthquake has been witnessed every 25 years. Smaller earthquakes and tremors are recorded instrumentally every year, and at least once every few years these movements are strong enough to be noticed without instrumentation. (We experienced two unrelated tremors within a ten-day period in July 2001.) Small tremors have little noticeable effect on local ecology.

The average recurrence interval of 25 years represents high-resolution variation overlaying a longer-term tectonic cycle of several centuries. Gilpin's (1995; see also Plafker et al., 1992) data indicate that great earthquakes recur at intervals of approximately 500 to 800 years on Kodiak. Data in Fitzhugh (1996) may indicate even longer intervals between major seismic events in the past. This longer-term periodicity in earthquake cycles suggests that there were long intervals (several centuries) during which Kodiak societies survived without having to deal with the effects of a major earthquake (e.g., landslides and tsunamis). A more enduring effect of larger earthquakes is the crustal deformation they can produce during or after an event, lifting or lowering segments of Kodiak by as much as a meter (Fitzhugh, 1996; Gilpin, 1995). These processes have their most profound effect on nearshore environments, where, for example, shellfish occupy vertically narrow habitat zones based on the amount of tidal exposure they can tolerate. As an example, O'Clair and Zimmerman (1986:327) report how the 1964 earthquake had a damaging effect on several shellfish colonies.

Great earthquakes often lead to broad zones of uplift on the shelf with subsidence (sinking of the land) of the area adjacent to the volcanic arc. These great earthquakes repeat approximately every 500–1300 years along the eastern Aleutian Arc (Gilpin, 1995; see also Plafker et al., 1992). Gilpin (1995) suggests that the cumulative effect of these great earthquakes is a subtle northwest down-tilt of the archipelago. While the axis of tilt (zero isobar) fluctuates from quake to quake, Gilpin's paleoseismic research has demonstrated that overall patterns throughout most of the Holocene have raised the southeast coast relative to sea level but left the northwest coasts more or less constant (after initial intervals of earthquake-induced submergence and post-quake rebound balance out). As a result, many

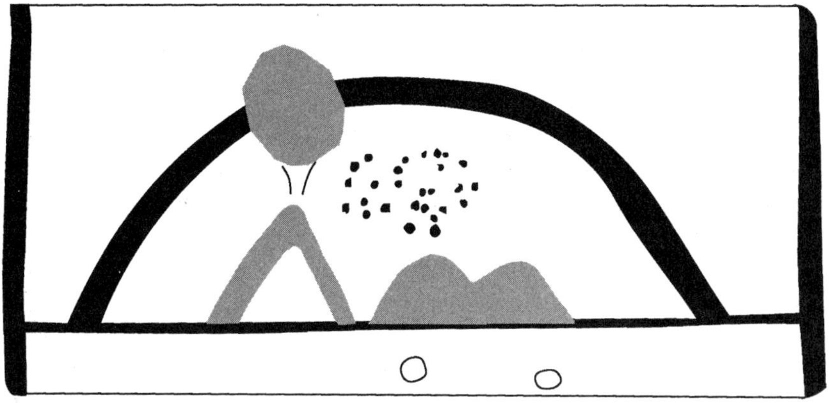

Figure 2.12. Painted wooden box panel from the New Karluk (KAR 001) site. (Drawing courtesy of Koniag, Inc. see Steffian et al. 1996)

archaeological sites on the southeast side of the islands are elevated and stranded inland, while those on the north and west sides are at the shoreline and more frequently subjected to modern wave erosion.

Volcanic activity is relatively common on the Alaska Peninsula adjacent to Kodiak, and "smoking" vents are often seen from the western shores of the archipelago (Figure 2.12). Actual eruptions are less frequent. During the past two centuries only four eruptions have been recorded for the central Alaska Peninsula region (Kienle and Swanson, 1983; data cited in Jacob, 1986:figure 33). Longer-term estimates of volcanic frequency are underdeveloped (Jacob, 1986:174); however, volcanic ash (tephra) evidence from around Kodiak suggests that as many as eleven eruptions blanketed parts of Kodiak with significant tephra deposits since the beginning of the Holocene (Gilpin, 1995:177, 216). Bounding radiocarbon dates from the peat layers in three stratigraphic sections spread through the archipelago (Big Bay on Shuyak Island and Narrow Cape and Three Saints Bay on Kodiak Island) provide minimal estimates of the frequency of ash-falls affecting different parts of the island network. The Big Bay record suggests that at least seven discrete ash-falls occurred during the 3000 year span from 9000 to 6000 BP.[4] Only two tephra layers correspond to this chronological interval on the east and southeast coasts of Kodiak. Three tephras are preserved in a Three Saints Bay section dating between approximately 6000 and 3000 BP, while a maximum of two tephras are recorded for the last 3000 years.[5] Even assuming discontinuities in each stratigraphic section, the evidence strongly suggests that significant ash-falls were fairly infrequent (perhaps once every 300–500 years).

Ash-falls would have choked lagoons, streams, and marshes. They would have produced respiratory difficulties, polluted drinking water, and altered ecosystems. Fortunately for many mobile populations (humans and other species), ash-falls tend to be confined by wind patterns and thus leave some areas in the archipelago unaffected. Human migration to the territories of friends and relatives would only need to last a few months to a few years; these migrations, however, could severely tax local resource productivity and social relationships.

Alaska Peninsula volcanoes can have indirect effects on the people of Kodiak. Large eruptions can generate landslides, local tsunami waves, and thick ash-fall on the Peninsula coast. Proximity to the volcanic arc mean these hazards are experienced much more frequently than on Kodiak. Occupants are sometimes forced to vacate these regions to seek refuge on Kodiak or elsewhere, as they did after the massive 1912 eruption of Mt Katmai.

2.4.4. Long Scale Variation

Our understanding of Holocene paleoclimate is driven by palynological reconstruction (e.g., Brubaker, Anderson, and Hu, 2001; Edwards et al., 2000; Heusser et al., 1985; Nelson and Jordan, 1988), evidence of glacial advance and retreat (Barclay et al., 2001; Calkin, Wiles and Barclay, 2001; Crowell and Mann, 1996; Mann and Peteet, 1994; Mann et al., 1998; Wiles, 1992), and tree ring sequences (Wiles, 1992; Wiles et al., 1999). Holocene climate can be characterized by two dominant climatic phases, the Hypsithermal from 9000 to 6000 bp and Neoglaciation from 5000–6000 bp to present. Neoglaciation, in turn, has three significant warm intervals (Figure 2.13). In the Gulf of Alaska, temperature and

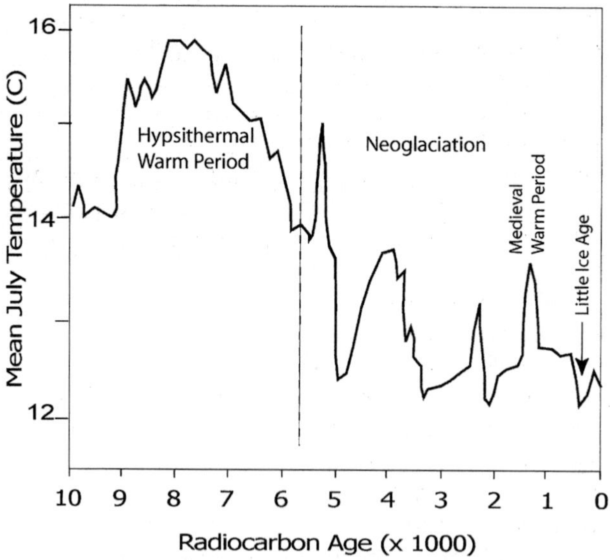

Figure 2.13. Climate variation in the Gulf of Alaska (compiled by and redrawn from Mann et al. 1998).

precipitation tend to be inversely correlated due to the connection between temperature and weather patterns across the region (Wiles, 1992). As a result the cold periods dominating the Neoglacial from 6000 bp to the present were periods of increased precipitation (Knecht, 1995:107; Mann et al., 1998; Wiles, 1992). Surprisingly, the only pollen core taken from Kodiak itself shows vegetational stability in the last 4000 years (the complete record of the core), which is often interpreted as evidence of climatic stability (Nelson and Jordan, 1988). This may indicate that vegetation communities on Kodiak were relatively insensitive to Little Ice Age amplitude cold phases or that the sample resolution was too coarse to capture the changes that did occur in the terrestrial biota. Whether the marine ecosystem was equally stable remains an unresolved issue. Nonetheless, low temperatures and higher amounts of precipitation would have increased the difficulty of maritime subsistence throughout the winter and spring months. Boat travel would have become more dangerous due to increased incidence of violent storms with hazardous winds, snow, and fog. Icing on skin boats would have been a significant concern for those caught in the open during a storm.

Unfortunately, little is known about the ecological consequences of warming versus cooling conditions (Partlow, 2000:120–121). Salmon have received more extensive research than most other species due to their economic importance in modern fisheries. References cited in Partlow (2000:117–121) suggest that salmon populations often suffer under reduced temperatures on decadal time scales, though the correlation is only partial (Mann et al., 1998). Spawning behavior and spawn/fry survival are sensitive to temperature and to stream turbidity, which would tend to increase in wetter cold intervals. On the other hand, warmer and drier intervals have their own potential impacts on salmon. Lower stream volumes can make it difficult for fish to enter spawning streams. Droughts impact adult survival in spawning streams and on the open ocean (Partlow, 2000:117–118). Some scholars see cold climates as potentially beneficial for salmon productivity (Chatters et al., 1995). We can conclude that climatic variation had a significant impact on resource availability, but the complex effects of climate on the North Pacific coastal ecosystem, at scales from decades to centuries, remain to be clarified by further research.

Chapter 3

A Historical Framework

3.1. OVERVIEW

The analyses presented in this book rely on the periodization established for Kodiak culture history. Developed over the past 60 years, the framework was first roughed out by Hrdlička (1944; see Heizer, 1956; de Laguna, 1946) and was further refined in the 1960s with the University of Wisconsin's Aleut-Koniag Project and related research (Clark, 1956, 1960, 1966a, 1966b, 1970a, 1970b, 1974a, 1974b; Laughlin and Reeder, 1962; Workman, 1966). The framework has seen no significant changes since Clark first documented the Ocean Bay period in 1963 (Clark, 1966a; 1974b). In various publications over the past 40 years, he has refined the culture history, characterizing the material culture, subsistence economy, and geographic variability of each period and phase (Clark, 1979, 1980, 1982, 1984b, 1985, 1986, 1987, 1988, 1992a, 1992b, 1994, 1996, 1997).[1] In the past two decades, Clark and other researchers have targeted the cultural and ethnic significance of transitions between periods (Clark, 1992b, 1997; Dumond, 1988a, 1994; Jordan and Knecht, 1988; Knecht, 1995). Also, a few studies have begun to consider issues of hunter-gatherer lifeways at different times in the past (Donta, 1993, 1994; Saltonstall 1995, 1996, 1997; Simon and Steffian 1994; Steffian 1992a, 1992b, Steffian and Saltonstall 2001; Steffian et al 1998).

As currently understood, Kodiak prehistory is divided into three periods each subdivided into two phases, as shown in Table 3.1. This sequence has clear affinities with archaeological patterning in the surrounding areas of the northern Gulf of Alaska. In addition, various typological relationships have been proposed to link Kodiak and the Eastern Aleutians, the Alaska Peninsula, Outer Cook Inlet (especially Kachemak Bay), and Prince William Sound. In this review, I focus on the Kodiak sequence, referring to related traditions in the surrounding area as necessary. More extensive reviews are found in Clark (1984b; 1992a, 1994), Crowell (1988), Dumond (1984a, 1984b, 1987), and McCartney (1984).

Table 3.1. Culture history of the Kodiak Archipelago

Period/tradition	Phase	Calendar age
Alutiiq	American	AD 1867–present
	Russian	AD 1784–1867
Koniag	Developed Koniag	AD 1400–1764
	Early Koniag	AD 1200–1400
Kachemak	Late Kachemak/ Three Saints	500 BC–AD 1200
	Early Kachemak/ Old Kiavak	1200–500 BC
Ocean Bay	Ocean Bay II	2500–1500 BC
	Ocean Bay I	5500?–2500 BC
Paleo-Arctic	(unconfirmed for Kodiak)	before 5500 BC

3.2. OCEAN BAY I AND II (7500–3500 BP)

Late in the summer of 1963, members of the University of Wisconsin Aleut-Koniag Project discovered flaked and ground-slate artifacts in a road-cut on a fossil beach terrace overlooking majestic Ocean Bay on Sitkalidak Island in the southeastern Kodiak Archipelago (Figure 3.1). This Roadcut site (KOD 119) helped

Figure 3.1. The majestic sand beach at Ocean Bay on the outer coast of Sitkalidak Island. Photo by author.

Clark define a new period in Kodiak culture history: the Ocean Bay period. We now recognize this period to extend from at least 7500 to 3500 years ago and divide it into two phases, the first an essentially pre-ground-slate and the second a ground-slate phase.

Subsequent to the discovery of the Sitkalidak Roadcut site, Ocean Bay period components and closely related assemblages have been recognized throughout the Koniag Archipelago, on the Alaska Peninsula, and in Kachemak Bay. Because of the antiquity and generally small size of these sites, they are often deeply buried and difficult to find. Their discovery, as at the Roadcut site, has usually been fortuitous and unsystematic. Due to tectonic changes in relative sea level, Ocean Bay period sites have frequently migrated inland (where coasts have emerged) or have been lost to coastal erosion (where coasts have sometimes submerged). As a result, reconstructions of Ocean Bay period adaptations and settlement history have been impeded.

Significant Ocean Bay period sites are shown in Figure 3.2. Prior to the Sitkalidak Archaeological Survey (which provides the core data for this book), only 11 sites were reported and entered into the Alaska State historical database (AHRS) across the Kodiak Archipelago (Clark, 1966a, 1979; Haggarty et al., 1991: table 18; Hausler-Knecht, 1991; Jordan, 1992; Jordan and Knecht, 1988; Knecht, 1995; Nowak, 1978).[2] The Sitkalidak Archaeological Survey (SAS) added 10 definitive and five probable Ocean Bay sites (Appendix A). Subsequent research conducted by staff of the Alutiiq Museum in Kodiak identified four additional Ocean Bay sites in recent years (Steffian et al., 1998; Steffian et al., in prep; Patrick Saltonstall and Robert Kopperl, personal communication 2001, 2002). Pending the addition of sites recently found but not yet reported, Kodiak boasts between 32 and 42 Ocean Bay period sites. Related components were identified in two locations on the Alaska Peninsula (Clark, 1984b, 1992a:112), on Takli Island (G. Clark, 1977) and at Kaflia (Oswalt, 1955). In Kachemak Bay, Workman and colleagues (1993) recently identified Ocean Bay II components. These show expansion of the Ocean Bay complex beyond Kodiak and the adjacent coast of the Alaska Peninsula approximately 4000 years ago. Additionally, possible late Ocean Bay deposits have been recovered from the Prince William Sound region (Linda Yarborough, personal communication 1995). Although the Ocean Bay Tradition has gained prominence as a cover term (see Dumond, 1987), terminology varies slightly across this region. For example, on the Alaska Peninsula, the Takli Alder and Takli Birch phases (defined in the Takli site) correspond to Ocean Bay I and II, respectively, and share strong technological and stylistic resemblance (Clark, 1984b; G. Clark, 1977).

A transition in technology can be discerned within the Ocean Bay sequence. Designated the Ocean Bay I phase, the earlier deposits include microcore and microblade assemblages, chert bifaces (points and knives), stone oil lamps, abraders, end and side scrapers, cobble scrapers, and hammerstones (Figure 3.3). Several of the earliest sites include a macroblade industry, which disappears early in the

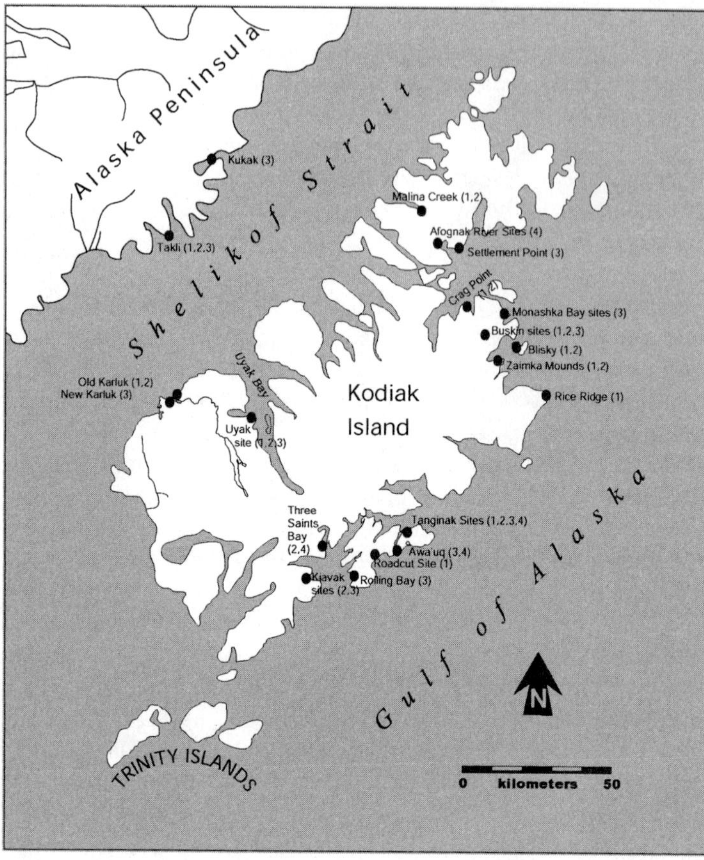

Figure 3.2. Archaeological sites (tested or excavated) that contribute substantially to present understanding of Kodiak archaeological history. Numbers indicate primary culture historical period represented (1 = Ocean Bay, 2 = Kachemak, 3 = Koniag, 4 = post Russian contact).

phase (between 7000 and 6000 BP). Throughout the Ocean Bay I phase we see evidence of experimentation with slate grinding, starting with the grinding of small slate rods and eventually leading to a full-fledged ground-slate point and knife industry. The emergence of a ground-slate hunting and food processing technology signals the transition to the Ocean Bay II phase about 5000 years ago (Figure 3.4). Ocean Bay II phase assemblages often include ground-slate projectile points (with or without notches cut into the blades), grooved cobble weights, cobble scrapers, hammerstones, and large flensing knifes and smaller knives with serrated stems. Chipped stone technologies (biface and flake tools) continue through this and all other phases of occupation, but are later eclipsed by the ground stone

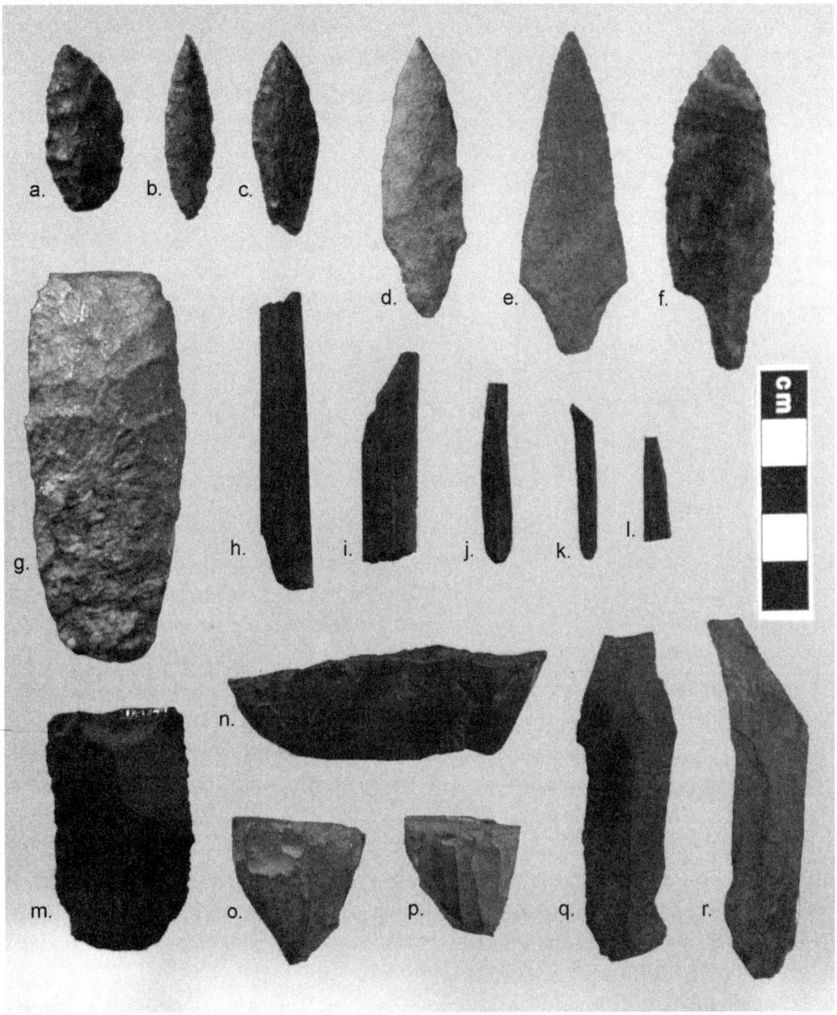

Figure 3.3a. Ocean Bay I phase artifacts: a.–f. bifacial points and knives; g. bifacial and polished adze bit; h.–i. ground slate rod fragments; j.–l. microblades; m. unifacially flaked macroblade; n.–p. microcores; q.–r. macroblades. a.–e., h.–r. Tanginak Spring site (KOD 481); f. Bear Island site (KOD 564). Photo by J. Taylor.

technology. Ground stone tools were produced by sawing and snapping slate preforms and then scraping and grinding the finished product (Clark, 1982). In some cases, preforms were also produced by flaking slate before scraping and grinding (Plate 3.4, d).

Figure 3.3b. Ocean Bay I ocher grinding stone (a.) and pecked stone lamps (b.–c.) from the Tanginak Spring site (KOD 481). Photo by J. Taylor.

The discovery of a stratified Ocean Bay period site with organic preservation at Rice Ridge (KOD 363) led to the first direct evidence of Ocean Bay period organic technology and subsistence. The Rice Ridge site contains both Ocean Bay I and II components and documents a subsistence adaptation broadly similar to later periods (Philomena Hausler-Knecht, personal communication 1993; Kopperl, 2002; Steffian et al., in prep). Organic artifacts include bilaterally-barbed bone harpoons with twin line-guards that are similar to later variants found in the North Pacific. They demonstrate the (often assumed) presence of a fully developed maritime-adapted technology at least 6000 years ago. Composite fish-hooks, needles, and wedges of bone further demonstrate the similarity of this period with later times. Faunal data demonstrate a remarkably diverse set of marine, littoral, and terrestrial resources; these include seal, sea lion, sea otter, whale, a variety of shellfish species, bear, fox, and caribou (Kopperl, 2002; Knecht, 1995:105).

The paucity of available information on Ocean Bay dwellings has led to current investigations at the Tanginak Spring site (KOD 481). Studies reveal Ocean Bay I houses to be oval or sub-rectangular in form, with little or no depression. Ocean Bay I floors are commonly coated with red ocher, a practice that ceases during the Ocean Bay II phase. Small post-holes, found at the margin of Ocean Bay I floors, likely supported tent frames, in contrast to later larger internal posts found within semisubterranean houses (Haggarty et al., 1991:120–121). Although pit-houses may be present in a couple of earlier deposits (Jordan and Knecht, 1988; Jordan, 1992), indisputable semisubterranean house depressions only appear in the later Ocean Bay II phase (Clark, 1979:138; Haggarty et al., 1991:120–121; Knecht, 1995:106).[3]

Figure 3.4. Ocean Bay II phase artifacts from the SAS survey: a. and e. ground slate, serrated-stemmed knives (a. KOD 089; e. KOD 564); b., c., f. ground slate lance fragments (b. KOD 379, 34:1004; c. KOD 564, 49:1070; f. KOD 564, 49:1074); d. flaked and scraped slate point (KOD 530); g. serrated ground slate fragment (KOD 564); h. barbed bone harpoon point with bilateral line guards (KOD 384); i. red chert dart point (KOD 486). Photo by J. Taylor.

Published evidence leads to several conclusions about the Ocean Bay period. First, this tradition was almost fully maritime in orientation. Sophisticated watercraft and navigational skills were a prerequisite to colonizing the Kodiak Archipelago. Sometime prior to the Kodiak colonization, such skills clearly developed on a mainland coast, along with a competent marine hunting and fishing technology (see Chapter 5). Second, initial colonists of the archipelago (either in the Ocean Bay period or during a yet undefined earlier phase) maintained a core and blade technology generally similar to that used throughout the interiors of Alaska and Siberia since the terminal Pleistocene, and by contemporaries around the North Pacific rim from the Aleutians to Vancouver Island, British Columbia.

Uniquely maritime stone-tool technologies developed on Kodiak and nearby regions only about 5000 years ago with the development of the ground-slate industry (Clark 1982). However, specialized organic tools such as bone harpoons and fishhooks were clearly developed earlier. The Ocean Bay diet was diverse, and the limited faunal evidence suggests that a full array of resources common to later traditions was already consumed to some degree. Finally, the dearth of identified Ocean Bay sites could indicate either low population levels or poor archaeological preservation, detection, and recovery. The Sitkalidak Archaeological Survey's success rate in identifying several Ocean Bay period sites (Chapter 7) suggests that research history and poor site preservation in other areas were responsible for the low number previously reported.

3.3. OCEAN BAY TO KACHEMAK TRANSITION

The character of the Ocean Bay to Early Kachemak transition remains poorly defined. Similarities in artifacts between Ocean Bay II and the Early Kachemak on Kodiak support the continuity between these cultural traditions. Clark (1996, 1997) recently documented a transitional assemblage at a site on Afognak (AFG 088) supporting cultural continuity from Ocean Bay to Kachemak. While uncorrected radiocarbon dates have left the appearance of a gap in the record of occupation dates, Clark (1997:36) also notes that this is not the case when statistical error is factored into the analysis. The Early Kachemak and Ocean Bay II phases contain sufficient technological overlap to warrant the assumption of cultural continuity between them (Clark, 1996, 1997). In a recent analysis of Early Kachemak sites from southeast Kodiak (Old Kiavak) and Afognak River, Clark (1997) shows the transition to the Early Kachemak technological tradition to be fairly rapid, spanning a period of less than 500 years (Clark, 1997:82–83).

While evidence strongly supports cultural continuity between the Ocean Bay and Kachemak periods, there are a couple of archaeological sites around Kodiak dating to about 4000 years ago that may reflect a brief incursion by outsiders. Tools diagnostically similar to those from Arctic Small Tool tradition (ASTt) sites north of the Alaska Peninsula have been noted in the terminal deposits of the Rice Ridge site, in basal deposits of the *Aw'uq* refuge site, and in Kachemak Bay (Clark, 1997:83).

If these assemblages are truly related to ASTt sites in the Aleutians and north of the Alaska Peninsula, they suggest an intrusion to an otherwise local sequence of developments. Clark points out that ASTt characteristics do not appear in Early Kachemak assemblages.[4]

3.4. EARLY KACHEMAK/OLD KIAVAK (3200–2500 BP)

The Kachemak tradition was first defined by de Laguna (1934) from her pioneering work at Yukon Island in Kachemak Bay. In 1963, the Aleut-Koniag project identified an early Kachemak deposit below a Koniag level at the Old Kiavak site (KOD 100) in Kiavak Bay. This site became the type site for the Kodiak expression of this phase (Clark, 1992a) and remains the best published representative of the Early Kachemak on Kodiak (Clark, 1996, 1997). Subsequent research has resulted in the publication of Early Kachemak dates for approximately seven additional sites (see Erlandson et al., 1992; Haggarty et al., 1991: table 18, 222–224; Mills, 1994: table 1). Among them, Early Kachemak components at the Crag Point (KOD 044) and Old Karluk (KAR 031) sites are the best reported. Not enough detail has been published, however, to generalize about the broader patterns of Early Kachemak on Kodiak. Recent research at the Outlet (KOD 562), Blisky (KOD 211) and Zaimka Mounds (KOD 013) sites promises to add to new insights to our understanding of Early Kachemak lifeways (Steffian et al., 1998; Amy Steffian, Patrick Saltonstall, and Robert Kopperl, personal communications, 2002). On the Alaska Peninsula, the Cottonwood and Beach phases contain affiliated assemblages (Dumond, 1987; Haggarty et al., 1991:121).

Throughout the northern Gulf of Alaska, the Early Kachemak is characterized by the presence of plummet style, end-grooved cobbles and large quantities of bilateral notched stones (sometimes called notched shingles; Figure 3.5d–g). The cobbles are identified as line sinkers for deep-water fishing or anchor stones for the ends of nets, and the shingles were probably used to weigh down nets for fish, birds, or sea mammals (Clark, 1997; de Laguna, 1934). The semi-lunar, ground-slate *ulu* appears at this time (Figure 3.5a–c), and the double-edged, serrated stemmed, ground-slate knife also continues well into this phase. Importantly, the plummet style, end-grooved cobbles, the ubiquitous notched shingles and the semi-lunar *ulus* co-occur in the earliest Early Kachemak deposit on the Afognak River (AFG 088; Clark, 1997). Unfired clay-lined pits, basins, hearths and floor patches are documented at the Old Kiavak site from the base of the occupation (Clark, 1997:58–63), suggesting some form of storage and/or processing. Similarly, massive smoking pits have been found at the Outlet site suggesting intensive fish processing (Patrick Saltonstall, personal communication, 2002). These technological changes appear to be related and have important implications for changes in

Figure 3.5. Ground slate ulus (a. KOD 575; b. KOD 474; c. KOD 101) and notched stone sinkers (d., f., g. KOD 506; e. KOD 510) from the SAS survey. Photo by J. Taylor.

human adaptation and social organization in the Kachemak Period. These implications will be explored in subsequent chapters.

One curiosity of the Early Kachemak is the prevalence of dark gray/black, carbon-rich sediment in many sites. Clark (1970; 1997) dubbed this the "Black Kachemak" on the basis of his observations at Old Kiavak and Crag Point. This characteristic may relate to decomposition of organic matter resulting from temporary sea level rise (Stein, 1984a, 1984b, 1992, but see Yesner, 1998) or to some unique cultural practice (such as the rendering and/or burning of large quantities of sea mammal oil and/or intensive fish processing). I suspect that the explanation of the "Black Kachemak" deposits will relate directly to the low number of recorded transitional sites bridging the late Ocean Bay and Early Kachemak phases.

3.5. LATE KACHEMAK/THREE SAINTS PHASE (2500–800 BP)

By about 2500 BP the Kachemak culture is substantially represented on Kodiak and the first villages are clearly identified (Crowell, 1986; Haggarty et al., 1991:123; Jordan and Knecht, 1988:131–132). The first Kachemak occupation recorded in the archipelago was found below Russian deposits at the Three Saints Bay site discovered by the University of Wisconsin team in 1962. The team recognized the similarities between this assemblage and the lower levels of the Uyak site excavated by Hrdlička in the 1930s and reported by Heizer (1956; Clark 1992a) and to the Kachemak sequence defined by de Laguna (1934) for Kachemak Bay. The phase was subsequently named the Three Saints phase of the Kachemak Tradition to distinguish it from the older Old Kiavak phase assemblage found in the 1963 excavation. The Three Saints phase is usually called the Late Kachemak by Kodiak archaeologists today. A diagnostic lance point, the Three Saints point (a finely honed ground-slate point, long or short with a squared-off stem and distinctly barbed shoulders) is common in Late Kachemak assemblages. This point is frequently found with stylized markings, often interpreted as maker's/owner's marks, although some of these markings may also indicate cultural affiliation (see Figure 3.6). Notched stone sinkers continue to dominate assemblages in this phase, and slate *ulus*, grooved cobble line weights (but no longer plummet-shaped ones), greenstone adzes, and cobble scrapers are common (Figure 3.7). Organic artifacts include barbed bone harpoon and dart points, fish and bird leister prongs, shovels, awls, needles, and decorated spoons. Toggling harpoons (efficient sea-hunting tools) are first found in the Early Kachemak and apparently came to the Kodiak Archipelago through trade with more northerly groups.

Elaborate decorative art (at least on durable artifacts) reached a peak in the Kachemak period (especially in the Late Kachemak phase). It was first observed by

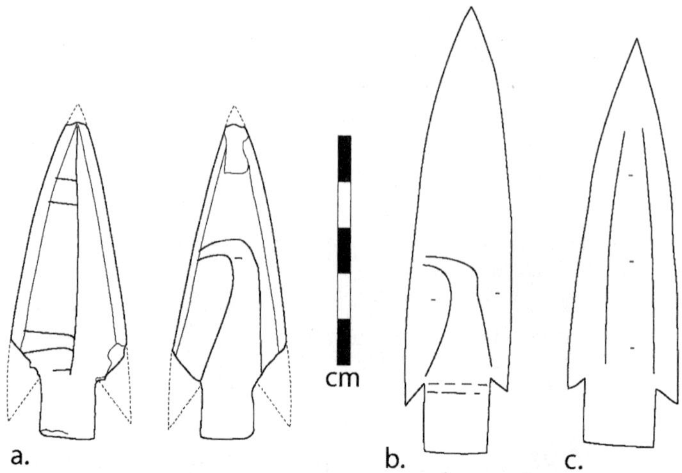

Figure 3.6. Examples of Three Saints Phase (Late Kachemak) ground slate points from the Sitkalidak region, showing "ownership marks" and similar bird-like engravings: a. Two views of a damaged slate point from the Tanginak Bluff site (KOD 384). b. & c. Two points from the Three Saints Site (KOD 083) (D. Clark, 1970a:fig 3).

Hrdlička (1944), who saw it as an expression of the highest cultural achievement in Kodiak prehistory. Large pecked stone lamps from the period contain relief carvings of whales, breasts, humans, and other stylized depictions (Figure 3.8). Coal labrets (lip-plugs) make their appearance at this time; their variety of stylized forms may relate to clan or lineage affiliations and competitive status displays (Steffian and Saltonstall, 2001). Throughout the Late Kachemak phase, evidence of complex mortuary treatment includes cremations, disarticulated burials, and apparently altered or curated human bones with drill holes and other modifications (Crowell, 1988; Simon, 1992; Simon and Steffian, 1994; Urcid, 1994; Workman, 1992). Some of these perforations appear to have been used to hold bodies together after death, suggesting the development of mummification and ancestor worship (Simon and Steffian, 1994). Other bones appear to have been drilled more or less randomly, perhaps for suspension during ritual performances. Still other burials are found lacking head and/or hands (Workman, 1992), perhaps signifying ritual torture or punishment.

Faunal remains indicate a continuation of the previous subsistence practices, although variation in dietary emphasis is seen in different regions (Amorosi, 1987, 1988; Yesner, 1992). Given sufficient samples for comparison, we might expect this for all time periods. Salmon fishing is a major seasonal attraction by this time, as evident in midden deposits and placement of settlements well up major salmon streams (Amorosi, 1987, 1988; Jordan and Knecht, 1988; Yesner, 1989). Away

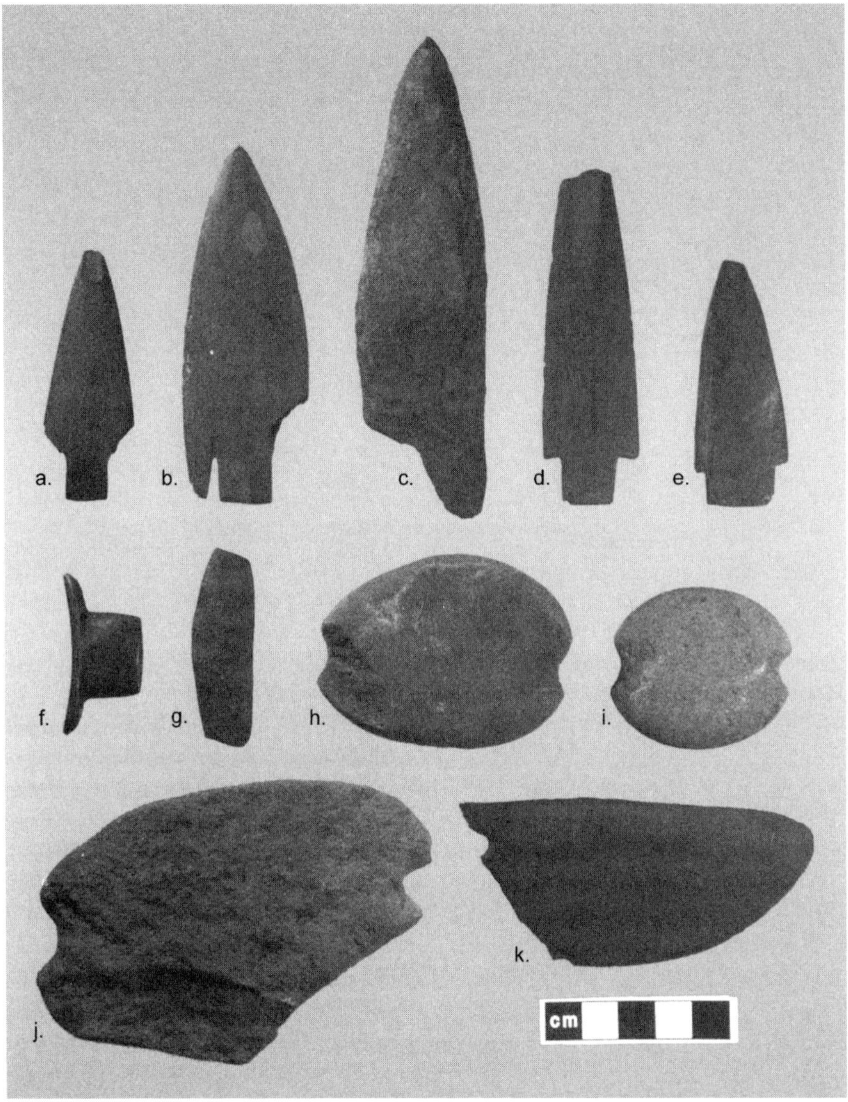

Figure 3.7. Kachemak period artifacts from Sitkalidak Archaeological Survey: a. Three Saints style point with bird engraving (KOD 384); b., d., e., ground slate lance points (b. KOD 564; d. KOD 384; e. KOD 384); c. ground tipped slate blank (KOD 486); f. polished slate labret (KOD 550); g. chipped and ground-tipped burin-like-tool (KOD 115); h.–j. notched stone sinkers (h. KOD 384; i. KOD 578; j. KOD 384); k. ground slate ulu fragment (KOD 384). Photo by J. Taylor.

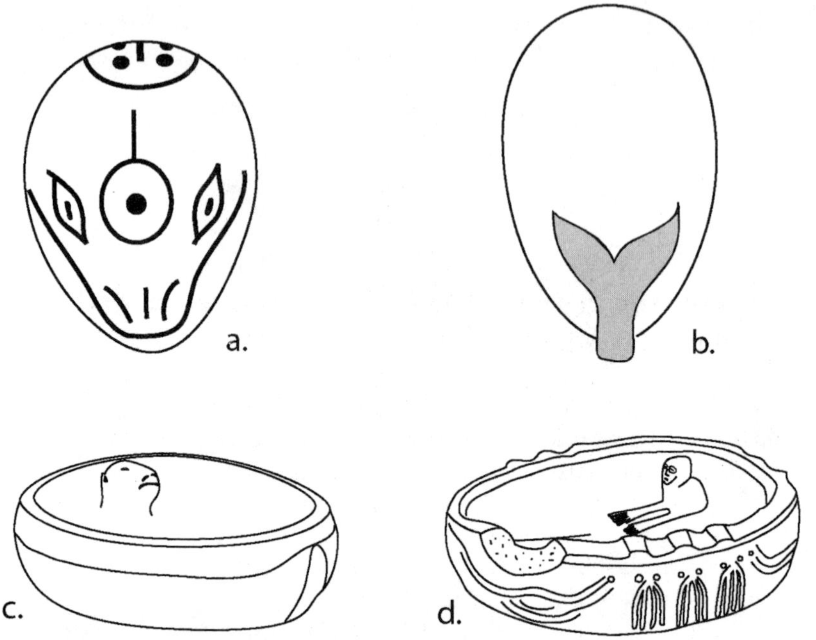

Figure 3.8. Examples of Late Kachemak decorated stone oil lamps. The top two images show the undersides of lamps. The bottom two images are oblique upright views of lamps with three dimensional relief carvings protruding from two lamp bowls. a. and b. redrawn from Heizer (1956: 32, fig 24d,i), c. bear head lamp drawn from photograph (Fitzhugh and Kaplan 1988:133, plate 156), d. seated figure lamp drawn from photograph (Crowell et al. 2001:37, plate 25).

from these rivers, greater focus is placed on codfish, sea mammals, shellfish, and other marine resources.

Late Kachemak semisubterranean houses are often seen on the modern ground surface. They average around 16 to 35 m^2 and tend to be square to rectangular, with or without an entrance passage slightly lower than the living floor in a style reminiscent of the cold-trap entrance passages of more northerly regions (Jordan and Knecht, 1988; Knecht, 1995:108; Saltonstall, 1996, 1997; Steffian, 1992b). Sub-floor clay-lined pits, raised earth benches along the walls, and stone slab hearths are commonly associated with these dwellings. Late in the phase, small corner alcoves are sometimes appended to the insides of these houses (Jordan and Knecht, 1988; Steffian, 1992b).

Five Late Kachemak sites on Kodiak have been excavated and their associated publications provide comparative data on the phase in this region. The Three Saints Bay site (KOD 083) has already been mentioned. Additional Late Kachemak assemblages and structural data come from the Crag Point site (KOD 044) tested in 1964 and substantially excavated in 1986; the Old Karluk site (KAR 031), partially excavated in 1985; the Uyak site (KOD 153) first excavated by Hrdlička (1944; Heizer, 1956) and later by Steffian (1992b); and the Outlet site on the Buskin River, excavated between 1999 and 2001 by the Alutiiq Museum community archaeology program (Robert Kopperl, Patrick Saltonstall, and Amy Steffian, personal communications, 1999, 2001, 2002).

3.6. KACHEMAK-KONIAG: TRANSITION, DISCONTINUITY, OR REPLACEMENT?

One legacy of Hrdlička's work on Kodiak has been a dispute over the significance of the Kachemak to Koniag transition. He was convinced that the "pre-Koniag" deposits of the Uyak site contained evidence of a biological and cultural stock completely separate from the following Koniag layers. The biological basis for Hrdlička's argument has been refuted (Scott, 1991, 1992, 1994) and appears to have been based on shifts in head-binding practices adopted later in time, as well as an increase in cross-cultural marriage and genetic mixture during the Koniag period. The lack of systematic excavation by Hrdlička at the Uyak site further undermines any serious attempt to correlate cultural and biological variation with the stratigraphic sequence (Heizer, 1956; see also de Laguna, 1946, 1956). Hrdlička's cultural attributions necessarily relied on sorting out perceived differences within the whole collection, and were only mildly controlled by the gross separation of upper, intermediate, and lower levels (Heizer, 1956). Unfortunately, Steffian's subsequent excavations at the Uyak site cannot help resolve this issue as she only found intact Late Kachemak phase deposits.

Dumond (1987, 1988a, 1988b) has argued in favor of Hrdlička's position, basing his studies on accumulated evidence from the Aleut-Koniag project and his own research on the Alaska Peninsula. He suggests that two factors strongly argue for a cultural replacement around AD 1000–1200: (1) the appearance of several features of more northerly Norton and Thule cultures at the inception of the Koniag period on Kodiak, and (2) the relatedness of modern Alutiiq (Sugtestun) and Yupik speech variants.

Recent archaeological work on Kodiak (in particular at the Old Karluk (KAR 031) and New Karluk (KAR-001) sites) has moved the Koniag period "start date" up to around AD 1250 (Knecht, 1995; Steffian, 1992b). This same research casts doubt on the replacement model by showing continuity in occupation, house construction, and most artifacts through the transition period (AD 900–1400: the terminal Late Kachemak and Early Koniag). Changes appear to be evolutionary not revolutionary with different novelties appearing at different times in these sites (Jordan and Knecht, 1988; Knecht, 1995). Many of the archaeologists familiar with the Karluk sequence maintain that Late Kachemak population was healthy and growing by AD 1100 and an invasion or abandonment followed by replacement seems unlikely. Knecht (1995) suggests that the changes in artifacts and house construction that mark the development of the Koniag were *in situ* adaptive adjustments to climatic deterioration during the Little Ice Age.

A third view suggested by Clark (1992b) describes continuity in Kodiak culture mixed with immigration from the Bering Sea and Alaskan Peninsula. In Clark's model, Koniag pottery, sweat baths, Yupik language, and other "northern" traits were incorporated into the evolving Kodiak sequence. How a minority of immigrants could influence such changes in language remains to be demonstrated. The model, however, is appealing for its incorporation of evidence of both continuity and mixing. Because we have no evidence of the language that was spoken on Kodiak prior to the Koniag period, and no model of time elapsed since Yupik-Alutiiq divergence is directly testable, all linguistically related arguments remain speculative.

3.7. KONIAG (800–200 BP)

Regardless of how the changes came about, the Koniag period contains evidence for major changes in social organization, political integration, ceremonialism, and economy. Substantial changes are seen in the size and density of villages and in house design. New technologies are introduced or invented, including steam-sweat bathing and, in some locations, coarse gravel-tempered pottery (Figure 3.9). The New Karluk site witnessed the most extensive (controlled) excavation in the Kodiak Archipelago to date and is the single most detailed source of information relating to the Koniag period (Figure 3.10). The site's excellent organic preservation along with its sequence of house floors and associated refuse extend from approximately AD 1250 into the Russian contact phase. Coastal erosion, unfortunately, destroyed unexcavated portions of this unique site during the winter following the 1995 excavations (Steffian, 1996).

Knecht (1995; and see Jordan and Knecht, 1988) has presented a detailed analysis of the artifact assemblage from the 10 superimposed house floor deposits excavated by the Kodiak Archaeological Project between 1983 and 1985 at New Karluk. His detailed descriptions and analysis suggest that major changes associated with the Koniag tradition only became established after about AD 1400.[5] Continuity in most artifact types is seen from the base of the site at AD 1250, with parallels extending back into Late Kachemak deposits across the lagoon at KAR-31 and in other Late Kachemak and so-called "transitional Koniag" sites around the Kodiak Archipelago. The Koniag period is therefore divided into "Early" and "Developed" phases at AD 1400. Common implements recovered from New Karluk include ulus, labrets, bentwood bowls, composite fishing harpoon points,

Figure 3.9. Koniag period artifacts from the SAS survey: a., d. ground slate spatula scraper (a. KOD 116; d. KOD 86); b. blunted-edged ground slate point (KOD 499); c. diamond cross-section ground slate lance (KOD 384); e. painted pebble (KOD 101); f. polished greenstone adze (KOD 116); g. ground slate lance fragment with "owners mark" (FOX:1001); h.–i. pottery rim fragments (h., i. KOD 116); j.–k. ground slate ulu fragments (j., k. KOD 384); l. ground slate axe (KOD 122); m. small bone bow fragment, possible toy or bow drill (KOD 116); n. bone wedge (KOD 474). Photo by J. Taylor.

wooden tongs and paddles, bow and arrow shaft fragments, wood slat armor, shields, gaming balls, gaming disks, tally sticks, boat parts, and various ceremonial items, such as drum fragments, masks and mask bangles, puffin beak rattles, and anthropomorphic figurines.

Knecht's (1995) analysis suggests that transitional (Early Koniag) forms include long medially ridged slate points, stemmed ulus (continuing from Kachemak

Figure 3.10. The New Karluk site (KOD 001) was located between beach and small pond. Abandoned cannery building stands on the beach in the distance, mostly abandoned 20th century structures in left foreground. Photo by author.

phases), scarfed composite fishhooks, and incised pebbles (see also Donta, 1993, 1994; Jordan and Knecht, 1988; Saltonstall, 1995, 1996, 1997). Donta (1993, 1995) analyzed collections of incised pebbles from an Early Koniag site in Monashka Bay (KOD-026) and New Karluk (KAR-001) and suggested that this narrowly dated artifact horizon reflects the emergence of rank-based social organization in the Early Koniag phase between AD 1300 and 1450. Analyzing ethnographic rituals and Koniag period ceremonial paraphernalia, Donta (1993) believes status hierarchy was already an important part of Early Koniag ritual as well as possibly Late Kachemak phase practices.

Houses during the Early Koniag phase include elaboration on corner alcoves and the addition of one or more side rooms to the Kachemak house forms. House size also expands at this time. Some confusion exists in the literature about the exact timing of the shift from single- to multiple-roomed houses (see Partlow, 2000:99–109). Recent excavations at the Settlement Point site (AFG-015) are helping to clarify this issue. At least one large multi-roomed house was constructed as early as AD 1300 and coexisted with classic Early Koniag diagnostics, including incised pebbles and medially ridged slate points (Partlow, 2000; Saltonstall, 1995, 1996, 1997). Saltonstall (1995) suggests that the earliest house (House 1) at Settlement Point may have been a very early multi-roomed house, created by joining several single-room houses together to enclose what was formerly a common "courtyard." Partlow (2000), in an informative study of Koniag salmon intensification, found evidence that increased specialization and storage of salmon paralleled the shift from single-room to multi-room houses during the Early Kachemak phase.

Changes occurring around AD 1400, based on Knecht's (1995) analysis, include: elimination of medial ridged points in favor of flatter stemmed points, higher incidence of notched stone sinkers (although these are often associated with Kachemak phases in other sites), replacement of scarfed fish harpoons with socketed variants, larger bentwood vessels, more woodworking tools, abandonment of stemmed ulus for perforated hafting techniques, increased frequency of trade goods from beyond the Kodiak archipelago, and more gaming and ceremonial paraphernalia. In southern Kodiak, gravel-tempered pottery is also common in Koniag assemblages (Clark, 1956, 1974a). Houses throughout the "Developed Koniag" phase have multiple rooms and increase in size to an average around 80 m^2. These structures normally have multiple side rooms extending from a large central room. At the New Karluk site, these houses are constructed with sizable driftwood supports and beams. Floors are constructed over the collapsed roof of previous structures, sometimes mimicking the outline and internal layout of the previous floor (Knecht, 1995) and suggesting unbroken continuity of occupation through the entire Koniag sequence at this site. Drainage channels lie below each floor (in the roof sods of the previous occupation) to prevent the wet conditions from affecting the living surface; the channels are covered with split planks, forming the appearance of patterned wood walkways within the houses (Figure 3.11 and Figure 9.2).

Large cobble-ringed hearths are situated in the center of the main rooms at the New Karluk site. Large wooden storage boxes are found inside the houses. Side rooms are reported to have been used as sleeping chambers, sauna rooms, and occasionally as burial crypts (Davydov, 1976). In other late Koniag sites, stone slab hearths and storage boxes are more common, possibly reflecting a difference in the availability of raw materials between sites (e.g., cobbles and driftwood are common at the New Karluk site while slabs are not).

The Koniag period witnessed an increased scale and intensity of warfare, as evident in armor and shields found in the New Karluk assemblage (especially in layers dating after AD 1400) and in the large defensive sites established on near-shore islets. Defensive fortresses have been identified for this period throughout the North Pacific coast as well as from the central Northwest Coast Tlingit, north and west all the way to the Aleutians (McCartney, 1984; Moss and Erlandson, 1992). On Kodiak, this form of defense was last used during the Russian attack on Sitkalidak in 1784 (Black, 1992).

In this study, I follow Knecht's use of the term Early Koniag to refer to the transitional period from AD 1200 to 1400, and I use "Developed Koniag" to cover the interval AD 1400 to 1784. The earlier phase is characterized by expanded

Figure 3.11. Excavation at the New Karluk site (KOD 001) in 1987. Dr. Richard Jordan inspects a House Floor 1 wood feature. Photo by author.

dwellings, few side rooms, incised pebbles, medial-ridged points, and little if any "Koniag pottery". The Developed Koniag phase has large multi-roomed houses, flat stemmed points with squared shoulders (unlike the barbed shoulders of the Late Kachemak Three-Saints Point type), gravel-tempered pottery (limited to Southeast Kodiak), and high proportions of fire-cracked rock (steam bath refuse).

3.8. ALUTIIQ/RUSSIAN-AMERICA (AD 1784–1864)

A major revolution in life and social organization was thrust upon the Kodiak inhabitants following the arrival of the first Russian fur hunters in 1763 (Black, 1992:165–166), and the conquest of Kodiak by Gregorii Shelikhov in 1784 (Black, 1992; Shelikhov, 1981). Documentary evidence of the Russian conquest and the subsequent Russian American period derives from scattered records written by the conquerors, their associates, missionaries, company men, and passing travelers (e.g., Black, 1989, 1992; Clark, 1984a; Davydov, 1976; Holmberg, 1985; Izmailov, 1790; Lisianskii, 1814; Merck, 1980; Shelikhov, 1981). With varying degrees of detail and objectivity, many of these Europeans made records of native life and organization. These records form the basis of ethnohistoric reconstructions of precontact Alutiiq/Koniag life-ways (see Chapter 4).

At this point I prefer to use the term "Alutiiq" to refer to the native inhabitants of Kodiak and the related cultures of the central Gulf of Alaska (previously designated Koniag and Chugach). I retain use of the term Koniag for the last prehistoric period as it distinguishes the final pre-Russian cultural patterns on Kodiak from the dramatically altered life-ways created in Alutiiq society after AD 1784.

Shelikhov justified his brutal attack on the Sitkalidak Refuge Rock site (*Awa'uq*—KOD 450) and his ten-year reign of terror against the Alutiiq of Kodiak and Prince William Sound as a response to the reportedly fierce resistance Russians met on three previous excursions to the archipelago in 1762–63, 1776, and 1779–1780 (Shelikhov, 1981). Black (1992) assembled various ethnohistoric reports of Shelikhov's attack and the early phase of conquest. According to her accounting, Shelikhov headed to Southeast Kodiak in the summer of AD 1784 with the assistance of 130 Russians and Siberians, 12 Fox Aleuts, and an elderly Kodiak Alutiiq slave named Qaspeq (serving as translator).[6] Searching for sea otter furs, Shelikhov was prepared for confrontation. He and his crew immediately established a fortified post in Three Saints Bay on the southeast side of Kodiak Island. After unsuccessful attempts to trade and command hostages, and after repelling an attack from the inhabitants of the area, Shelikhov demonstrated his strength by mounting a decisive offensive. Aware of impending trouble, as many as 4000 Alutiiq retreated to *Awa'uq*, their strongest and best concealed defensive refuge on the outer-coast of Sitkalidak Island (Black, 1992:171, citing Shelikhov, 1787:230).[7] Unfortunately for the native defenders, Shelikhov learned of the location of this hiding place and its vulnerable back approach (Black, 1992:172: Holmberg, 1985:59).

Acting as an intermediary between the Russians and the besieged Alutiiq, Qaspeq pleaded in vain with his relatives to surrender to Shelikhov. Finally, apparently fearing the arrival of reinforcements from other Alutiiq settlements, Shelikhov (1981) decided to attack. From a high spot on the spit on Sitklidak Island, the Russians fired a barrage of cannon balls on the exposed side of the cliff-faced refuge islet and then stormed the encampment (Figure 4.2). Mayhem broke out among the Alutiiq, and reports suggest that between 500–2000 Alutiiq were killed or drowned while attempting to escape. Women and children flung themselves over the cliffs in fear and desperation. So many people died in the attack that the location earned the name *Awa'uq* (to become numb) in the Alutiiq language of Sugtestun (Pullar and Knecht, 1995). Arsenti Aminak, a Native who survived the attack and later served as a primary informant for Holmberg (1985), reported that a year after the attack "the stench of corpses lying on the shore polluted the air so badly that none could stay there, and since then the island has been uninhabited" (Holmberg, 1985:59).

Following the attack of *Awa'uq*, Shelikhov claims to have captured over 1000 people, detaining 400 as hostages. (Numbers vary between accounts, and this

Figure 3.12. Shelikhov's settlement Three Saints was founded in 1784 just across the strait from Sitkalidak Island. Lithograph from Gavriil Sarychev's (1802) Atlas, courtesy of Alaska and Polar Regions Archives, Rasmuson Library, University of Alaska, Fairbanks (accession # rare C0015).

may be an inflated figure; Black, 1992 and included references). Of the captives brought back to the Three Saints outpost (Figure 3.12), several adult men were executed on Shelikhov's order. The number of executions reported varies from 6 to 200. Other captives were resettled a few kilometers from the Russian post under the care of a chief under Shelikhov's control. Shelikhov then began consolidating his grip over the island, demanding that each high-ranking man deliver a male child into his custody for insurance against malice. Numerous incidents of subsequent brutality by Shelikhov's militia occurred around Kodiak and as far as Prince William Sound (Black, 1992). By the time Shelikhov returned to Russia in 1786, the Alutiiq of Kodiak had been demoralized and decimated. Russian control was secure. Shelikhov continued to administer his fur company from Russia, attempting to gain an Imperial monopoly over the Alaskan colony he had founded by force. Alexander Baranov ran Shelikhov's company in Kodiak and the successor Russian-American Company in Sitka from 1790 until his removal in 1818. Like Shelikhov, Baranov was an officious organizer as well as ruthless towards the Alaskans under his influence (Fisher, 1990).

In 1792 the Russian colony at Three Saints Bay relocated to Saint Paul's Harbor (the location of the city of Kodiak today). From this outpost the Shelikhov Company, and later the Russian-American Company under Baranov, expanded its control over southern Alaska. It moved rapidly into the Tlingit area of the Northwest Coast and relocated the colonial capital to New Archangel (Sitka) in 1808. From there the company ultimately expanded Russian colonial domination as far south as northern California, where Baranov founded Fort Ross in AD 1812 (Pritchard, 1990:84).

The Alutiiq were forced to work for the colony, producing quotas of furs, wild foods, and labor (Clark 1984a). In pursuit of furs to fuel Russian-Chinese trade in sea otter and fur seal skins, able-bodied men were sent as far as the Kuril Islands south of Kamchatka and the Channel Islands in southern California (Crowell, 1997; Ogden, 1941, 1991; Shubin, 1994). Introduced disease and conscription soon reduced the native population of the archipelago; at the time of sustained contact in AD 1784, the estimated contact population was still more than 9000 (Clark, 1984a:187). By AD 1850 the population was 3000 (Clark, 1984a:187). Under Russian governance, native settlements were consolidated by colonial proclamation and by the logistical demands of diminished populations (Crowell and Lührmann, 2001:60). Subordinated exchange and tribute to the colonial officials replaced local and inter-regional political activities such as trade and warfare.

The 1794 arrival of Orthodox missionaries in Kodiak brought some relief from the brutal tactics of Shelikhov and Baranov. The missionaries' sympathies for native peoples, along with condemnation of the treatment they received from Baranov's forces, ultimately led to the Russian Tsar's declaration of humanitarian decrees for the protection of these new citizens of the Russian Empire (Black in Gideon, 1989:*vii*). Fathers Ioasaf, Herman, Gideon, and their contemporaries and successors sought to Christianize, "civilize," and protect the Alutiiq people. In the aftermath of conquest, disease, and depopulation, the surviving Alutiiq openly embraced the Russian Orthodox faith.[8] According to many spokespeople today, the Russian Orthodox religion and traditional native faith shared many features, and Russian missionaries emphasized the complementary aspects of these traditions. Many natives became priests and the Russian Orthodox Church became a mainstay of Alaskan Native life from the Aleutians to Southeast Alaska (Oleksa, 1992).

Most archaeological research on Kodiak focuses on pre-Russian periods. Nevertheless, several projects have examined contact and Russian-Alutiiq period sites. The Aleut-Koniag Project tested Shelikhov's Three Saints Bay post in 1962. Their investigations were diverted by the discovery of a deeper underlying Late Kachemak deposit (Clark, 1974a, 1992a). Later, in 1990 and 1991, Crowell (1997) led a crew that expanded excavation at this site in an effort to understand the economic, social, and political dynamics of early colonial life as well as the relations between the Russians and Alutiiq under their charge. The Aleut-Koniag

Project researchers also conducted surveys that identified many contact era Alutiiq villages reported in early historic accounts (Lisianksii, 1814); these provided Clark (1987) with the material for the first comprehensive settlement pattern analysis for the region, focusing on eastern Kodiak.

In 1984 Rick Knecht excavated the first complete Alutiiq sod house (*ciq'lluak* in Alutiiq or *barabara* in Russian) in the abandoned village of Nunakakhnak on the southern shore of the Karluk Lagoon. The excavation of this Russian phase dwelling provided data on traditional (including pre-Russian) house construction techniques and a host of artifacts attesting to the degree of cultural change brought about by the Russian conquest (Knecht and Jordan, 1985). Funded by the Old Harbor Native Corporation, Knecht directed a testing and excavation program on the *Awa'uq* refuge site in 1992. A few other excavations have been conducted at sites of this historic phase, sponsored by the City of Kodiak (Shinkwin and Andrews 1979) and the Kodiak Historical Society with the Alaska State Office of History and Archaeology (Heritage, 1983). Supported by the Afognak Native Corporation's Dig Afognak program, research by Woodhouse-Beyer (2001) focused on the Russian American phase on Afognak.

In general, the archaeology of Alutiiq sites from the Russian phase illustrates the violence of the Russian conquest and its impact on traditional life-ways.

Figure 3.13. Sod houses and drying salmon in the village of Old Harbor ca. 1888–1889, Photo courtesy of the Alaska State Library (PC-186-40).

However, while new implements (e.g., metal knives and axes) replaced traditional ones, many tools, house construction techniques, and subsistence practices endured; this despite drastic population reductions and consolidation of several hundred settlements into only seven by the time of the American purchase of Alaska in 1867. Photographs of Alutiiq dwellings from Karluk and Old Harbor in the 1880s show the continuation of traditional sod house construction well into the American phase (Figure 3.13).

3.9. ALUTIIQ/US AMERICA (AD 1867–PRESENT)

In 1867 Russia gave up control of its American colony to the United States. Factors contributing to this realignment include: the decline of sea otter populations and trade, growing competition with the Hudson Bay Company, pressure from the Spanish American colony and then the United States, and domestic troubles closer to home. The Alaskan Territory was nevertheless soon recognized as a land of great economic potential. When gold was found in central Alaska, just as the 1849 California gold rush began to decline, prospectors inundated the territory. Alaskan natives were an afterthought to American expansion.

Following precedents established in its incursion into the southern part of the North American continent, the US treated Native Alaskans as conquered tribes subject to minimal rights and ignored the natives' interests in all administrative decisions covering the territory. While the Russian Empire had granted natives full citizenship by the end of the Russian period, that status did not transfer to the United States. The Bureau of Indian Affairs (BIA) assumed stewardship for Alutiiq and other Native Alaskans and facilitated their assimilation into English-American culture. They established schools where native language was discouraged (sometimes even violently prohibited) and where Euro-American history was taught as national history without regard for the past of the Alaskans themselves. Soon feelings of low self-esteem (common among groups treated as conquered people) became pervasive among the Alutiiq and other native groups (Pullar, 1992; see also Brody, 2001).[9] Alcoholism, encouraged by both good-willed and manipulative outsiders, quickly grew to crippling proportions.

In the early years of the American period, sea otter and fur seal hunting accelerated without regard for the sustainability of the harvest. Already dwindling populations of sea otters were forced to near extinction from California through the Aleutians, and recent research suggests that there may have been fur seal rookeries as well throughout the North Pacific Rim that were exterminated (Etnier, 2002). The search for other viable industries soon led to the rich stocks of salmon in Alaskan waters. Kodiak quickly became the hub of the growing Alaskan salmon fishing industry (see Roppel, 1986). In 1882 the Alaska Commercial Company constructed the first Kodiak cannery, locating it on the spit fronting Karluk Lagoon

Figure 3.14. Salmon canneries on the Karluk Spit in the late 19th century. Photo by "T.H.B.," courtesy of the U.S. National Archives (22-FFA-1145).

(Campbell 1992:63). Within seven years, four more canneries lined this same spit. They all took advantage of the immense volume of salmon entering the Karluk system until its eventual catastrophic decline (Figure 3.14). At the same time, other canneries opened in the Afognak, Olga, Moser, and Larsen Bays. Early in the development of the fishing industry, the United States Congress became concerned about the potential for over-fishing and mandated the construction of hatcheries in commercial salmon streams. One of the first hatcheries was built in Karluk (Roppel 1982). Nevertheless, salmon populations in many of the larger streams around the archipelago were soon reduced due to a lack of sophisticated hatching techniques, pressure from in-stream fish-traps installed by canneries, and the development of intensive seine fishing from commercial fishing vessels.

Competition and profitable markets spurred a high influx of non-Natives in search of economic opportunities. Cannery companies from outside Alaska led the way, non-Native fishermen followed. From the outset, Alutiiq laborers were employed in canneries throughout the region, though often as second-class laborers. Soon many Alutiiq realized they could do better by applying their intimate knowledge of the regional waterways and fish behavior to the operation or crewing of fishing boats. Since then other workers became attracted to the cannery jobs, including immigrant Chinese, Germans, Scandinavians, Italians, and Filipinos.

In 1912 the Katmai eruption blanketed much of Kodiak in ash and forced a temporary evacuation (Figure 3.15). While several communities on the Alaska Peninsula were permanently abandoned, Kodiak recovered relatively quickly.

Figure 3.15. The town of Kodiak under 30 cm of volcanic ash from the 1912 eruption of Mount Katmai (Novarupta). Photo courtesy of the University of Washington [Thwaites collection # 1252].

In 1964 the Good Friday earthquake leveled the Native villages of Akhiok, Kaguyak, Old Harbor, and Afognak, and wreaked havoc on other communities. Federal assistance helped most communities rebuild, but Kaguyak and Afognak remained vacant. Fortunately for those with fishing boats that survived the tsunami of 1964, the king crab fishery boomed throughout the 1960s and into the 1970s, providing income and stimulating the Kodiak economy to unprecedented levels. This boon was not without its drawbacks for Alutiiq people. Like the gold rushes of the previous century, the king crab fishery attracted large numbers of outsiders, many with little care for, or interest in, the welfare of the original inhabitants of the Archipelago. Money, alcohol, and drugs flowed like water into the swelling town of Kodiak. As a boom-and-bust industry, fishing, like the tsunami, left some devastation in its wake. When the crab-fishing industry declined in the 1980s, permanent residents had to live with the consequences while outsiders moved on to better prospects. Fortunately, the salmon, cod, herring and halibut fisheries remained viable and continue to provide the primary economic support for Kodiak residents. These, however, are also becoming less reliable for primary support. In 1989 the Exxon Valdez oil-spill coated Kodiak's shores, damaging the ecology and threatening fish stocks (Richardson and Cline, 2001).

Throughout these changes many Alutiiq people have tried to hold on to their culture and traditions. Traditional subsistence dominated the local economies well into the 20th century. Even today, communities more distant from the city of Kodiak supplement wild food with store-bought food rather than the other way around (Davis, 1978). Attachment to the land and sea is strong. Native communities attempt to maintain self-sufficiency and control their fate, with increased success in recent years. In 1971 Native Alaskans won a land claims victory and were subsequently empowered to administer large tracts of productive land under business incorporation. Lacking business training and opposed by powerful competing interests, many of these corporations have suffered. On Kodiak, however, Native corporations have been reasonably successful. Today they lead in the promotion and funding of projects to rediscover and revitalize the history (including archaeological history) of their vibrant Alutiiq heritage. The research reported in this book was partially funded by the Old Harbor Native Corporation and logistically supported by the Kodiak Area Native Association through the Alutiiq Culture Center (now the Alutiiq Museum and Archaeological Repository). These groups have been interested in identifying the Alutiiq archaeological heritage and rekindling knowledge about the rich history suppressed by 200 years of colonial cultural repression.

3.10. SUMMARY

The cultural sequence of Kodiak provides a consistent schematization for the study of temporal change in cultural characteristics. Further refinements are, however, necessary. One of the most pressing needs is a refined chronology that escapes the limitations of a typologically driven period/stage approach. Currently published evidence is insufficient to break the phases apart and closely examine the spatial and temporal variability underlying changes in artifacts, residence patterns, or economic and social practices. Until we can work within a chronology of finer resolution, we remain constrained to view changes as if they occurred in punctuated fashion (at phase boundaries) and that many occurred more or less simultaneously at these junctures. By adding the requisite resolution to the Koniag period, the New Karluk site analysis shows how hazardous these assumptions can be for interpretations of cultural importance (Jordan and Knecht, 1988).

This chapter outlined the cultural history that forms the foundation for the archaeological research and synthesis that follow in the next several chapters. Even though Kodiak has a better understood cultural sequence than many other areas of Alaska, little attention has been systematically directed to explaining the underlying cultural structures and processes of change. After 30 years of continuous archaeological investigation, we are now in a position to begin addressing these issues.

Chapter 4

Complex Hunter-Gatherers on the Kodiak Archipelago

4.1. INTRODUCTION

In the late 18th century Alutiiq society was the product of thousands of years of struggles and triumphs, the aggregate effect of billions of decisions and actions, some in perilous circumstances and others of seemingly trivial importance. In this chapter we explore the result of this evolutionary history, the product of at least 7500 years of development prior to substantial influence from Eurasian and Euro-American intervention. In the next chapter we will establish the context and character of the initial Kodiak occupation. These two chapters, then, mark the end points, at least 7500 years apart, of a time line. The rest of the book connects these points through evolutionary modeling and archaeological analysis.

The primary documentary record for 18th and 19th century Kodiak is found in ethnohistoric accounts written by early fur traders (*promishleniks*), Russian colonial officials, and missionaries, supplemented by much later ethnographic accounts. In the natural historical tradition of the time, some observers paid close attention to Alutiiq (Koniag) economic, social, political, and ceremonial practices; these accounts give us our best insights into traditional Alutiiq lifestyles. Because Kodiak was peripheral to the major ethnographic studies of the late 19th and early 20th centuries, we find it difficult to compare Alutiiq culture with that of the Northwest Coast, where extensive anthropological ethnography paints a rich picture. Kodiak Alutiiq's early incorporation into the Russian colonial system may partially explain this relative ethnographic neglect.

By anthropological convention, the Alutiiq (of Kodiak and the greater Gulf of Alaska region) have been most closely associated with their cousins, the Yup'ik Eskimo of the Bering Sea (e.g., Lantis, 1947), who live in distinctly different environments. There is little question that the Alutiiq and Yup'ik share deep cultural, linguistic, and biological roots. But scholars note that the Alutiiq also share strong cultural parallels with other groups of the North Pacific Rim from the Aleuts to the Northwest Coast Indians (e.g., Donta, 1993; Townsend, 1980). Donta (1993), in particular, makes a compelling argument that the emphasis of social status (rank and hierarchy) in Alutiiq ceremonies is much more similar to other North Pacific groups than to the Yup'ik. Because the societies of the northern Northwest Coast are considered classic cases of complex hunter-gatherers, we will examine the Alutiiq documentary evidence in comparison to Tlingit ethnography for developing a picture of the Alutiiq cultural complexity at the dawn of Eurasian contact and colonialism. The Tlingit make a worthwhile comparison because Alutiiq people engaged with them in military competition and trade at the time of Russian contact.

The Kodiak data allow us to infer that, like the Tlingit and other northern Northwest Coast peoples, the Alutiiq organized themselves into semi-sedentary villages, held regular seasonal potlatch festivals, developed rank status differentiation, and engaged in extensive intertribal warfare and slavery. Each discussed factor of economic and social life illustrates a dimension of the complexity of the pre-contact Kodiak Alutiiq. In addition, each factor is related to variables highlighted in various models for the development of emergent complexity (see Burch and Ellana, 1994).

4.2. FEAST AND FAMINE FOR THE KODIAK ALUTIIQ

Many scholars argue that subsistence is a key factor in the emergence of social complexity. To some, resource abundance is decisive as it provides latitude for social competition through modes of accumulation, craft production, trade, and status display (see Clark and Blake, 1994; Hayden, 1995). According to this view, food surpluses give people the leisure to invest their efforts in activities less directly related to the food quest or to facilitate population growth above some threshold that inspires political competition. Social complexity is then seen as an almost inevitable result of the release of subsistence constraints, when individuals with dominating personalities are free to accumulate resources and, in turn, subordinate others as they compete for prestige and power among their peers. Others see social complexity as a result of organizational change for the purpose of dealing with recurrent resource shortages or of the difficulties associated with managing large cooperative work forces (e.g., Ames, 1985). For these scholars, social changes are the intended and/or unintended consequences of attempts to adapt to hardships, not a result of the absence of hardship. These positions are perhaps less contradictory than often assumed. (See Chapter 6 for elaboration). The North Pacific Rim plays an important role in the development of these arguments, as it supports highly productive maritime and littoral ecozones. This ecosystem, though often characterized as "rich" or "abundant," is actually spatially uneven and subject to variable productivity, punctuated by intervals of both predictable and unpredictable resource shortfalls (see Chapter 2).

Like all ethnographically-known North Pacific cultures, the Alutiiq traditionally employed strategies that helped them deal with cycles of abundance and scarcity. Seasonal changes were the most prevalent cycle (see Chapter 2). Fish (especially salmon) and sea mammals played an important role in the Alutiiq economy, as they did in other North Pacific groups. Unlike most mainland residents, however, Kodiak inhabitants lacked access to significant supplies of terrestrial game. Besides the famous Kodiak brown bear and a few smaller land mammals, subsistence came almost exclusively from the ocean, rivers, and littoral zones and from birds drawn to these areas.

It is often assumed that the combination of food storage and seasonal abundance meant the population would rarely, if ever, suffer food shortage; however, archaeological indicators suggest prehistoric occupants did in fact suffer from occasional metabolic stress, probably related to seasonal dietary insufficiency. For example, a study of skeletal samples from the Crag Point site reveals a high frequency of Harris lines in the long bones of Late Kachemak people (Steffian and Simon, 1994). Ethnohistoric accounts and oral tradition provide insight into how traditional methods were employed to defend against hard winters. According to these accounts, each family built up its own stocks of fish and foodstuffs for winter which typically lasted to the end of January, the close of the festival season (Merck, 1980:100).

During the lean winters, people hunted and gathered resident resources farther from the village whenever the weather permitted them to go to sea (Davydov, 1976:15). During winter months, shellfish from close shorelines became critical dietary staples. Although some anthropologists argue that shellfish could have played a significant role as a dietary staple, at least in times of low protein sufficiency (Erlandson, 1988; Erlandson and Moss, 2001; Moss 1993), native informants characterize shellfish as "starvation food." Additionally, diets rich in shellfish have significant drawbacks including protein starvation and paralytic shellfish poisoning (Fitzhugh, 1995). The fact that shellfish were eaten in great numbers suggests that higher quality foods were frequently unavailable.

Stone weirs and fish harpoons were used in many rivers to capitalize on large fish runs during the summer and fall. Bone harpoons helped hunters secure sea mammals in the water, and lances tipped with slate endblades dipped in paralyzing poison helped kill small and medium-sized whales (Crowell, 1994). As discussed in Chapter 2, halibut, cod, and other bottom fish were caught with large barbed hooks and stone weights (Figure 4.1). Flesh of fish, sea mammals, and shellfish was eaten fresh or dried and smoked for preservation. Edible roots, stems, berries, and leaves were collected during the summer and stored in oil made from whale, seal, sea lion, or fish fat (Holmberg, 1985:41–42). This "train oil" was eaten with most other foods. Indeed, "without [oil] they could barely exist and would never be happy" (Davydov, 1976:16). This addiction to oil can be understood as a biological response to the need for fat in an overwhelmingly protein-biased environment (cf. Speth and Spielmann, 1983).

Figure 4.1. Fishing for cod in the Eastern Aleutians as illustrated by Henry W. Elliot. Reproduction courtesy of the National Anthropological Archives (SI-NAA 7119–13).

Storage was clearly an important aspect of the Alutiiq economy, as islanders could postpone consumption of fish, meat, fat, and plant products. Salmon "are stored in the home, piled high along the walls, and if the room is not sufficient they are also piled high on the floor and then eaten" (Holmberg, 1985:41). Due to their reduced fat content, late-season salmon were easier to preserve; evidence from the Northwest Coast suggests these supplies only kept for six months or so before spoiling (Jorganson, 1980:128, cited in Mitchell and Donald, 1988:320). Holmberg (1985:41) suggests that food stores were insufficient to last the winter and that shellfish often served to tide Alutiiq families through to spring and the return of migratory resources. In spite of this (or perhaps because of the limited shelf-life of dried and smoked foods), stored foodstuffs were often expended in winter festivals or potlatch feasts between November and January (Davydov, 1976:18).

4.3. POTLATCH FEASTING

Feasting is a social activity receiving considerable anthropological attention (Blitz, 1993; Donta, 1993; Jordan, 1994; Junker, 1999; McAllister, 1990; Miller, 1984; Potter, 2000). Feasts are formal and ritualized events that bring people together to eat and celebrate. Feasts embody multiple dimensions of cultural life that make them useful (and often idealized) reflections of society. Feasts often both dramatize and institutionalize structural relationships, and thus help to reproduce social order, provide outlets for social competition, and sometimes serve as venues for dissention and sanctioned protest. As ritualized events, feasts typically are culturally conservative and resist changes that pervade daily life (Rappaport, 1978). At the same time, feasts can become vehicles for sanctioning changes already pervading society.

Feasting is relevant to the evolution of social complexity in direct and indirect ways. For example, the competitive Northwest Coast-type potlatch feasts may have arisen to provide avenues for small-scale competition between families and lineage groups. These competitions may then have propelled increased production of foodstuffs and valuables (e.g., blankets, fine clothing, adornments), and stimulated craftsmanship. Amplification (or positive feedback) in such a socially motivated economy could then have led to changes in the social structure. Alternatively, competitive feasting may have arisen as a means of sanctioning competition that was already developing for other reasons. Regardless of the cause-and-effect relationship between feasting and social change, ethnographic and ethnohistoric accounts of feasting provide useful measures of social organization.

December and January were festival months on Kodiak (Davydov, 1976:18). During this time neighbors from nearby villages would be invited for days or weeks of feasting, gambling, and dancing. During these potlatch events people congregated in a large house or in the village *kazhim* (ceremonial house; also known as a *khasu* or *qásxiq*; Donta, 1993:168; Merck, 1980:100) to boast about their ancestors and to show off through various acts of conspicuous consumption, self-mutilation, gambling, gift-giving, and oratory (Davydov, 1976:6–8, 15, 18). Potlatches brought neighboring villages into reciprocal relationships as hosts and guests, allowing social bonds to grow between otherwise potentially hostile groups (see discussion of warfare, below). "The headmen of the villages sometimes only return visits [as guests/hosts] after a lapse of two to three years, and then they are made great with ceremony" (Davydov, 1976:15).

Women often spent months making headdresses or costumes to be worn during the festivals (Davydov, 1976:10–11). At the end of the potlatch, these adornments were destroyed as a means of thanking guests for the honor of their presence (Holmberg, 1985:54). Even the *kazhim* and other paraphernalia were sometimes destroyed at the conclusion of each feasting season (Davydov, 1976:18; Holmberg, 1985:54). As an act of honor, men occasionally cut up their finest garments and distributed the pieces to all guests at the conclusion of a potlatch. This same practice was required of a father before his children were first admitted to the potlatch (Holmberg, 1985:54).

Potlatch feasts often memorialized the dead, and according to Donta (1993), these were among the most important Alutiiq ceremonies. Large funerary feasts were thrown after the death of a wealthy man (Holmberg, 1985:53). "When the Islanders celebrate the memory of their relatives all those present are given presents at the end of the ceremony. During this ritual one person will give a speech praising the dead man, and for this he will be given expensive presents" (Davydov, 1976:18). "During festivals they will relate deeds performed by their fellow countrymen in far-off times" (Davydov, 1976:6). Donta (1993) points out that in addition to feasts of the dead, feasting was an important part of many Alutiiq ceremonies such as rituals before and after warfare and whale hunting, after an elite birth, and following marriages involving an elite or rich person. In all cases, ceremonial rules emphasized social position in seating, service, and speaking privileges. Donta (1993) notes that the Alutiiq emphasis on personal status was distinctly different from the Yup'ik practices described by Lantis (1947).

Potlatch feasts on Kodiak were similar in many ways to the better-documented customs of the Tlingit of Southeast Alaska. While differences exist, it is instructive to consider the way that Tlingit potlatching served to institutionalize social hierarchy and ranking. Limited analogies can then be drawn between the two neighboring groups.

The Tlingit potlatch season was November and December, after the seasonal stores had been established (de Laguna, 1990:206). Considerable effort was expended amassing large amounts of consumable resources and wealth (prestige goods) by the host lineage or "House." Larger and more elaborate potlatches would take several years to plan. All members of the lineage or clan would be asked to contribute food and wealth objects in order to throw a large and respectable potlatch. Often the host chief's spouse would request donations from her/his matrilineal House in order to increase the amount of wealth displayed at a potlatch. Such donations from a guest House would be paid back with interest (de Laguna, 1990:220). Preparations for a potlatch would include the production of valuable wealth items to give away at the end of the potlatch. Some of these might be produced by elite craftspeople such as the Chilkat blanket-makers. (It took a full year for a woman to produce one of these highly valued blankets.) Other items, such as canoes, carvings, exotic shells, ivory, and slaves, were procured through raiding of, or trade with, neighboring groups. While potlatches could only be planned by lineage chiefs, the entire host lineage gained prestige through the event; their crest was displayed and honored, and their social debts were zeroed. During the potlatch, hosts gained prestige by the number of guest Houses they could entertain, the amount of food they could serve (dictating the length of the potlatch), and the amount of wealth objects they could assemble and give away or destroy publicly.

As with Alutiiq groups, Tlingit potlatches were often organized to honor the dead (de Laguna, 1983:76). Sometimes potlatches were designed to repay funeral debts to opposite moiety House members who had prepared bodies for cremation and administered wake ceremonies since the last potlatch season. Larger potlatches celebrated the ancestors ("Memorial Potlatch") and might be called to celebrate the reconstruction of a house structure by a wealthy lineage. These potlatches constituted major events, and took several years to organize. Often the Memorial Potlatch was thrown by a chief to honor the passing of a former leader and to demonstrate the viability of the new lineage leadership (de Laguna, 1990:220).

Another type of Tlingit potlatch was designed to give high-ranking lineage members or heads ("nobles") the opportunity to honor and "ennoble" their children (a practice required for them to be recognized as "nobles" in their own right). During this ceremony the children of high-ranking members would receive an ear piercing (for both genders) and hand tattoos (for girls only) from their close maternal kin. Adults of truly high-ranking parents might be able to show-off as many as eight ear holes bearing witness to the same number of ceremonial potlatches thrown in their honor as children (de Laguna, 1990:250).

The Tlingit ennobling ceremony has parallels to the less well-documented Alutiiq practice of destroying fine clothing prior to introducing children to the Kodiak festival for the first time. It is unclear if the Alutiiq practice led to the segregation of people into "noble" and "commoner" groups as it apparently did for the Tlingit; however, the destruction of fine clothing (such as guillemot bird skin coats) would certainly be recognized as an act of grandeur, conferring honor on the children being introduced. While the Alutiiq frequently wore jewelry in their lips, ears, and nasal septum, I have found no documentation relating piercings to the winter festivals or potlatches, although Holmberg (1985:38) indicates that the piercings were done by close kin. The noted correspondence between number and size of labrets and rank (Holmberg 1985:38) suggests that piercings were controlled by social sanctions and open only to individuals recognized by a larger group as deserving honor. Potlatch ceremonies were likely contexts for such ennobling.

The aforementioned parallels between Tlingit and Alutiiq winter potlatches should not be accepted as evidence of cultural symmetry between the two areas. Several differences are apparent. Davydov claims that the *kazhim* where festivities occur belonged to the whole village (Davydov, 1976:18; Holmberg, 1985:54). Gideon (in Black, 1977:93) implies that these ceremonial structures were owned and that non-owners could use them if they paid gifts to the owner for the privilege. Frequently, however, villages did not have separate *kazhims*. Gideon (1989:40) notes that he "only saw three such *kazhims* on the whole of Kad'iak." Where dedicated *kazhims* were not constructed, festivities were held "in the foreroom or in a large sleeping room" (Gideon in Black, 1977:93).

Tlingit potlatches, in contrast, were held in the center of lineage houses, and independent Houses in the same village often competed for honors by hosting the best potlatches. Tlingit potlatching was fully incorporated into the highly structured kinship system of lineage and moiety. Whereas Kodiak potlatch facilities were often "shared" between closely-related families (with payments made to the owner for the privilege), Tlingit facilities were part of the competitive expression of the hosting lineage or clan House. Kodiak villages were organized into extended-family kin groups, but there is no evidence of the highly integrated lineage system that structured Northwest Coast social life. These apparent differences, if in fact real, suggest that Alutiiq villages were somewhat less structured and less internally competitive than Tlingit villages, at least in post-contact times. Russian interference is known to have forced the consolidation of Kodiak villages, led to great reductions in population due to disease, and caused the deterioration of lineage cohesion, all of which would have de-emphasized intra-village competition in favor of village unity. Davydov's suggestion of jointly-owned *kazhims* may refer to the post-contact period when structured intra-village political competition would have been less feasible.[1] In another potential contrast, Davydov (1976:15) suggests that gifts exchanged at Kodiak festivals had to be of equal value so that no one remained in debt. It is probable that this sentiment more accurately reflected the ideal, that after a series of potlatch exchanges over several years, guests and hosts would be approximately "equal" in social reciprocation, as on the Northwest Coast.

Both in and out of festival situations, Kodiak Islanders distinguished themselves according to rank and wealth by the clothing, jewelry, and tattoos they wore. Women could wear from two to six labrets in their cheeks. The number and size of labrets would testify to their rank and respectability (Holmberg, 1985:38). Only women wore tattoos and high-ranking women sported them on their backs as well as chins and arms (Merck, 1980:103). Men generally wore a single lower lip labret. Beads and bangles were inset into labrets, earrings, and nose ornaments. Amber was highly valued due to its rarity (Holmberg, 1985:38). A pearl-like shell (probably dentalium shell) traded from the Queen Charlotte Islands (via the Tlingit) was also considered extremely valuable on Kodiak (Davydov, 1976:3).

4.4. GENDER RELATIONS

It is commonly noted that truly egalitarian societies do not exist because even among societies with simple organizations, asymmetries based on age and gender prevail (Flanagan, 1989; Kelly, 1995; Lee, 1982). Yet, gender dynamics often change with the emergence of social complexity. There is considerable evidence that women often lose relative status and power over their labor and reproductive rights with the emergence of increasingly structured society (e.g., Brody, 2001:249–254; Collier and Rosaldo, 1981; Gailey, 1987). As they permeate social relations and dynamics in all societies, gender relations are necessarily involved in all social changes. These social changes, in turn, influence gender dynamics. While the archaeological study of gender is difficult, it is important to recognize that the relations of men and women are implicated at every level (Bacus et al., 1993; see Conkey, 2001; Conkey and Gero, 1991). For this reason, and because gender asymmetry is often perceived as the foundation of broader scales of institutional inequality (Kelly, 1995:302), it is important to consider the position of men and women in Alutiiq ethnohistory.

The status of women and men among the contact era Alutiiq is obscured by somewhat contradictory statements in the literature (as it is for the Tlingit: compare, for example, de Laguna (1990) and Moss (1993)). According to Davydov (1976:11), while Alutiiq women had considerable power and were often able to choose their husbands (or at least accept or refuse an offer of marriage), they could not attend public councils and could not eat with the men. Also, men would sometimes "share" their wives and daughters with visitors. On the other hand, women would sometimes participate in war raids (Davydov, 1976:22), and, at least on the Northwest Coast, sometimes gained reputations as fierce warriors and chiefs (de Laguna, 1990).

The Alutiiq had female puberty rights similar to those of the Tlingit. Pubescent girls would be secluded for six months in small, dark huts. Limited space restricted their positions and kept them from straightening their backs. After this six-month period the huts were enlarged so that the girls could straighten out their backs. After another six months, the initiates were accepted into society and a 'rebirth' ceremony was held (Holmberg, 1985:53). Upon completion of their seclusion, the young women were honored at a feast and received tattoos (Donta, 1993:142 and references). Among the Tlingit, girls might spend up to two years in confinement (de Laguna, 1990:217). In both cases, the initiate would emerge with a "translucent complexion," (and weakened condition) culturally suitable for marriage. At other times, Alutiiq women would wash in water mixed with urine to whiten and clarify the skin (Davydov, 1976:15). During menstruation, women were again confined to small shacks away from the main living quarters of the village (Davydov, 1976:14). During birth, and for up to three weeks following birth, mothers and babies were confined to small huts or side rooms of houses (see Donta, 1993:139 and references).

Very little information has been recorded about male initiation rites. Donta (1993:144) suggests that initiation into men's secret societies may have been one of the primary forms. These societies maintained secret knowledge, songs, and dances that gave them the power to emulate evil spirits, devils, and demons in rituals. It is widely reported that the performances of these secret societies were designed to invoke fear in women and children. Observing Aleut customs in the early 19th century, Veniaminov (1984:314–316, cited in Donta, 1993:161) believed that their secret societies were "used as the most emphatic means of keeping their wives in submission and fidelity and for teaching their daughters the virtues proper to them." This may have also been the case on Kodiak.

Open only to men, another secret society was the more exclusive whaling cult (see Donta, 1993:145, 177–186). Whalers were a revered class of individuals who inherited the rights, songs, and paraphernalia to hunt whales and who were inducted into the role through "a long and arduous period of instruction in which the young whaler learned the ritual preparations, songs, etc." (Lantis, 1938:440, cited in Donta, 1993:145). Whalers were steeped in secret magic designed to assist in the hunt. They maintained secret rites and rituals, including stealing, preserving, and consuming the bodies of deceased whalers. They slept apart from their families during the hunting season and were revered and feared for the danger of their magic.

According to both Merck (1980:108) and Davydov (1976:11), Alutiiq marriage was not a highly formalized event. Women could choose their husbands and, although bride service and bride wealth were often involved (Davydov, 1976:19), we have little evidence that marriage initiated a significant economic relationship between the parents or kin groups of the married couple. Nevertheless, marriage between lineage leaders or wealthy families had both economic and political implications. Rich men and chiefs were known to have as many as eight wives, and women did on occasion take up to three husbands (Davydov, 1976:11; Iosaf in Black, 1977:85; Lisianskii, 1814:198; Merck, 1980:108). In both cases the extra spouses provided additional productive labor and expanded social and political integration. Donta (1993:146) notes that marriages were occasionally arranged by the parents of the bride and groom for political reasons (Gideon, 1989:50). For richmen and chiefs, security came in the form of alliances forged through marriage. Donta (1993:146) notes comparability to Aleut customs described by Veniaminov (1984:194):

> . . . for political reasons, such as avoidance of conflict or establishment of friendly bands [alliances], wives were taken from distant villages, and not just from one [but from several], in the conviction that no one would dare to insult, or even aggrieve [in any manner] a man who has large affinal connections.

4.5. LEADERSHIP

Political organization is an important dimension of social inequality and complexity. Leaders in all societies play a significant role in reproducing the social structure in which they operate. Among small-scale societies, leaders obtain limited power from their ability to make fair or wise decisions; followers will often chose to follow only to the extent that they feel the leader makes decisions effectively for the greater good of the group (see Boehm, 1993; Lee, 1972, 1979; Lee and Daly, 1999). In more complex societies, leadership often falls to individuals who can compel followers to help fulfill the leader's personal goals. Many tribal societies exhibit a dynamic tension between these two extremes, when, for example, leaders seek to attract followers through acts of generosity while they use the followers' labor to increase their own wealth, power, and prestige (Hayden, 1995; Sahlins, 1972).

Alutiiq leadership is described inconsistently in the ethnohistoric literature. Some observers have suggested that village headmen had no real power "because no one can command any but his relatives" (Davydov, 1976:7; Merck, 1980:108). Each "Koniaga" (Kodiak Alutiiq) had an "equal share in matters of general concern. In his personal affairs, each follows his own will" (Davydov, 1976: 24), not unlike the people of the Northwest Coast (see Ames, 1995). Other observers, however, claim that the Alutiiq maintained "ordinary and hereditary chieftains" (Holmberg, 1985:36; see also Gideon in Black, 1977:91).

Some accounts suggest that leaders (*anayugak*) had considerable power in some arenas. According to Davydov (1976:23), "some of the chiefs rule over many settlements and are all descended from one tribe." Davydov (1976:12) also recounts a story of a greatly respected and feared leader early after Russian conquest. Challenged to rid Kodiak of the Russians, this leader realized these actions would ultimately lead to the murder of a great many Alutiiq women and children. He vowed to preserve peace with the Russians and was able to ensure it for as long as he lived.

These accounts become coherent when we recognize (1) command over one's relatives can be a significant source of power in tribal societies, and (2) leaders of large kin groups can use threat or deed to exercise considerable influence over the course of events affecting smaller lineage groups. These influences would have also reached neighboring communities where kin might reside after marriage or where local leaders owed debts to the chiefs in question. Marriage, especially to multiple partners from different neighboring and distant villages, helped to extend kinship and with it greater leverage in social negotiations with kin and non-kin alike (Veniaminov, 1984:194 in Donta, 1993:146).

A leader's power beyond his/her kin group ("family, fosterlings, and *kalgas* [prisoners]" Gideon in Black [1977:91]) was contingent upon the leader's ability to marshal wealth and wage war. Leaders acquired high status through mobilizing labor and capturing the resources of enemies and slaves. However, positions of leadership and the supporting conditions of wealth (material goods, rights to resource locations, slaves and kin groups) were inherited. According to Gideon (in Black, 1977:91):

> [The *Anayugak's*] office was hereditary for his children and kin: eligible for succession to his office were his son, brother, uncle, nephew, or son-in-law. The heir assumed the office gradually. The *anayugak* designated his successor well before his anticipated death. On such an occasion he called everyone into the *kazhim*, then he escorted the designated successor thither himself, having dressed the candidate in his best clothing. In the presence of all he declared that the individual was his choice for a successor; the latter had to treat all assembled to food and drink.

Upon death, individuals of high status would be buried in the side rooms of their houses (Merck, 1980:108). In some cases, the entire house was collapsed and another one was built at this time (Holmberg, 1985:53). Particularly well-respected leaders and whale hunters were sometimes mummified and kept in secret caves, or they were cut up, dried and distributed between living whale hunters as talismans of great power to be carried on the hunt (Holmberg, 1985:49). Rich men's deaths would often be celebrated with large potlatch feasts thrown by the widow of the deceased. Non-elites would generally be buried some distance from houses (Merck, 1980:108).

One reason for confusion over the character of power relations and indigenous governance involves the Russian practice of seeking a chief (*toion*) as an intermediary between each village and the local Russian administration. Rather than using the existing political structure, the Russians both intentionally and unwittingly changed the nature of political rewards, attracting some into positions of access to European material goods and killing uncooperative or resistant leaders to prove Russian domination (see Black, 1992, with references). Prior to the Russian conquest, warriors were highly regarded for their bravery and success in battle. High-ranking leaders (*anayugak*) were fierce warriors as well as skilled traders and politicians. Their status was generated by their ability to attract followers, to accumulate wealth through the labor of those followers, and to garner reputations as both fierce warriors and generous hosts. Resistance of Russian invaders was a logical extension of Alutiiq political economy, and the most successful *anayugaks* initially had the most to lose in supporting Russian interests. After the Russian conquest, leaders' most valued characteristics (from the Russian perspective) were the ability to work as subordinates to Russian administrators and to assist in the pacification and taxation of the Alutiiq people. As a result, individuals who took advantage of alliance with the Russians were more successful at gaining the material advantages that benefited them in political positions in their native villages. The village chiefs (*toions*) instituted by Russians in place of traditional "bigmen" or "richmen" (*anayugak*) were often puppets of the Russian administration, backed up in political competition by their privileged access to Russian resources that probably included force. Not surprising then, Davydov (1976:23) reports "the power of the chief of a village is very limited. The islanders frequently give more respect and obedience to some rich person or good hunter."

Along the Northwest Coast, leadership was also fairly confined. De Laguna (1983:80) states, "the Europeans and Americans who met the Tlingit in the 18th century expected to encounter tribes organized under 'head chiefs.' As sea captains and officials, they were used to dealing with the heads of hierarchical bodies. When they did not find the chiefs of their expectations, they created them, consciously or unwittingly." As Ames (1995) suggests, with the exception of *power over* slaves, Northwest Coast chiefs really only had *power to* organize potlatches and other festivals, make war, and signal the opening and closing of fishing and hunting seasons. Their power was contingent on an appearance of generosity and an ability to unify large lineages. Disgruntled junior kinsmen could defect to more successful lineages, and potlatches provided opportunities for "richmen" and "richwomen" to compete for supporters (see also Ferguson, 1983, 1984).

The key to respect and leadership throughout southern Alaska (from the Northwest coast to the Aleutians) was inherited status at the top of a large kin group or lineage. Successful reproduction of status for lineage heads depended on the ability to invest their children with symbols of status through potlatches; this required the ability to produce large surpluses, negotiate wealth exchanges and political alliances with other elites, and engage in raids and warfare with enemies. Ultimately lineage leaders gained their support by creating a climate of advantage for junior kin. As Townsend (1980: 133–134) suggests, "members clung to the household not only because of kinship but also because of personal advantages which might accrue from such membership."

4.6. PROPERTY OWNERSHIP

Property ownership is generally an alien concept in small-scale, egalitarian societies; it is, however, an important, if not central, concept in complex societies. In important ways, legitimized rights to exclusively control things (e.g., land, goods, structures, people, symbols, stories, and ideas) lies at the core of what it means to be non-egalitarian. In studies of small-scale hunter-gatherer groups around the world (such as the !Kung, Ju/'hoansi, Hadza, and Inuit), ethnographers have been impressed not only by the lack of a concept for property, but by its active suppression (Biesele and Kxao Royal – /o/oo, 1999:208; Brody, 2001; Lee, 1979; Woodburn, 1979, 1980, 1982; see also Ingold, 1999). The ability to own something and exclude others from its use or consumption, to accumulate goods for personal purposes, to acknowledge superiority in the hunt or talent in a craft, all are perceived as great social dangers in small-scale egalitarian societies. Possible reasons for this cross-cultural distinction are detailed in Chapter 6. In sum, a pervasive ethic of sharing lies at the core of egalitarianism and property ownership is antithetical to uninhibited sharing.[2] Here we consider the extent to which the concept of property was integral to 18th century Alutiiq society.

Territorial ownership is poorly documented for the Alutiiq in comparison with peoples of the Northwest Coast area, although it is likely that it was significant. We find no unequivocal claims concerning the ownership of resource extraction locales close to villages, and while this may indicate a lesser emphasis on exclusive ownership among the Alutiiq, it may be better understood within the context of the violence Russians used when installing themselves on Kodiak. By taking hostages, removing male hunters from their family roles, introducing diseases, consolidating villages, and reorienting economic interaction to move goods directly between the villagers and the Russian forts, the Russian colonial officials quickly altered traditional patterns of resource territoriality.

Holmberg (1985:59–61) provides one reference to individual ownership of a resource extraction location. Prior to the Russian conquest, Arsenti Aminak's father received the uninhabited island of Chirikov from a relative who discovered it while lost in a storm. Chirikov Island was renowned as a source of sea mammals and valuable amber. While elements of this account may be true, the story may also describe a myth developed over time to justify control over a valuable piece of property and to explain an ancestor's wealth. Whether true in detail or not, this story shows that resource extraction locations could be controlled and that some individuals were considered legitimately qualified to own them. The relative who found the island and its riches could have claimed the property for herself or himself; instead the relative chose to "give" it to Aminak's father who subsequently became a very rich chief.

I define wealth as material goods with exchange value that can be accumulated; and, the accumulation is only meaningful in comparison to what other people accumulate. While wealth draws its significance from a climate of social competition, it often bears a close relationship to economic well-being, both through the potential to consume wealth directly (e.g., amassed food) and to leverage social and material assistance indirectly (e.g., through debt obligations or through trade). Compared to territoriality, accumulation of wealth is a well-documented aspect of the Alutiiq political economy. Given the relationship between territorial control and the production of wealth (e.g., the aforementioned control over the amber source of Chirikov Island), this evidence suggests the existence of pervasive territoriality despite its near invisibility in the ethnohistoric accounts. One of the most thorough descriptions of Kodiak Alutiiq wealth is given by Father Gideon, who visited Kodiak between AD 1804 and 1807:

Among the Aleuts [here meaning Kodiak Alutiiq; see Black, 1977:90, footnote 45] that person was deemed wealthy who, besides having the members of his family clad, possessed the following items: for himself, one sea otter parka, five *evrashechiei* [ground squirrel], five marmot (*tarbagan*) deer parkas, 5 bird skin (*urile*) parkas, 5 bird skin (*toporok*) parkas, a twenty-oar *baidara* [*anyak* or open skin boat], three *baidarkas* [*qayaks* or kayaks], many *lavtaks*, that is sea lion and seal skins from which the hair has been removed by steaming, plenty of whale meat, two sea lion bladders of whale oil, two casks for water, 5 *kaliukakh* (baskets made of bent wood . . .) filled by *shiksha* [crow berry] in oil and the same quantity with *sarana* [Kamchatka lily], a sea lion bladder filled with red fish caviar, 10 large bundles of *yukola* [dried fish, in this case salmon] and the same quantities of halibut and cod yukola. In addition, leaving aside darts and other tools and utensils which everyone needs, he should have for adornment the following items: Firstly, four or more amber pieces, which they sometimes find themselves on the island of Ukamok [Chirikov]; they value the light colored, almost white variety much more than the yellow amber; secondly, they had to have *aiminak* (*sukli* . . . [spiral shell-like ornaments, possibly dentalium shell]) that is thin small bones, resembling a worm in length more or less and a half *vershok* [1 vershok is approximately 2 inches]; thirdly, they had to have *kuikak*, thin small plates, about one vershok in length and width, cut out from the very middle of the mother-of-pearl-bearing mollusk; and fourthly, a substantial quantity of various beads. (Gideon in Black, 1977:91)

Owning a kayak made a man "wealthy" because he could then use the resources of the sea and coast without depending on others (Davydov, 1976:12). More significantly, wealth was a product of the labor one could control for generating surplus to invest in winter festivities, warfare, and trade. Women's labor was vital for the production of skin boat covers, utilitarian clothing, and prestige garments. Bird skin/feather coats made of murre, puffin, and guillemot were highly prized. The finest of these were made from the necks of guillemots killed in the spring, using 140 birds to manufacture a single coat (Davydov, 1976:3–4). From a social perspective, the most significant dimension of property ownership was the practice of slave ownership, discussed later in this chapter.

Prior to Russian contact, property ownership (territoriality and accumulated wealth) was clearly an important component of Alutiiq political ideology. Wealth differentials existed and "rich" people were able to claim control over certain resource extraction locations. Possibly documentary evidence understates the degree of economic differentiation of the pre-contact Alutiiq. Almost everywhere else around the North Pacific Rim, from the eastern Aleutians to the southern Northwest Coast, elites were able to claim access to productive resource locations. These claims were based on the ability to defend resource locations and punish unsanctioned use.

4.7. TRADE

Many accounts of emergent complexity include long-distance trade as a variable. Though trade is common among most societies, differences exist between small-scale hunting and gathering societies and more complex groups in at least two dimensions. The first distinguishes societies with egocentric trade networks (in which most individuals are able to forge their own alliances with trading partners [Wiessner, 1977]), from those with socio-centric trade networks (in which a subset of the population has exclusive rights for trade with their counterparts among other groups [Blanton et al. 1996; Clark and Blake 1994]). The second dimension divides societies into those where exchange is primarily a form of "risk-buffering" in the subsistence economy (e.g., Weissner, 1982) and those where the economy is based significantly on the exchange of non-consumable valuables (a "prestige economy"). Prestige economies emerge from the development of property ownership and competitive displays of wealth. Trade in non-consumable prestige-goods contributes to a self-reinforcing cycle, where increasing amounts of wealth and displays of wealth are avenues of social mobility.

Early accounts often describe the Alutiiq as skilled traders, knowledgeable in the economics of supply and demand (Davydov, 1976:4). They maintained a brisk trade in exotic goods, acquiring Bering Sea ivory from trade contacts in the Aleutian Islands and the Alaska Peninsula. Sometimes this ivory was traded along with sea mammal oil to Prince William Sound and the northern Northwest Coast in return for pearl-like shells (probably dentalium; Davydov, 1976:3) and craft goods (see Burch, 1988a:236–237; de Laguna, 1990:209). Amber (found occasionally on Kodiak beaches, but more frequently on the isolated island of Chirikov), was highly valued (Gideon in Black, 1977:91; Holmberg, 1985:38) and constituted an important export for trade with the natives of the Alaska Peninsula (Burch, 1988a:236–237), from whom the Alutiiq obtained spun caribou-gut thread and caribou-skin parkas. Bear parkas were acquired in trade with peoples of the Kenai Peninsula (Tenaina) and Prince William Sound (Chugach) (Davydov, 1976:4).

While ethnohistoric records reflect nothing on the issue of Alutiiq trading politics, ethnographic patterns on the Northwest Coast suggest that trade was initiated by wealthy individuals who were able to free themselves of productive tasks during the summer months because of large numbers of junior kin and slaves (see Mitchell and Donald, 1988:324–328). Establishing peaceful partnerships with like-minded elites in neighboring regions, these individuals would capitalize on the profit and prestige generated by access to exotic trade goods. According to de Laguna (1990:208), Tlingit traders operated through "gift-exchanges" with partners in different clans. Trade along the Northwest coast was vigorous and cross-cut "tribal" affiliations and language groups. Traded commodities included dentalium, blankets, walrus ivory and hide, canoes, carvings, and slaves (de Laguna, 1990:209), as well as various subsistence commodities and raw materials (Mitchell and Donald, 1988:326). Alutiiq traders similarly exchanged with each other and with partners on adjacent mainland regions.

Trade with mainland partners involved dangerous ocean crossings. This enabled powerful individuals some control over "crossing points." The physical isolation of Kodiak set limits on the number of individuals with access to long-distance trade wealth and allies. These individuals needed the ability to fund trade expeditions (including the costs of boat building supplies and labor for construction), amass large factions of warriors to defend trading parties (because war and trade would have been strategic alternatives in many inter-group interactions), and impose sanctions on local competitors. As reported earlier, the conditions of wealth include the means and products of vigorous inter-coastal trade networks (e.g., *anyaks*—open skin boats, kayaks, and exotic goods). In the Russian-American period, Tlingit traders fiercely guarded the "grease-trails" to the interior, trading with interior Athabascan groups for copper and other goods, while preventing these groups access to the coast where they might trade directly with others (de Laguna, 1990:208; Legros, 1985). Likely, trade between Kodiak and the mainland was similar because the tools of trade (in this case large skin boats), were preferentially available to rich individuals. With larger kin networks, these rich individuals likely had greater control over labor needed for hunting sea mammals, preparing and sewing skin covers, and defending trade missions.[3]

4.8. WARFARE

Warfare occupies an ambiguous position in anthropological thought about emerging complexity (Ferguson, 1983, 1984). Interpersonal violence characterizes all societies. Organized aggression and defense, however, require a degree of social cohesion that is lacking when disputes can be resolved through relocation. Endemic warfare is characterized by ongoing, regularized, and often ritualized military engagements between established enemy groups. Such regular cycles of aggression may often have less to do with resolving political disputes than with reinforcing dominance, intimidating rivals, and encouraging political support among allies and subordinates. Thus, endemic warfare is a condition of political competition under conditions of social inequality. Warfare and trade are important dimensions of inter-group political dynamics, promoted in part by intra-group social and economic organization.

Intertribal warfare has been well documented along the Northwest Coast (e.g., de Laguna, 1983, 1990; Emmons, 1991; Mitchell, 1983). Warfare is also evident around the Gulf of Alaska, from the Aleutians to Kodiak, and into Prince William Sound (e.g., Maschner and Reedy-Maschner, 1998; Townsend, 1980).[4] Before the Russians arrived, the Alutiiq were "constantly at war amongst themselves" (Davydov, 1976:8). While warfare apparently did occur between Kodiak Alutiiq groups, it was even more common for Kodiak Alutiiq to skirmish with the natives of the Kenai (Chugach, Alutiiq, and Tenaina) and Alaska Peninsulas (Aglegmiut) (Davydov, 1976:8). Occasionally longer-distance attacks were mounted on Fox Aleut and Tlingit groups (Davydov, 1976:22–23). Because warring raids were common between the Fox Aleuts and Kodiak Alutiiq, both groups considered each other unconditional enemies and had war-captive slaves from each other's group when the Russians first arrived (Holmberg, 1985:57–59).

Davydov (1976:8) claims that Alutiiq conducted warfare because of their desire for wealth and slaves, deep-seated hatred grounded in revenge for prior insult, injury, or attack, and hunger. Warriors included both men and women (as among the Tlingit: de Laguna, 1983:74) and death was considered more honorable than capture and the resulting torture (Davydov, 1976:8). During the summer raiding season, villages would move to defensive islets, "lonely and inaccessible rocks from which people were lowered by ropes even to go and fetch water" (Davydov, 1976:14). It was on such a "rock" that Shelikhov conquered the people

Figure 4.2. Awa'uq Refuge Rock, scene of Shelikhov's devastating attack of the Aluttiq in AD 1784. Photo by author.

of southeast Kodiak in August of AD 1784 (see Chapter 3; Black, 1992:172; Holmberg, 1985:59).

4.9. SLAVERY

Slave capture motivated warfare across the North Pacific Rim (e.g., Ames, 2001; Donald, 1983; Mitchell, 1984; Townsend, 1983). Used as laborers and as tradable commodities, slaves increased the productivity of corporate kin groups, freed elites for non-foraging activities (trading and raiding) during the summer months, and could be converted themselves into other forms of wealth through ransom and trade. Mitchell (1984:table 1) provides equivalence values for a slave in trade: Tlingit (2 sea otter skins); Tsimshian (15 elk skins); Haida (150–200 blankets). On Kodiak, a slave could be exchanged for 20 (dentalium) shells or their equivalent (Merck, 1980:109).

Forming a separate class, slaves were forced to labor, deprived of privilege, and could be traded or killed at will (e.g., to bring prestige to their owners). Children of slaves also became slaves. Even if a slave was freed, her or his children remained stigmatized for generations (de Laguna, 1983:75). Orphans were raised as slaves until at least adulthood (Davydov, 1976:12; Merck, 1980:109; see also Emmons, 1991:46; Townsend, 1983:124). Most slaves captured in war were women and children because men were usually killed, sometimes after gruesome torture (Holmberg, 1985:36). When kept alive, men were the most highly valued slaves. Most male slaves were captured as children or acquired through trade (Holmberg, 1985:36).

Among many of the Northwest Coast groups, slaves were sometimes executed at the funeral of a wealthy lineage head (e.g., Emmons, 1991:271; Ruyle, 1973 with references). Based on his ethnographic interviews in AD 1790, Merck (1980:108) notes that "rich" Alutiiq, like Tlingit, were sometimes buried with slaves who were killed to accompany them into the afterlife. Davydov (1976:18), who visited Kodiak almost fifteen years later, denies that slaves were put to death in this fashion. Merck's earlier observation holds more weight since it was made only six years after conquest and before the Russian-American Company and Orthodox priests had significantly influenced Alutiiq tradition and ideology. The Orthodox mission, begun on Kodiak in AD 1794, brought the Alutiiq people relief from the oppressive treatment of the fur-hungry Russian-American Company. Wishing not to diminish their "victim status" in the eyes of the priests, Alutiiq informants would have been disinclined to emphasize their own harsh treatment of slaves.

4.10. SUMMARY

The Kodiak ethnohistoric record provides sufficient detail to reconstruct a picture of contact era Alutiiq socio-political organization. While the political structure was modified more rapidly after contact than the subsistence economy, the available data suggest that the Alutiiq had social, economic, and political organizations similar in many ways to their Northwest Coast and Aleut neighbors. Alutiiq society of the 18th century was complex relative to egalitarian hunter-gatherer bands documented around the world. Factors associated with this complexity include:

- an emphasis on hunting and gathering (primarily maritime)
- feasting for demonstration of generosity, competitive display, and ennobling of high-ranking individuals
- securing status and power through achievement and inheritance
- property ownership
- wealth accumulation
- prestige trade
- endemic warfare
- institutionalized slavery

The Alutiiq were full participants in a North Pacific-wide cultural pattern noted for its relative complexity. The rest of this book examines how this cultural pattern evolved on the Kodiak Archipelago.

Chapter 5

Colonization

5.1. BACKGROUND

We have already looked at the ethnohistoric record to establish the character of social and political complexity at the end of the archaeological sequence. Here our goal is to set a baseline, or starting point, at the beginning of that sequence. All starting and ending points in historical/evolutionary analysis are arbitrary. Ours are defined by the earliest evidence of occupation on the Kodiak Archipelago and its precedents around the North Pacific Rim (this chapter), and by the Russian conquest of the archipelago in the later half of the 18th century (last chapter). So bracketed, in the next chapter we will ask the central question of this work—by what processes of change did Kodiak hunting and gathering societies evolve into the ethnohistorically documented pattern? This chapter examines the background of, as well as the arguments and evidence for, the initial colonization of the Kodiak Archipelago. This chapter also looks at the character of early to mid-Holocene coastal hunter-gatherer lifeways around the coast of the Gulf of Alaska.

Detailed examination of the archaeological record of the early Holocene is beyond the scope of this book. It is sufficient to note, however, that most evidence points to the movement of peoples from the Alaskan interior to the coast sometime between 9500 and 7500 BP (Yesner, 1998). These people appear to have brought interior technologies to the coast, where these technologies gradually developed a more coastal character (Jordan, 1992). Some have suggested coastal migration from Asia as an alternative source for some of these assemblages (e.g., Laughlin and Aigner, 1966; see also Aigner, 1970, 1976a; Aigner and Del Bene, 1982; Laughlin, 1975).

5.2. MARITIME ADAPTATION

Prerequisites for Kodiak colonization included skills in boat construction and navigation as well as an economy focused on maritime resource exploitation.

Questions surrounding the timing and processes that give rise to maritime adaptations in various parts of the world have received considerable attention over the last 30 years; answers, however, remain elusive (Erlandson, 2001; Fitzhugh, 1972, 1975; McCartney, 1974; Workman and McCartney, 1998; Yesner, 1981). As Erlandson (2001) notes, aquatic and maritime adaptations are prominently featured in models of the evolution of social complexity (e.g., Ames and Maschner, 1999; Binford, 1968; Matson and Coupland, 1995) despite ambiguity in both theoretical and empirical referents (Erlandson, 2001, Erlandson and Moss, 2001). Maritime adaptations involve the overlapping, but distinct, dimensions of seafood subsistence capabilities (from littoral to open ocean), seafaring technologies and skills, and coastal settlement. Evidence for maritime adaptations of all three dimensions extends well back into the Pleistocene (Erlandson, 2001), and these adaptive strategies probably developed independently several times on separate continents and shorelines. Here I will only consider the North Pacific case, recognizing that maritime resource use and seafaring have been identified in other parts of the world. The earliest coastal sites on the North Pacific Rim are currently located along the Northwest Coast of North America (Figure 5.1). These include several sites from British Columbia and Southeast Alaska that date from 10,000 bp forward. All indicate the use of marine resources and probably boat navigation from earliest times (Ames, 1998:79).

A seminal issue of *Arctic Anthropology* (1998, v. 35, no. 1) documents the accumulated evidence for the development of maritime adaptations around the North Pacific. Articles summarize developments from northern California to northern Japan, including the Bering Sea and Sea of Okhotsk, and range from the earliest

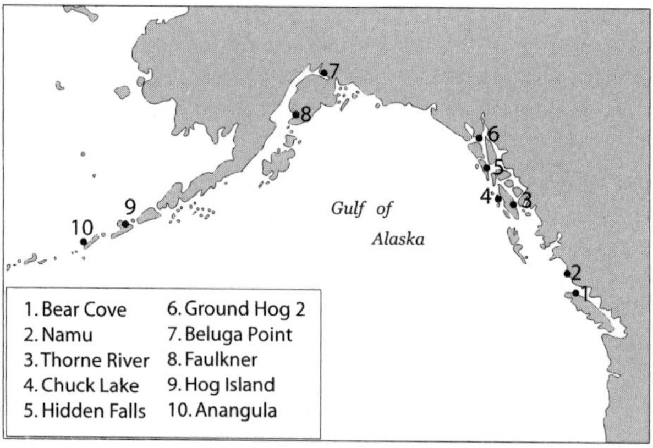

Figure 5.1. Early Holocene archaeological sites around the North Pacific Rim.

coastal site components through the Late Holocene. Sites of the Northwest Coast region provide the best evidence for early sites with some degree of maritime subsistence adaptations. These include On Your Knees Cave of 9700 to 9900 bp (Dixon et al., 1997), Namu beginning 9720 bp (Carlson, 1998), Chuck Lake Locality 1 at 8200 bp (Ackerman, 1988; Ackerman et al., 1985; Carlson, 1998; Moss, 1998; Okada et al., 1989), and Bear Cove beginning 8200 bp (Carlson, 1998:27). Other early sites without faunal preservation that presumably supported maritime hunter-gatherer groups include Kasta and Lawn Point in the Queen Charlotte Islands dating between 8000–6000 bp (Ames, 1998; Fladmark et al., 1990), Thorne River on Prince of Wales Island (Holmes et al., 1989), Hidden Falls on Baranoff Island (Davis, 1989, 1996) and Ground Hog Bay 2 on Icy Strait (Ackerman, 1968; Moss, 1998).

Early Holocene coastal sites are rare north of Southeast Alaska. Beluga Point in inner Cook Inlet is associated typologically with early Holocene Paleoarctic sites in the Alaskan interior (Reger, 1977, 1981). Beluga Point's location on the narrow coast of Turnagain Arm suggests an estuarine adaptation (Reger, 1998:162). Sites on the Kenai and Alaska Peninsulas from 9000–7000 bp apparently represent interior and riverine occupations. These include the Round Mountain site on the Kenai River, Ugashik Narrows on the upper Ugashik River (Henn, 1978), Koggiung near the mouth of the Kvichak River (Dumond, 1981), and possibly sites from Lake Illiamna and the Chignik river region (Dumond, 1998:190). The Faulkner site in Kachemak Bay has been dated to the early Holocene and may be another early maritime site (Table 5.1; Janet Klein, personal communication, 2002). Being located well into an island chain with few (current and probably ancient) terrestrial options, sites from the eastern Aleutian Islands are also strongly suggestive of maritime subsistence orientation. These include the Anangula Blade site and the Hog Island sites (Russian Spruce site and Oiled Blade site) of ca 8000 bp (Dumond and Knecht, 2001; Knecht and Davis, 2001; Knecht, Davis, and Carver, 2001).

While limited knowledge of complex sea-level histories hinders the ability to make inferences, island sites, such as On Your Knees Cave, Chuck Lake, Thorne River, Hidden Falls, and especially Anangula, suggest ocean-faring as far back as 10,000–8,000 bp (Black, 1974). Thus it is reasonable to believe that early Holocene occupants of the North Pacific were able to make and use boats. There is, however, no necessary reason to expect boat use or open ocean navigation in all cases, a point emphasized by Ames (1998:79). Certainly by 8000 bp, when Anangula Tradition sites were established, seafaring supported occasional inter-island navigation and probably some form of boat-based hunting and fishing. Engaging in both activities, inhabitants of the Eastern Aleutians were pre-adapted for the longer open ocean crossings (across the Shelikof Strait) and the focus on marine and littoral foods that would sustain Kodiak inhabitants through the millennia to follow (Figure 5.2).

Figure 5.2. Two hatch kayak and residents on beach in front of the village of Old Harbor, 1888–1889. Photo courtesy of the U.S. National Archives (22-FFA-268).

5.3. EVIDENCE FOR THE EARLIEST OCCUPANTS OF KODIAK

Assuming that early Kodiak sites preserve clues to the origin of colonizing ancestors, comparison of the archaeological record from these and earlier sites in adjacent regions may help identify the source of colonizing populations. Unfortunately, early Holocene sites from coastal South Alaska are rare and poorly preserved. A second difficulty emerges from the fact that, from Siberia to Alaska, early Holocene technologies were largely similar, reducing the specificity of typological comparisons. Additionally, many aspects of technological variation may only relate to the specific functional requirements of unique micro-environments and available raw materials (see below).

The Alaska Peninsula, perhaps hosting an expansion from the Eastern Aleutians, remains one of two likely departure points for Kodiak's colonization. Possibly the archipelago's first occupants came from this region, bringing with them technology and subsistence adaptations that they had developed and modified there. Alternatively, colonists may have come from the Kenai Peninsula, by way of the Barren Islands. Unfortunately the archaeological record of outer Kenai Peninsula has an extremely shallow time depth, limited by violent tectonic reconfiguration of the coastline and glacial advance during the Little Ice Age (Crowell and Mann, 1996). The Faulkner site, could represent a source for Kodiak colonists, even though it has just one radiocarbon date only 500 years older than the earliest reliable Kodiak dates (Table 5.1). Other sites with potential clues about the Kenai ancestry of Kodiak's first colonists may eventually be found. Evidence presented here, though, suggests that the Alaska Peninsula is the most likely staging area for migration to Kodiak.

Early radiocarbon dates from sites of the Aleutians, the mainland coast of the Gulf of Alaska, and Kodiak (Table 5.1) suggest that the colonization of the Kodiak Archipelago proceeded from the Aleutian and Alaska Peninsula region. This suggestion is at least partially supported by the typological similarity between the early sites of the Anangula Tradition, the Upper Alaska Peninsula Takli Alder phase (G. Clark, 1977; Dumond, 1984b, 1987; Dumond and Knecht, 2001; McCartney, 1984), and Kodiak's early Ocean Bay I phase (e.g., the lowest occupation of the Tanginak Spring site: see Chapter 7 and Steffian et al., n.d.). Recent investigations of the Pacific coast of Katmai National Park identified several sites with assemblages similar to those of early Ocean Bay I (e.g., blades and red ocher grinders; Patrick. Saltonstall, personal communication, 1995; Mann and Crowell, 1996). These sites are comparable in age and character to the Takli Alder phase sites from Takli Island previously reported for the area (G. Clark, 1977). Yet, the earliest dates reported for this region post-date those from Kodiak by approximately 500 years. Poor site preservation or recovery may account for the discrepancy (Crowell and Mann, 1996).

Examination of the lithic raw materials used by early Kodiak people strengthens the inference of affiliation between the earliest Kodiak sites and those of the Alaska Peninsula and eastern Aleutian area. Though a careful analysis of raw materials from different early Kodiak sites is beyond the scope of this book, raw materials from the Tanginak Spring site (KOD 481) suggest a connection to the volcanic regions of the Alaska Peninsula and/or the Aleutian Islands, rather than the Kenai Peninsula (see below).

5.3.1. Estimating the Timing of Colonization

We can assume that the initial peopling of Kodiak resulted from an expansion or relocation of coastal and maritime adapted groups living around the Gulf of Alaska in the early Holocene (10,000 to 7000 BP). It is also reasonable to expect that the earliest colonists to Kodiak had some degree of technological, economic, and cultural affinities with their source populations. That leaves us to question the timing of colonization (including the processes underlying it) and its direction (identifying cultural links between antecedents and early Kodiak assemblages).

It is rarely possible to unequivocally determine the timing of an area's human colonization on the basis of archaeological remains (Cherry, 1984, 1992). Securely dated archaeological deposits in a region, at best, represent a *terminus ad quem*

Table 5.1. Early radiocarbon dates for archaeological sites in regions surrounding the Kodiak Archipelago. (The oldest Kodiak dates are included for comparison.)

Region	Site/Location	Uncalibrated date (bp)	Calibrated age range (BP)	References
Aleutian Islands:	Anangula Blade	various	8750–8250[a]	Aigner & Del Bene 1982: figure 2.8b
Lower Alaska Peninsula:	XCB 029	not available	5500–3500[b]	Maschner 1996
	Johnson's Point	5460 ± 220	6739–5739[b]	Mills 1994; Okada et al. 1984
Upper Alaska Peninsula/ Ugashik Region:	UGA 001	8995 ± 295	11067–9434[c]	Mills 1994; Henn 1978
		8425 ± 115	9595–9092[c]	
		7675 ± 260	9237–7961[c]	
Upper Alaska Peninsula/ Naknek Region:	NAK 018	7895 ± 95	9012–8431[c]	Mills 1994; Dumond 1981
		7765 ± 95	8976–8883[c]	
		7475 ± 60	8390–8169[c]	
Shumagin Islands:	XSB 029	not available	4510 (MASCA corrected date)[b]	Winslow & Johnson 1989: Table 2
Alaska Peninsula (Pacific Coast):	XMK 006a Kukak Bay	5830 ± 120	6980–6356[b]	Clark 1977; Mills 1994
	XMK 030	5990 ± 85	7147–6640[c]	Dekin et al. 1993
	Takli Island	5815 ± 85	6847–6409[c]	
	XMK 018 Takli site	5650 ± 115	6729–6199[b]	G. Clark 1977; Mills 1994
	XMK 106	5630 ± 60	6530–6300[b]	Crowell & Mann 1996: Table 2
	XMK 056	5060 ± 90	5970–5610[b]	Crowell & Mann 1996: Table 2
	Russian Anchorage	4940 ± 60	5870–5590[b]	
Cook Inlet	SEW 214	4795 ± 165	5919–5049[b]	Mills 1994; Reger and Boraas 1993
		4640 ± 150	5724–4869[b]	

Kachemak Bay:	SEL 009 Faulkner site	7150 ± 100	8172–7754[c]	Zollars and Klein 2002
		7070 ± 90	8108–7682[c]	
		6830 ± 90	7835–7509[c]	
		6790 ± 70	7784–7509[c]	
		6670 ± 90	7677–7423[c]	
	SEL 245 Silva Site	4440 ± 90	5319–4849[b]	Mills 1994; Reger and Boraas 1993; Workman 1993
Outer Kenai Peninsula & Prince William Sound:	SEW 430	4440 ± 70	5299–4859[b]	Haggarty et al. 1991; Mills 1994
	SEL 188 McArthur Pass	1960 ± 140	2310–1347[c]	Crowell & Mann 1995:Table 1
		1710 ± 120		
Kodiak:	KOD 363 Rice Ridge	6080 ± 90	7229–6723[c]	Knecht 1995:33
	KOD 013 Zaimka	6390 ± 70	7430–6163[c]	Amy Steffian, p.c. 2002
		6250 ± 70	7313–6950[c]	
		5530 ± 130	6635–5992[c]	
	KOD 324 Kizhuyak	6620 ± 60	7579–7374[b]	Crozier 1986, 1987; Mills 1994
	KOD 044 Crag Point	7790 ± 620	9988–7393[b]	Jordan 1992; Mills 1994
	KOD 224 Uganik	6220 ± 70	7274–6897[b]	Clark 1984; Mills 1994
	KOD 481 Tanginak Spring	6580 ± 60	7574–7340[c]	Table 7.1
		6250 ± 70	7314–6951[c]	
	KOD 555	6340 ± 100	7395–7005[c]	Table 7.1

[a] It is unclear if this is a calibrated or uncalibrated age range from the publications consulted.
[b] Calibration range derived from published source.
[c] Calibration range derived by author, using Calib 4.3 (Stuiver et al. 1998a, 1998b; Stuiver and Braziunas 1993).

Figure 5.3. The environment of the Tanginak Spring site (KOD 481). Dr. Donald Clark crosses the salt marsh in the foreground. Photo by author.

(latest possible) date for colonization of that region. Earlier assemblages always may remain undiscovered or lost to the ravages of time. In some unique cases, the paleontological and paleobotanical records can provide regional signals of human colonization due to dramatic effects of human-induced environmental modification. This is particularly true for environments first colonized by agricultural peoples. Collected from non-archaeological contexts, pollen, charcoal, and faunal records often show rapid change following the colonization of islands (e.g., Burney, 1997; Dewar, 1997; Kirch and Ellison, 1994; Steadman, 1989). Introduction of domesticated plants and animals, slash-and-burn clearing, or even hunting pressure can lead to the extinction and replacement of indigenous species. Unfortunately, the paleoenvironmental signals corresponding with island colonization by subarctic maritime hunting and gathering peoples—as on Kodiak—have yet to be identified. Thus, we are forced to focus exclusively on the archaeological record to estimate the timing of colonization. A logical place to start, then, is the current chronology of early Kodiak archaeological sites.

All of the oldest reported Kodiak dates (Table 5.1), fall within two standard deviations of 7400 BP (calibrated) or younger. With the exception of the Crag Point date, the average probability (calibration intercept) for the oldest dates of each of the six sites lies between 7100 and 7500 BP. The accuracy of the seventh date (7790 ± 620 bp, from the Crag Point site) is disputed (Knecht, 1995; Mills, 1994:table 1d; Steffian et al., n.d.). Though Jordan (1992) used this date to suggest a Paleoarctic occupation on Kodiak, the date is problematic because of its large standard error.[1] Because the younger end of the probability range for this date overlaps the range of the other dates reported here, the Crag Point sample may not be significantly older than the others.

The tight clustering of early dates suggests a pattern of colonization and rapid population expansion (through immigration or reproductive growth). The relatively sharp temporal boundary also could result from a regional taphonomic process that selectively removed or concealed evidence of sites older than 7500 BP. The rising post-glacial sea level is the most likely process that could have obscured early Holocene coastal sites. As illustrated in Figure 5.4, global (eustatic) sea level rise began to stabilize around 3 000 bp, well after the first Ocean Bay I sites appeared around the Kodiak Archipelago. But coastal sea level history in this region, unfortunately, is complicated by isostatic and tectonic processes. Mann and colleagues (1998:120) suggest that Southeast Kodiak may have a similar profile to that of Hecate Strait in the Queen Charlotte Islands (Figure 5.4). If so, we would expect the archaeological record to be preserved somewhat farther back than under pure eustatic conditions. Tectonic processes in this region favor coastal

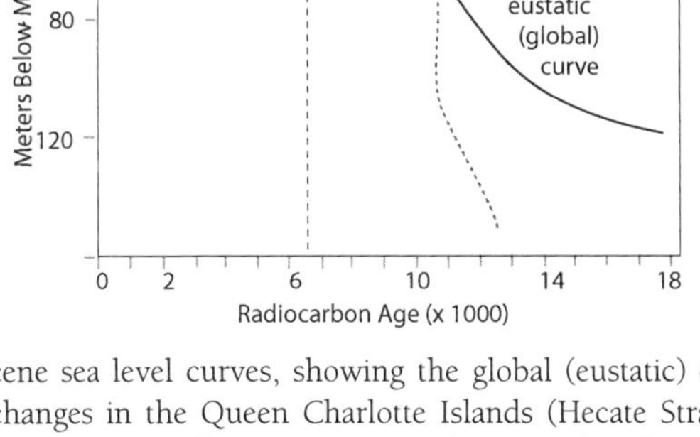

Figure 5.4. Holocene sea level curves, showing the global (eustatic) estimate and local relative sea level changes in the Queen Charlotte Islands (Hecate Strait) and Southeast Alaska (Icy Strait). Mann and colleagues (1998:120) expect that the Hecate Strait curve could approximate the Kodiak sea level history. (Redrawn from Mann et al., 1998; data from Bard et al., 1996; Fairbanks, 1989; Josenhans, 1997; Mann et al., 1998). Variation in curves is due to differential effects of glacio-isostatic load and rebound conditions on different parts of the shelf margin with different glacial loading histories. Tectonic processes in the Kodiak region complicate the sea level picture further.

emergence, at least in the late Holocene (Fitzhugh, 1996:264–277; Gilpin, 1995), and would serve also to extend the archaeological record further back in time. The presence of a substantial archaeological record prior to stabilization may be explained by this complex process. But even if future research shows that relative sea level stabilized roughly 7500 years ago, we still could not rule out the possibility that people colonized the archipelago at roughly the same time.

As it is currently impossible to disprove the claim that inundation has removed archaeological sites older than 7500 to 8000 BP, we are left to ponder the significance of negative evidence. An alternative involves developing testable hypotheses about the positive archaeological evidence we could expect for colonization and comparing those expectations to a truncated sequence of occupational continuity in the earliest assemblages already discovered. Ongoing research addresses this problem using raw material data from the Tanginak Spring site (KOD 481). While detailed analysis will be published elsewhere, the operational principle is as follows.

Colonizing populations from off-archipelago would likely arrive without detailed knowledge of archipelago-wide raw material distributions and their relative qualities. They would continue to use high quality materials from their prior homelands to the extent that transported supplies could be maintained through careful curation, return travel and/or trade—and to the extent that newly discovered alternatives are technologically inferior or otherwise considered of lower quality. With time, quality sources in the new territory—should they exist—would be discovered and replace more costly alternatives from farther afield.

The raw material used for stone tools at the Tanginak Spring site provides just such a clue about the antiquity of colonization of Kodiak from the mainland and subsequent technological innovation and change on Kodiak. It appears that large blades (macroblades), or the raw material from which they were made, originated on the Alaska Peninsula or the Aleutians, where fine-grained volcanic rocks are common.[2] The macroblades from the basal component at this site were made of basalt, silicified slate, and obsidian. These may have arrived at the site in blade form, as no macroblade cores have been found. In comparison, microblades and microcores were typically made of a lower quality, regionally-derived cherts. Between the time of the first occupation and subsequent reoccupations, the population appears to have lost access to, or interest in, off-archipelago sources of raw materials as well as macroblades. Figure 5.5b illustrates the raw material composition of blades (macro to micro) from the 1994 "Trench A" excavation grouped by stratigraphic zone (see Chapter 7). If the site's occupants were in the process of adapting to a newly colonized archipelago, we would expect increased use of red chert at the expense of fine-grained igneous (non-archipelago) rocks. This is just what the blade data indicate. A larger sample of formal tools from "Trench A" shows a less definitive pattern but remains consistent overall with the colonization hypothesis.[3] Because of its close proximity, Tanginak chert

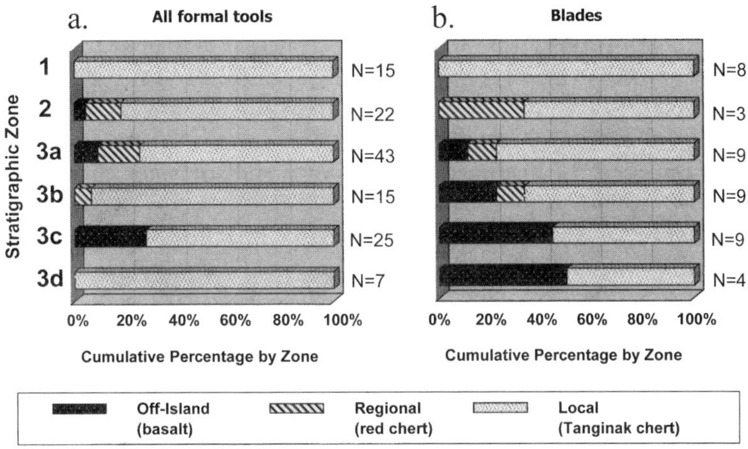

Figure 5.5. Frequencies of select raw material types by stratigraphic zone in the Trench A excavations at the Tanginak Spring site (KOD 481): (a.) all formal tool classes; (b.) microblades and macroblades only.

was always easy to get and could be used expediently with little conservation. For this reason it dominates the proportional measures through the occupational sequence.

Transitions in technology, within and following the occupation of the Tanginak Spring site, may relate to a change in the focus of adaptation, with macroblades becoming less essential than microblades, bifacial points and knives. On Kodiak, poor raw materials may have eventually forced this change in stone tool technology. Bifacial flaking is more forgiving of imperfection in raw materials than blade manufacture; possibly, this explains the increase in bifacial tool production at the expense of core and blade tools. Within two or three thousand years of experimentation on Kodiak, people seemed to have settled on slate grinding as the most effective and economic technology given the archipelago's available raw materials (Clark, 1980, 1982). The only evidence of slate grinding in the Tanginak Spring assemblage comes from ground slate rods.

Assuming hazardous journeys across the dangerous Shelikof Strait (between Kodiak and the Alaska mainland) were avoided when possible, the shift from basalt and obsidian in the lowest levels of the Tanginak Spring site to red chert in upper levels suggests that early Tanginak occupants were recent arrivals. Apparently they did not have a large social network for sharing information and resources around the archipelago. In addition they either curated lithic materials carried over during colonization from the Aleutian/Alaska Peninsular volcanic arc, or they made infrequent journeys to the Peninsula to trade or mine for raw materials, maintaining critical social networks in the process.

The Tanginak Spring site data need to be compared with findings from other early Kodiak occupations such as the Ocean Bay and Early Kachemak Zaimka Mound site in Woman's Bay (Steffian et al., 1998). Other sites, with different positions in respect to local and regional raw materials, are expected, of course, to have different colonization signatures. We would, for example, expect to find an earlier uptake of red chert and a more rapid fall-off of Alaska Peninsula materials closer to the red chert source. Furthermore, the island-wide pattern of raw material use would depend on where colonization occurred on the archipelago, how often colonization occurred, how long it took populations to explore/expand through the islands, and how often quality materials were traded between groups spread across the archipelago. Though many of these factors are archaeologically discernable, insufficient data are currently available to pursue this here.

From this information, we conclude: (1) that people arrived on Kodiak no later than 7500 BP, as evident in the radiocarbon dates from several sites (Table 5.1), (2) that earlier sites may have been lost or obscured through inundation due to local sea level rise, and (3) that the data from one site suggests a colonization not much before 7500 BP as an alternative to a missing early Holocene record. The earliest *explorers* of Kodiak likely preceded the earliest *colonists*, perhaps by many generations (see Spiess et al., 1998). The earliest attempts at colonization probably had been tentative and tenuous affairs. Isolation is unfavorable for demographic survival, and initial forays to the archipelago by small groups could easily have failed, possibly several times. Island biogeography suggests that local extinction is more common among small colonizing groups cut off from larger breeding populations (MacArthur and Wilson, 1967; MacArthur, 1972; Wobst, 1974). Successful colonization requires suitably large founding populations coupled with, or offset by, continued social and reproductive connections to populations in the source area. Apparently it took 5500 years for people to expand across the Aleutian Islands, moving from the Fox Islands in the east to the Near Islands in the west (contributions in Dumond, 2001a). This biogeographical principle underscores the possibility that the earliest available archaeological evidence from Kodiak represents colonization shortly before 7500 BP, despite the earlier evidence of coastal occupation in the eastern Aleutians and northern Northwest Coast. While Clark (1998) may have correctly anticipated human visitation to the archipelago prior to 8000 BP, there are good reasons to suggest that successful colonization was delayed. At least this hypothesis can be disproved relatively easily if older deposits are discovered. The claim for older sites yet to be discovered can never be sufficiently disproved.

While room remains for disagreement, I tentatively propose that Kodiak colonists arrived from the Alaska Peninsula region sometime between 8000 and 7500 BP. They came with technologies best suited to the raw materials available in the volcanic landscape of their former residence, and they were soon forced to adjust their technologies to those dictated or facilitated by the properties of local and regional materials. We now have a reasonable estimate for the timing and source of Kodiak human colonization. What evidence exists of the lifeways and social organization of these groups?

5.4. LIFEWAYS OF EARLY HOLOCENE COASTAL PEOPLES

Available data provide some evidence to characterize the economic, demographic, and social variability of early Holocene coastal occupations around the North Pacific Rim. Summarizing the evidence for the northern British Columbia coast, Ames (1998:80) argues that "the most parsimonious hypothesis is that the coast's earliest occupants were generalized foragers exploiting a range of habitats from terrestrial to littoral, if not marine." This pattern is consistent with most evidence from the Northwest Coast to the Alaska Peninsula, where sites of the earliest time periods are generally small, relatively thin, and rarely densely deposited. We have sufficient evidence to suspect economies fully dependent on maritime subsistence, seafaring, and some degree of open-water hunting and fishing technology by the time of Kodiak colonization.

Most coastal sites known from the early Holocene around the North Pacific are small, would not have supported large populations, and seem to have been abandoned frequently (although occupation duration and frequencies of reoccupation are currently unknown). In the next chapter, I present an ecological argument to support the expectation of periodic residential mobility among small maritime hunting and gathering peoples in the Early and Middle Holocene North Pacific. In most cases we would not expect large aggregations or permanent (sedentary) occupations. Evidence for at least a few semisubterranean pit structures and occupation over a large area at the Anangula Blade site (Aigner, 1976b; Aigner and Del Bene, 1982) has led Maschner (1996) to interpret this site as a 9000 year old aggregated and sedentary community. His interpretation is open to question. But there is no *a priori* reason why early Holocene populations could not aggregate and settle around isolated oases of bountiful productivity, such as those provided by the passes between the Aleutians. Under these very unique natural conditions, large population aggregations and sedentism can be expected. In most cases, however, long histories of economic and demographic change are necessary to establish the ecological structures that support aggregation and settlement.

The archaeological record provides less direct evidence for inferring aspects of social organization, political structure, and warfare. For the time being, we must rely on conceptual and analogical arguments to establish believable parameters. Most early Holocene North Pacific hunter-gathering sites are small and would have supported no more than three to five small family groups. Most site assemblages are thin and have limited artifact assemblages suggesting impermanence and periodic residential mobility. From these attributes, we infer that group membership would have been flexible, with families banding together for periods of time and shifting to join other families. Factors contributing to fluid group membership would include: changing ecological patterns that cause certain regions to gain and lose economic potential, formation and dissolution of social alliances, marriage, divorce, and relocation of children. This kind of social structure, with freedom of alliance and movement within the family (or at the individual level), is relatively intolerant of social inequity beyond dimensions of age and gender (Boehm, 1993; Kelly, 1995), although gender inequalities can be marked in the most egalitarian of societies (Collier and Rosaldo, 1981). While violence occurs in all societies (including relatively egalitarian hunter-gather societies), institutionalized aggression such as organized warfare is unlikely where group membership is fluid, egalitarianism prevails, and rewards for aggression (territory, production, labor) are transient. These expectations will be justified empirically in examinations of the earliest archaeological evidence from Kodiak presented in Chapters 8 and 9. The point is raised here to facilitate the modeling effort in the next chapter.

Chapter **6**

Modeling Emergent Complexity on the North Pacific

6.1. INTRODUCTION

In this chapter, I develop a model of the processes expected to generate systemic change in the lifestyles of Kodiak hunter-gatherers, drawing from the insights, assumptions, and biases of evolutionary ecology and related theoretical frameworks. This model illustrates how Kodiak social, technological, and economic life may have moved from its initial (or at least early Ocean Bay period) character to the more complex organization observed in ethnographic contexts.

6.1.1. Modeling

Models idealize and simplify conceptualizations of phenomena. While there are a number of different kinds of models developed for different purposes (e.g., maps are representational models), scientific models are tools developed to help formulate testable hypotheses (see Clarke, 1972; Winterhalder, 2002). In essence, they illustrate abstract theoretical concepts in a way that makes sense of specific events, processes, or states of being.

The following model is narrative, synthetic, and processual. As it examines a long-term developmental sequence, it is an evolutionary or historical model. Explanatory models in archaeology range from formal mathematical or graphical models (demonstrating deterministic or probabilistic relationships between specific classes of variables), to narrative models (relating variables to hypothetical scenarios). Formal models reflect a high degree of specification between theory and prediction; these models are most powerful when developed to predict the relationship between a small number of variables (e.g., which prey foods would provide the greatest caloric returns), and they tend to lose utility as the complexity of the modeling task increases (Winterhalder, 2002). (In this case modeling complexity refers to the number of variables and the potential kinds of dynamic feedback common in the systems of concern.) Ideally, it would be preferable to develop complex formal models to account for evolutionary questions. Even though complex systems theory (complex adaptive systems, non-linear systems theory, chaos theory) may be advancing along these lines (e.g., Bentley and Maschner, 2001; Bourque, 1995), this approach is still in the early stages of development. In the meantime, narrative models serve as effective tools for organizing theoretical expectations of complex phenomena; they are, indeed, the primary locus of concept development and often a prerequisite to more formal modeling.

In modeling the complex interplay of social, technological, economic, and political variability in a long-term evolutionary perspective, no single formal model is likely to generate useful predictions. However, a number of formal models have been developed to deal with the subsets of phenomena under consideration. A synthesis of a number of these formal models drawn from evolutionary (behavioral) ecology (see Kelly, 1995; contributions in Smith and Winterhalder, 1992) guides the following narrative. The behavioral ecological models utilized in this account include foraging theory models, such as the prey-selection/diet-breadth model and the patch-choice model (Kaplan and Hill, 1992), models of reproductive decision making with implications for demographic variability (Rogers, 1992; Voland, 1998), models of sharing and tolerated theft (Blurton-Jones, 1983, 1987b; Goland, 1991; Hawkes, 1992; Smith, 1988; Winterhalder, 1986, 1996), egalitarian versus non-egalitarian group formation (Boone, 1992; Smith, 1981; Vehrencamp, 1983), territoriality and scramble/context competition models (Boone, 1992; Dyson Hudson and Smith, 1978), and overlapping issues of risk-sensitivity in adaptive decision-making (Fitzhugh, 2001; Hawkes, 1990; Winterhalder, 1986; Winterhalder and Leslie, 2002; Winterhalder, Lu, and Tucker, 1999).

Behavioral ecological theory is defined by the assumption that individual behaviors are guided by cognitive decision-rules that have been shaped by a history of natural selection favoring reproductive fitness-seeking behaviors. Behavioral ecologists operate under the assumption that natural selection encourages individual self-interest over self-sacrificing altruism as a primary decision rule. Behavioral ecologists also recognize that fitness-optimal behaviors differ by environmental context, making flexible behavioral strategies critical to survival and reproduction (the fundamental principles of adaptation). Importantly, cognitive and behavioral flexibility foster the ability to solve problems and to react to circumstances in a way that changes those circumstances. Through the emergence of symbolic communication, humans have developed a capacity to adapt and change environments in much more complex ways than other animals. Permeating every aspect of adaptive decision-making, language has, in some ways, enabled the emergence of an entirely new level of environment (culture). In efforts to understand human behavior, anthropologists often focus on the cultural environment and sometimes neglect the biological and ecological foundations of behavior. But living in a culturally structured environment does not release humans from the influence of natural selection on cognitive capacities; humans continue to pursue goals that enhance the probability of survival and successful reproduction within the context of the total environment (i.e. physical, social, and cultural). Interested readers are referred to the seminal literature in human behavioral ecology for details and justifications of the approach (e.g., Boone and Smith, 1998; Smith and Winterhalder, 1992).

I find it useful to view social evolution as a process in which human propensities for survival and reproduction condition individual motivations and behaviors in response to specific environmental circumstances. These motivations and behaviors combine to produce social organization and propel social and cultural changes. The specific strategies individuals use to adjust to perceived adaptive problems will be conditioned by their position in the socioecological (and ideological) environment. In any specific case, then, it is the socioecological environment that must be modeled to locate the impetus for cultural change. When we know the symbolic/ideological contexts in which people make adaptive decisions, we expect to better understand how decisions are adaptively designed, and when they are not. More proximate models make use of this logic by supplying conscious or culturally-driven motivations for behavior (e.g., desire for social acceptance, perpetuation of tradition, wealth, prestige, security, survival, power, etc.).

While the assumptions and mechanisms underlying the following model are drawn from behavioral ecology, the form of the model is processual in its attempt to consider the dynamic development of long-term change in human socio-cultural organization. In this regard, I follow an established tradition in ecological archaeology tracing back to the early work of Binford (1968), Flannery (1972), Ford (1977), and Thomas (1983). The following model is non-traditional in that it uses formal behavioral ecological models to guide the construction of the processual narrative. Importantly, I reject some of the key assumptions of many processual models (e.g., culture as an extrasomatic means of adaptation: Binford, 1962), and replace them with what I consider to be more useful working assumptions drawn from behavioral ecology. Formal modeling underlies this shift. Many of the assumptions underlying processual ecological anthropology have been called into question by attempts to make more formal models of ecological variability (see Grayson and Cannon, 1999; Kelly, 1995; Smith and Winterhalder, 1992).

The use of reductionist biological logic to account for evolutionary change in hunter-gatherer societies invites the criticism (or worse, the belief) that hunter-gatherers are somehow less cultural or less human than agriculturalists. How else could a theoretical framework developed to explain variation in nature be more appropriate to the explanation of hunter-gatherers than to other human groups? I have two responses. First, while human behavioral ecology is commonly applied to hunter-gatherer contexts for logistical reasons (the most developed theory applies primarily to foraging behaviors), it should not be inferred that hunter-gatherers are any less evolved than agriculturalists. (This point has already been made, in Chapter One, with respect to misconceptions about the term "social evolution.") Second, if evolutionary ecological theory applies at all to hunter-gatherers in the way suggested here, then it must apply to all humans. This may seem to strain logic for those who see complex societies as obviously inexplicable with a theory based on evolution by natural selection. The critical point here is that cultural environments (be they simple or complex, rural or urban, egalitarian or hierarchical, traditional or modern, and based on hunting and gathering, farming, or techno-industrialism) form a significant part of the surroundings to which humans must constantly adapt. If those cultural environments emerge through the iterative behaviors of fitness-seeking individuals who share linguistic modes of interaction, coordination, and manipulation, the cultural contexts should be just as explicable by evolutionary logic as the behaviors found within them.

6.1.2. Outline of Theoretical Orientation

The following eight items summarize the theoretical orientation and working assumptions of this model:

- The dynamics of long term social, economic, technological, political and demographic change are interrelated and productively understood with reference to ecological models.
- Formal evolutionary ecological and economic models that seek ultimate causes for evolutionary changes explicate complex and dynamic processes of social evolution. Ultimate (evolutionary) causes are those that condition or underlie more proximate (ideological, cultural) actions, and, in the context of human behavior, relate to the postulate of fitness-maximization. In many cases the formal models support or qualify a host of qualitative (and more proximate) models developed under theoretical paradigms such as cultural ecology, political economics, neo-Marxism, and practice theory (Fitzhugh, 2000).[1]
- At an archaeologically observable time-scale, social and cultural evolution is a product of behavioral variability. In biological parlay, "macroevolution is microevolution extended." This does not necessarily mean macroevolutionary change is gradual. But it does imply that the mechanisms of macroevolutionary change lie in the contextualized interactions of individuals embedded in populations. Thus, it forces us to consider the contexts of behavioral variability.
- Behavior is guided, in part, by an evolved propensity to pursue salient self-interests before altruistic acts. Salient self-interest prioritizes survival, well-being, and reproduction (all with direct implications for evolutionary fitness) over other goals.

- Behavioral variability arises from decisions made by individuals who need to adapt to variable environmental circumstances. Thus, individuals are the appropriate units of analysis for evolutionary modeling. As in organic evolution, population level effects of evolution are expected to be the consequences of individual level characteristics and actions.
- Adaptive decision-making addresses any condition affecting survival, welfare, and reproduction (as perceived by the decision-maker). While many external parameters precipitate adaptive responses all adaptive decisions involve social predicates and consequences that are included in behavioral calculations.
- Environment is a composite of physical, social, and cultural constraints and opportunities. That is, fitness is affected by both parametric (external, non-dynamic) and strategic (interactive, reflexive, dynamic) variability (Winterhalder and Smith, 1992:8–9).
- Lacking insight into symbolic or ideological dimensions of the cultural environment (that would provide greater specificity for predictions about individual adaptation to emergent and culturally contingent socioecological contexts in the archaeological past), it is sufficient to take a biologically reductionist perspective in deriving predictions about variation in adaptive behavior. In practice, the reductionist point of reference is used as a foundation for building more context-dependent predictions and interpretations. (See for example the discussion of shamanism in Chapter 10).

6.2. MODELING KODIAK SOCIAL EVOLUTION

The following model is presented in four "Acts" or stages. The theatrical analogy is chosen to reflect the fact that the subdivisions are somewhat arbitrary, and to emphasize that this is a hypothetical model, not a statement of empirical reality. Even though stage models are handy for articulating the conditions that are expected to lead to directional change in the long term, they potentially generate two problems.

First, they can channel unwary researchers to expect unilineal trends in social evolution. Recent views on social evolution tend to reject unilineal and stadial models of change that seem to march inevitably and progressively through a series of "stages" leading to the complex condition. In support of this critique, Feinman and Neitzel (1984) and O'Shea and Barker (1996) demonstrate that a considerable degree of variation in trajectory actually underlies social evolution. From the perspective of social complexity, any particular trajectory might be characterized by periods of increasing or decreasing complexity, or long periods of stasis. Given this, we need to remind ourselves that specific evolutionary trajectories, on Kodiak and elsewhere, are historical processes occurring in response to complex webs of changing environmental, social, and cultural circumstances. In this case, we know that the end result of 7500 years of change on Kodiak was increased complexity; but we do not assume that the path was unidirectional or that change was constant. Our goal is to make sense of the dynamic relationships between these circumstances while referring to principles drawn from general theory and expanding and modifying that theory in the process.

A second problem is that sequential models (of which stage models are a type) tend to bundle expectations into typological pigeonholes that may be too coarse to really evaluate the hypotheses under examination. Though finer-resolution models can be employed in specific circumstances, there are few alternatives to sequential models if the modeler's goal is to predict long-term change appropriate to multidimensional archaeological scale analyses. Sequence models, at least, provide a place to explore the theoretical implications of socio-environmental variability. Testable archaeological and paleoenvironmental predictions can then be developed to evaluate the utility of the modeled alternatives. Finer resolution data sets are always desirable in order to identify more precise timing and relationships between events; as methodological advances allow, archaeology moves in this direction.

Thus, the "Acts" in this model serve as logical steps in a sequence, rather than empirical ones. The model is calibrated to the Kodiak archaeological sequence using the general culture historical details already presented. In addition, series of predictions are developed for each Act that are evaluated in the following chapters.

6.2.1. Act I—A Beginning

The history of human occupation in a region begins with colonization, whether the known archaeological record extends to that event or not. Colonization has one of two outcomes: persistence (viability and population growth) or failure (local extinction/abandonment). Demographic models have convincingly shown that small founding populations (especially where disconnected from their parent population) are extremely vulnerable to extinction (Meindl and Russell, 1998; Wobst, 1974, 1975). Demographic extinction can occur even when resource productivity is high, resulting from random variability in reproductive rates and isolation from other breeding populations. To avoid demographic extinction, successful colonists often maintain social and reproductive ties to other populations during and soon after colonization. On Kodiak, that minimally entailed hazardous 20 km open-water crossings to and from the mainland. While this may not seem difficult by modern standards (or to highly capable maritime boaters such as the contact period Alutiiq), navigating this dangerous body of water likely would have been avoided if possible. Though we have no evidence of the kinds of boats used in the early Holocene, we expect that skills and technologies for open-water travel were under development. I suspect that a lack of reliable open-water technologies hindered colonization of Kodiak for millennia as compared to other regions of the North Pacific that lie on, or closer to, the mainland.

In relatively productive and previously unexploited environments, populations escaping demographic extinction grow steadily until they reach a point when resource return rates diminish substantially and/or crowding begins to limit mobility and the adaptive flexibility it provides (Erlandson et al., 1992; cf. Dumond, 1965; Rogers, 1992; Voland, 1998). Cohabitating groups remain small and population growth occurs as new groups bud off from parent groups, gradually exploring and filling in surrounding regions. Whenever co-residential groups outgrow local resources, stretch logistical constraints on labor and resource sharing, and foster social tensions that cannot be mitigated without social re-organization, group fissioning becomes the path of least resistance (see Kelly, 1995). This is a predictable process among egalitarian groups with regularly fluid group composition. The expansion phase for hunter-gatherers becomes relatively protracted where their fertility and mortality are conditioned by seasonal resource impoverishment, limited knowledge of the local geography and ecology, and the absence of storage strategies to extend resources through the lean winter season.

Optimal foraging theory offers a number of useful formal models of foraging behavior that generate expectations for economic and social patterns during and after the colonization phase. These models predict strategic variability in foraging activities related to fluctuations in resource productivity and resource distribution in space and time (Kaplan and Hill, 1992; Kelly, 1995; Stephens and Krebs, 1986). The prey-choice model (or "diet-breadth model"; Kelly, 1995:78–90), predicts that foragers concentrate on fewer high-return targets when they are doing well (when the return rates of the best resources are high) and increase the prey diversity when the best resources decline in productivity. When controlled for its rather restrictive simplifying assumptions, this model proves useful in both ethnographic and archaeological applications (e.g., Broughton, 1994; M. Cannon, 2000, 2002; Hill and Hawkes, 1983; Winterhalder, 1981).

Using the prey-choice model, all else being equal, one expects seasons of abundance to result in narrowly focused diets and lean seasons to result in more diverse diets. On the North Pacific Rim, summer yields peak resource productivity. Of the Kodiak resources discussed in Chapter 2, sea mammals such as seal and sea lion (given their predictability, aggregation, relatively large body size, and high caloric value as well as their hides and bones that provide raw materials for a variety of tools) are among the most attractive summer targets. Other reasonably high-return resources include salmon (in spawning streams) and large halibut (close to shore for the summer). As winter approaches each of these resources disperses, and the remaining resources are smaller, less concentrated, and more easily depleted. According to the prey-choice model, winter diets should expand to take in a number of lower-return resources like cod, flounder, rock fish, sculpin, and shellfish in addition to any stored food.

Applying the prey-choice model to Kodiak we expect a greater diversity of prey harvested in winter than summer. But resource distribution is as important as productivity in determining annual changes in foraging and residential mobility. Around Kodiak the spatial distribution of available species in winter is more even compared to high-yielding, highly concentrated prey in summer (Fitzhugh, 2002a). Using the central place foraging model, foraging theorists recognize that as search and transport difficulties/demands increase (e.g., with increased distance from camp), lower valued resources drop off the list of resources that can be harvested profitably (Kelly, 1991, 1995:133–141; Orians and Pearson, 1979). Thus, as distances from home increase, foragers more frequently limit their activity to high-return prey. "Effective foraging radius" refers to the range limit of profitability for any given resource. As return rates decline, so too does the effective foraging radius for that resource (Kelly, 1995:135). Coupling this with the seasonal variation in diet predicted for Kodiak, we expect longer distance targets associated with summer foraging, and a range contraction in winter. For Kodiak this latter pattern would be amplified—beyond the ecological predictions of subsistence—because of the increased hazard and difficulty of boat travel in the dark, stormy months of winter. Thus, as high-return taxa disappear and diets expand during winter seasons, the most effective foraging radius decreases, and people stay closer to camp. This prediction is universal because we expect foraging patterns in similar environments to follow this rule regardless of other social, cultural, or demographic factors; this prediction does, however, provide an important backdrop for change in residential mobility and storage practices.

All human societies practice some degree of resource storage. Storage allows groups to delay consumption of goods for a later time, either as a hedge against uncertainty in future returns or as a means for leveling predictable ebbs and flows in the availability of resources (Binford, 1980; Goland, 1991; Rowley Conwy and Zvelebil, 1989; Woodburn, 1980, 1982). But heavy reliance on storage has drawbacks that make it less desirable than alternatives. When resource fluctuations are spatially unsynchronized, mobility (either logistical or residential) often achieves better results than storage in moderating the effects of local fluctuations. Alternatively, if individual foragers each face equally unpredictable, out-of-phase returns on a daily or weekly basis, sharing becomes an effective method for minimizing individual variance in returns. Adaptive mechanisms possibly underlying the sharing approach to variance reduction include intentional risk pooling, reciprocity, and tolerated theft (Blurton Jones, 1983, 1987b; Speth, 1990; Smith, 1988; Wiessner, 1982; Winterhalder, 1986).

To varying degrees, storage implies a structural opposition to both mobility and sharing, although different ecological circumstances can make a given strategy preferable. Storage is indispensable when resources are periodically unavailable across a region. In such situations, moving from one unproductive location to another is ineffective and there is nothing to share. Noting Kodiak's seasonality

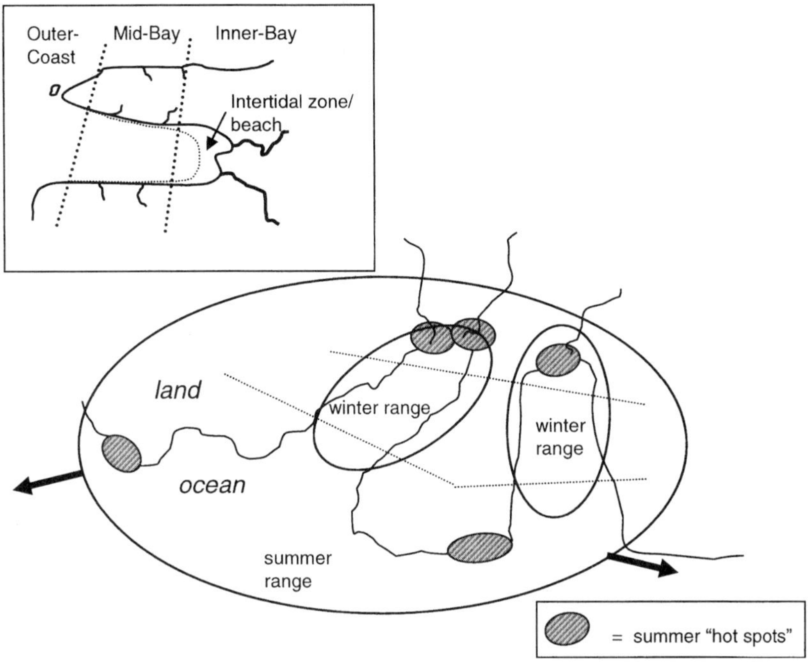

Figure 6.1. Show logistical ranges and predicted locations of camp moves (Fitzhugh 2002; reproduced with permission from Kluwer Academic/Plenum Publishers).

and ethnohistoric pattern of seasonal storage, some researchers view seasonal storage as a precondition to life there. With respect to the early phases of occupation on Kodiak, I disagree, for at least three reasons. First, some resources were available during winter as they are today. These resources were taken relatively close to camp because of their generally low effective foraging radii. Second, these "immediate-return" resources (*sensu* Woodburn, 1980, 1982) provided sufficient winter survival for the initial population densities. Third, occasional residential mobility (camp moves) could alleviate local harvesting pressure and allow these patches to "fallow" and recover.

Combining these factors with the pattern of seasonal variation in foraging ranges, we expect that in summer the earliest colonists foraged across wide areas and took fewer resource types, while in the winter they employed restricted foraging ranges and occasionally moved camp (Figure 6.1). We would not expect significant attention to seasonal storage because it required higher investments of seasonal labor (production and processing costs; Bettinger, 1999), greater coordination, and—importantly—greater social costs in protecting stores from less industrious individuals.

Considering the aforementioned factors, periodic residential moves likely occurred in winter when population densities were low. In summer, better weather and more profitable resource patches were exploited logistically by retaining a single camp and bringing high-yield resources home from a greater distance. This prediction reverses the ethnohistorically-documented pattern of winter sedentism and summer dispersal to foraging camps; this prediction also qualifies the argument that maritime hunter-gatherers always tend towards residential stability (Yesner, 1980; see also A. Cannon, 2002).

During the initial colonization and expansion phase, the most significant challenges to the colonizers and their descendents involved mastering the new environment and adapting or innovating technologies and strategies appropriate to it. These technologies were fairly generalized and capable of a wide range of functions befitting mobile foragers. Increased efficiencies through technological specialization were unlikely. As long as expansion into adjacent territory was relatively easy (e.g., ranges were close together and similar to each other), there was little leverage for evolutionary changes in the degree of economic, social, or political complexity. In anticipation of occasional moves, infrastructural investments were channeled towards portable facilities like tents and mobile tool kits. And, because of low population density, small group size, movement flexibility, and the absence of storage, a significant degree of egalitarianism prevailed in this initial phase of occupation.

In summary, Act I colonizers adapt to the new landscape, exploring and expanding by budding off daughter groups, and refining technologies as necessary to the new raw materials and environmental circumstances. They remain linked to their ancestral homeland through social and economic exchanges, but these ties weaken as colonist populations expand and the new homeland becomes more familiar. At this stage, subsistence is focused especially on large, productive resources in summer and small, locally available resources in winter. Subsistence difficulties are addressed primarily with sharing and residential mobility (especially in winter).

Box 6.1. Archaeological Predictions for Act I—the Colonizing Phase

- predominantly small campsites, distributed in central places with logistical access to a variety of productive summer patches and with close proximity to a variety of year-round resources
- proportionately few large, ephemeral (aggregation) sites located close to productive summer resource patches (as indicated by exclusively summer faunal remains, where preserved)
- low site densities and low variability in site functions as measured by landscape positioning, infrastructure, and artifact assemblages
- portable residential structures (investing in portability over location; tents were more likely than sod houses)
- low and normally distributed variation of tent/house sizes
- portable labor-intensive tools (expedient tools need not be portable)
- *relatively* unspecialized hunting, fishing and gathering tools (maritime hunting and fishing already imply a degree of specialization)

- lower proportion of processing tools in assemblages as compared to following phases
- decreased use of exotic raw materials, finished tools, and other products from distant locations as colonists increased familiarity with local environment (see Chapter 5)
- change in lithic technologies as adjustments were made for local and regional raw materials (Fitzhugh, 2001)
- *relatively* little diversity (low species richness and evenness) in summer faunal assemblages as compared to winter assemblages, where preserved. Seasonally-averaged assemblages emphasize larger bodied, K-selected sea mammals and fish as well as a wide variety of products available near camp. Resources demanding high processing costs or specialized facilities are present in low proportions, if at all.

6.2.2. Act II—Initial Effects of Circumscription

As we left our colonists, they were slowly expanding across the archipelago, exploring new ranges and incorporating local raw materials and resources into their technological and subsistence orientations. Eventually the archipelago fills, populations become circumscribed, and expansion is impossible. Three things could happen: (1) the economic system intensifies and group foraging ranges contract to pack more groups onto the landscape; (2) the economic system intensifies and foraging groups get larger; or (3) the economic system persists unchanged and population growth slows to replacement.

Circumscription reduces adaptive flexibility and the ability to absorb population growth. Individuals initially perceive circumscription as a reduction in mobility options when local conditions get tough and secondarily as a decrease in per capita foraging efficiency as more people forage the same ranges. Before population growth is curbed, however, population packing or crowding (Isaac, 1992) should stimulate more intense use of resources. Population growth should initially be accommodated by reducing territory sizes and packing more small groups onto the landscape. This is because larger groups would experience a greater number of unprecedented challenges than small ones. For example, larger groups would more quickly depress near-camp (winter) foraging ranges and have to move more frequently than small ones. In addition, social tension would arise from the increased stress on the sharing system; to accommodate the group, resources would be shared in smaller portions or more selectively. Either effect would lead more frequently to disagreements and anxiety. Smaller groups, while constrained to smaller ranges, would thus be preferable. For these reasons, on Kodiak we expect continued small co-residential groups but contracted ranges following filling of the archipelago. This should have led to frequent reoccupation of camp-sites as well as anticipation of regular return (marked by investment in stationary infrastructure over portability/flexibility).

Packing into smaller territories requires more intensive use of resident resources. Intensification could include increased labor effort (labor intensification) or efficiency (technological intensification) in the foraging of relatively high-return resources and/or in the capture and processing of lower-return resources. Foraging theorists demonstrate that, without mass harvesting technologies, return rates (calculated as benefits remaining after deducting search and handling costs) generally increase with prey body size (Bayham, 1979; Broughton, 1994). Thus, within the class of resources available to a given forager, larger game usually provide the highest returns and are thus most desirable (i.e., highest-ranked). Unfortunately, large game also tends to be the most vulnerable to over-predation or resource depression. This is because these species have relatively low reproductive rates (K-selected species; Hayden, 1981). As populations of large, slowly reproducing species are pushed into resource depression, they often become unstable (Belovsky, 1988; Stiner, 2002; Winterhalder et al., 1988). The result for predators (humans in this case) is lower average return rates as well as increased variance and unpredictability in returns. Resource depression should result in a reduction in the average body size of the depressed prey and increase in the capture of juveniles relative to adults. It could also lead to expansion of the diet breadth, if the addition of lower-ranked species would help to mitigate against the reduced returns of the higher-ranked prey.

Some prey populations escape harvest pressure by developing more effective defensive behaviors, such as relocation of colonies to more remote areas (Hildebrandt and Jones, 1992; Lyman, 1995). In this case, the greater distance to these hunting patches would increase the economic imperative to take only larger animals (Cannon, 2000; Hames and Vickars, 1982; Kelly, 1995:135; Orians and Pearson, 1979; but see Hildebrandt and McGuire, 2002). A combination of resource depression (with local extinctions) and relocation may result from intensified predation and the evolution of less vulnerable species (with a tendency to locate themselves in locations that are difficult to access). Both resource depression and relocation adversely affect foragers, exposing them to higher unpredictability and overall declines in resource returns, even though the sustainability of the prey population (and archaeological signatures of each mechanism) would be different. On Kodiak, seal and sea lion rookeries and breeding colonies would have been particularly vulnerable to hunting impacts.

Life-history models show that individuals often shift reproductive strategies when subsistence opportunities are limited, sometimes choosing to invest reproductive/parental resources in fewer offspring rather than reducing each offspring's chance of survival by having too many (Voland, 1998 and references; Winterhalder and Leslie, 2002). When resource depression generates increased nutritional stress, people experience decreased fertility and increased mortality (especially infants; Ellison, 1994; Kelly, 1995: 249–250; Wilmsen, 1982). The end result is retarded population growth, which may or may not stabilize at some equilibrium size (Rogers, 1992; Winterhalder et al., 1988; Wood, 1998).

As discussed in Act I, when high-ranking resources decline, diets expand to include resources of lower return rates; this change fosters an incremental shift to species that better withstand predation ("r-selected") but have higher individual search and/or processing costs (Bettinger, 1999; Hayden, 1981; Pianka, 1970). Without changes in harvesting and processing technology, these species are more expensive to harvest (after encounter) than higher-ranked resources; for this reason, expanding diet breadth cannot, by itself, relieve constraints on forager population growth. It can, however, help minimize vulnerability to risk or variability in returns of high-ranked resources.

As diet-breadth expands, the amount of per capita labor expended in subsistence pursuits increases to compensate for the lower rates of foraging returns (Boserup, 1965, 1981; Broughton, 1994; M. Cannon, 2002). This is the traditional meaning of the term "intensification" or what I call "labor intensification." Because labor intensification reduces the time and energy available for alternative pursuits (such as social networking, parental investment, and reproduction), it is expected only when environmental or human impacts on top-ranked resources result in periods of reduced return rates. Physical constraints on environmental productivity and conflicting demands on forager time allocation impose limits on the effectiveness and extent of labor intensification.

With greater variability and unpredictability in daily return rates, social tension and interpersonal conflict often increase. In ethnographic cases, such as among the !Kung and Hadza of Africa, individuals and families usually move from one residential group to another when social tensions flare (Kelly, 1995; Lee, 1972, 1990; Shostak, 1981; Woodburn, 1968, 1970). For our expanding Kodiak population, social conflicts could still be addressed by fluid group membership and movement between residential groups. The importance of sharing might also increase as a way to reduce daily or weekly scale variability in individual harvest rates (Blurton Jones, 1987b; Hawkes, 1992; Winterhalder, 1986, 1996). Both sharing and conflict-induced mobility work against the development of economic inequality at this stage. While economic asymmetries can emerge in low-density populations where stable, predictable, and productive resources are scarce, clumped, and defendable (Legros, 1985), the absence of methods (or motivation) to produce and/or extend productive resources for the winter also diminishes the utility of resource hoarding or defense. Occasional winter residential mobility (within the newly confined range) remains a better strategy for buffering spatio-temporal resource failure (Halstead and O'Shea, 1989). Barring major modifications of technology or environment, a density-dependent equilibrium emerges, which could persist indefinitely, engendering little change in subsistence economy, population densities, social integration or differentiation.

To summarize the developments of Act II, our colonists fill the archipelago to the density that their (generalized and non-intensive) foraging strategies support. Population densities increase by packing more small groups into the same space. This, in turn, leads to systemic resource depression of high-ranked resources. Diet expansion follows to include a greater number of species and increased effort in foraging. Eventually, the system stabilizes (density-dependence) until something triggers a change to Act III.

Box 6.2. Archaeological Predictions for Act II—Initial Effects of Conscription

- increase in contemporaneous site density compared to Act I.
- no significant increase in site sizes
- no increase in frequency of large, ephemeral (aggregation) sites (The increased proximity of other groups might even reduce the social imperative for seasonal aggregation, as social networking could be accomplished as/more easily by individuals and households.)
- no significant change in site function/land use variability
- more durable construction and thicker, denser site deposits (reflecting repetitive re-use of camps and foraging patches)
- increased use of non-portable, reusable technologies
- no significant evidence of technologies for mass-harvesting, processing, or storage
- no change in mean or standard deviation of habitation structure size (reflecting continued redundancy in household composition and status)
- no increase in frequency of military tools or defensive sites (reflecting minimal organized violence)
- no increase in frequency of violent trauma in mortuary contexts or catastrophic site destruction
- no change in frequency of elaborate ornamentation, monuments, or labor-intensive crafts (reflecting minimal prestige competition/political economy)
- increased taxonomic richness and evenness in faunal assemblages, with an increase, relative to Act I, in the dietary contribution of smaller prey of higher processing cost (Broughton, 1994; Grayson, 1984).
- decrease through time in average body size of larger bodied, K-selected resources with increased frequencies of subadult prey (if experiencing resource depression)
- increase or no reduction in average prey body size (if prey relocated to more distant and less accessible locations)

Calibrating Acts I and II to the Kodiak sequence, it is important to distinguish between evidence that calibrates and evidence that evaluates. Note, too, the model was formulated *before* I conducted my fieldwork. The results of the fieldwork, then, are appropriately saved for evaluation.

The Ocean Bay I and II phases are the best candidates for Act I and II, respectively. The only prior data suggesting this fit are (1) evidence for semisubterranean structures developing in Ocean Bay II as red ocher floor surfaces disappear; (2) lack of substantial, non-portable tools in either phase; and (3) absence of evidence for substantial mass-harvest technologies (net sinkers) until the following Early Kachemak phase.

The first piece of evidence is inconclusive. It is based on the argument that (a) pit depressions indicate investments in place and tents indicate investments in portability; (b) red ocher floor surfaces derive from hide preparation and hide tent use, and (c) the apparent lack of pit depressions in Ocean Bay I contexts is, in fact, real. Testing the lack of pit depressions in early Ocean Bay components is one of the research goals at the ongoing Tanginak Spring site (KOD 481) excavations. (A future publication will report these results in detail.) To date, no evidence exists for any sunken floor construction at that site (see Chapter 7). There may be at least one pit depression near the bottom of the similarly aged Rice Ridge site (KOD 363: Clark, written communication, 1996).

Stone lamps are one of the most labor-intensive pieces of material culture preserved from Ocean Bay times. Universally small (5–20 cm long and rarely more than one kilogram), Ocean Bay lamps could be carried in one hand and would not have significantly burdened boat or foot travelers. Beginning in the Kachemak period, stone lamps get much larger (25–40 cm long and up to 40 kg; Clark, 1994:141); they are much less portable, although somewhat portable by boat. The only stone implements exceeding the size of lamps in any period are grinding stone palettes used for making red ocher in the Ocean Bay I phase. These implements could be abandoned without great loss because they required little labor investment prior to use and because appropriate stones could be found at most locations.

The most striking evidence of mass-harvesting and intensive processing technology appears with the onset of the Kachemak. During that period we see, for the first time, notched stone (netsinkers) and semilunar (*ulu*) knives. Notched stones occur in very low frequency at Ocean Bay II sites, but they become a truly abundant artifact at Kachemak sites. Implications of these artifacts will be discussed after presenting Act III.

6.2.3. Act III—Eluding Malthus: Effects of Technological and Labor Intensification

Act I exhibited unconstrained expansion and Act II witnessed increased crowding and possibly increased difficulty in acquiring food, at least during winter months when locally available resources are more easily impacted. The specific range sizes, population densities, and diet breadths would vary from environment to environment and from group to group, depending on the balance of technological efficiency, resource diversity, and environmental productivity and stability. Among non-human populations, escaping this density-dependent equilibrium state requires evolutionary change in the behavioral and/or morphological adaptive capacity of future generations, or an easing of environmental parameters. For people, the evolutionary change can also be generated through cultural mechanisms such as technological invention or social reorganization (Richerson and Boyd, 1992).

As Boserup (1965) recognized, a more efficient adaptive strategy is needed. For individuals experiencing declining living standards, development of an effective new strategy is risky, and attempts likely fail more often than succeed. From a population-level perspective, on an archaeological time scale, we can expect successful changes to emerge and spread through the population more often in difficult times than in times of plenty. I have argued elsewhere that people more frequently display inventiveness when vulnerable to risk in subsistence returns (Fitzhugh, 2001). Factors associated with this vulnerability include: increasing population packing, diet expansion, and limitations of labor for intensified production. While many innovations fail, the increased inventiveness in times of stress often yields technological change. Technological change is critical for promoting further economic, social, and political complexity.

Two likely strategies of technological development are (1) increasing the yields of already high-ranked resources and (2) intensifying the production of previously low-ranked resources. Both fall under the category of "technological intensification." Efforts to increase return rates of high-ranked resources often create more efficient hunting and gathering technologies. For maritime foragers, prey wounded or killed on the water are lost easily, if not secured; thus, efforts to design new technologies might focus on better retrieval mechanisms. The introduction of toggling harpoon heads along with traditional barbed forms found in Kodiak sites from the beginning of the Early Kachemak phase suggest this dynamic. Even before that, during the Ocean Bay period, the development of ground slate points mark an improvement in marine hunting effectiveness. Ironically, increasing the effectiveness of marine mammal hunting would only increase the resource depression of these high-ranking species after a short time.

The second, and more enduring change, involves efforts to increase the productivity and efficiency of lower-ranked prey. Where these targets are more resilient to predation, technological changes in harvest efficiency *can* result in greater long-term productivity. To be profitable and thereby reduce the resource stress, costs must be lowered relative to the benefits involved in harvesting these products. Bulk capture and bulk processing technologies offer solutions to this problem, whether among agricultural or foraging groups (Madsen and Schmitt, 1998). As previously mentioned, notched stone net-sinkers and semi-lunar knives (*ulus*) of the Early Kachemak phase strongly indicate the paired development of mass-harvesting (nets) and processing technologies. Implementing these technologies increased local resource efficiencies and simultaneously facilitated the production of winter stores.

In the seasonal environments of the North Pacific Rim, schooling fish are some of the most profitable resources for intensification. Herring and a variety of anadromous salmon species are found in high abundance for short periods of time. Potentially, immediate surpluses could be generated and stored for use in the lean months of winter; this, however, requires available and timely labor and technologies because fish have a short shelf life prior to processing. If techniques are developed for capturing larger amounts of fish (*en masse*), the processing problem becomes even more intense. Unlike grains or nuts that can be stored prior to processing, fish need immediate attention. The efforts involved in later stages of processing for long-term storage (drying and smoking) only add to the logistical

Figure 6.2. Mary Haakanson splitting fish in 1993 for the National Geographic documentary "Island of the Giant Bears." Photo by author.

problem (Bettinger, 1999). Finding a place to keep the stores safe from animals and thieves poses yet another concern.

Supporting a tradition in economic and social evolutionary studies, this argument sets expectations for a significant economic reorganization hinging on technological change before registering further increases in population density, residential aggregation, or socio-political complexity. Embedded within the social relations that support it, technological change entails changes in the social context of economic activities (Dobres, 2001). Labor reorganization for intensified production of stores represents such a change. For example, implications of fish-harvesting intensification include factors associated with divisions of labor and gender dynamics (Figure 6.2). If women were responsible for fish processing in the Early Kachemak (as they were ethnographically), then nets (a technological solution for increasing the yield of capture without an equivalent increase in processing efficiency) should have prompted an increased demand for women's labor ("labor intensification"). Social consequences might include increased polygamy, reduction in female infanticide (if previously practiced), increased or decreased status for women, and demand for larger families to provide more processors.

Because this intense labor demand is seasonal and allows for more constant nutritional status throughout the year, fertility likely increases (Frisch, 1978). All else being equal, when technological and labor inputs for production are increased, populations grow, packing more people onto an already crowded landscape. This set of circumstances triggers other changes.

Within the context of shrinking home ranges, population growth would ultimately limit the potential for fissioning to maintain small residential groups. For the first time in the evolutionary sequence, residential group size and permanence (represented by the earliest use of villages) would increase. Domestic groups become more protective of stores and extraction locations, and households become increasingly self-sufficient through storage. Storage facilities become more concealed and defensible from other members of the same village. Obligate sharing declines, promoting changes in the nature of social interaction and organization.

Marriage-based residence transfers are a common mechanism for different residential groups to intermix. Marriage alliances in Act III, however, should become increasingly politicized, as access to labor and food resources become more integral to winter survival. The growing demand for processing labor might reduce individual freedom to choose marriage partners on equal terms, resulting in a concomitant loss of power for the spouse who leaves his or her cognates to join the other spouse's family. When women's labor is highly valued, bridewealth payments are often offered as compensation for a family's loss of a productive daughter. This situation can reduce young women's power in relation to their increased exchange value (Collier and Rosaldo, 1981).

Based on lines of marriage and kinship, individuals and families probably still move between villages to resolve disputes precipitated by enhanced competition. The effectiveness of this strategy is more limited, however, if recipient communities are especially protective of their claims on productive resources, which is more likely when a delayed-return system of storage replaces an immediate-return system of sharing (Woodburn, 1980, 1982). Leaving one's home village for another disadvantages the transient in political support and her/his access to the productive zones of the new village. Without a strong kin support network in the recipient village, immigrants are obliged to settle for the least productive foraging locations and are often seen as a drain on the village resource base.

Ownership claims over resources, tools, resource extraction locations, and facilities emerge as competition intensifies. Effective storage relies on the store owners controlling distribution, a situation taxing any "egalitarian ethos" (Boehm, 1993). Dyson Hudson and Smith (1978) argue that resource concentration and predictability are essential prerequisites for developing economic territoriality (defense of resources). Cashdan (1983) shows that intensity of resource competition is also an important factor. Combining these arguments, we would expect the most concentrated and predictable patches to be the first targets of defense under conditions of increased economic competition. Being the most concentrated and predictable, stores are carefully controlled. Methods of defense should also vary.

Cashdan (1983) suggests that the physical defense of territorial perimeters (especially large ones) is expensive and hazardous and would only occur under conditions of intense competition, when resources are predictable and concentrated. By contrast, social boundary defense is a less costly, and possibly less effective, option; under this practice, territorial claims are known and neighbors respect each others' rights to control territorial access, a common form of territoriality among roughly egalitarian hunter-gatherer groups. This practice works well when violation of established boundaries produces significant negative consequences for critical future social and economic relations. In the situation imagined for Act III, social-boundary defense remains the primary defensive mechanism for productive resource patches. Growing villages retain traditional first-use rights to productive patches within their ranges, while individuals and families within the villages develop rivalries for access to the best locations.

Territoriality is directly related to economic and social inequality. When productive patches are unpredictable or mobile from year to year, or when predictable patches are not easily defended, individuals find it difficult to gain leverage to control others. Boehm (1993) argues that humans in small-scale societies maintain relatively egalitarian interrelations because individuals resist, and actively intervene against, anyone seeking to gain disproportionate advantage. According to the logic of self-interest, people are more likely to collaborate when deposing an upstart than when supporting an emerging despot.

Violence is one means of punishing inappropriate or unfair behavior. In the absence of institutionalized inequality of power, prestige, and wealth, violence occasionally occurs within and between competing families or between neighboring villages. Factors precipitating such conflicts include competing claims over: productive resource locations, sexual jealousy, failure to repay social or economic obligations, and a group's attempts to maneuver unfair advantage in political networks. Act III depicts a liminal phase in the transition between egalitarian and ranked social organization. Small independent groups compete for resources, while households (though not as economically inter-dependent as in earlier stages) remain committed to preventing anyone from taking advantage of them. Additionally, economic interdependence increases in this phase because the social economy of matrimonial exchange alters households' interrelationships.

Dynamic equilibrium is achieved when localized competitiveness becomes endemic (especially in family rivalries), but no single group or individual establishes control or asserts dominance (or if they did, not for long). This situation persists as long as limited resources elude monopolization. When individuals (or, more likely, coalitions of kin) monopolize a disproportionate amount of resource production through territorial defense, those resources are leveraged into asymmetrical systems of distribution. Thus, inequality (the core feature of Act IV) begins to emerge.

Box 6.3. Archaeological Predictions for Act III: Technological and Labor Intensification

- technological change towards increased efficiencies of harvest and processing (including development/adoption of more lethal, higher-recovery technologies for big game and subsequent development of mass-harvesting and processing technologies for small game and plants)
- increase in the proportional representation of "r-selected" species relative to "K-selected" taxa
- increase in the proportion of processing tools in assemblages
- increase in the proportion of processing facilities found in excavations
- larger and more structurally developed sites ("villages") compared to earlier phases
- possible increase in mean house size but insignificant change in house size variability (reflecting rough equality in household population)
- evidence of internalized storage (e.g., pits or rooms inside the dwellings)
- increase in site infrastructure (more feature types)
- diversification of functional site types, including seasonal occupation of mass-harvesting locations
- increase in total site density (including sites of different functions)
- increase in artistic and exotic crafts designed to display success or importance of political and economic pursuits
- increase evidence of traumatic and/or lethal injury on skeletons (reflecting increased social conflict)
- addition of defensive sites

Several lines of evidence suggest that the Early Kachemak phase should correspond with Act III. As already noted, sinker stones and *ulus* of the Early Kachemak phase provide evidence of technological intensification. *Ulus* were more efficient processing tools than the elongated, ground-slate flensing knives or the bifacial and microblade tools of the Ocean Bay period. Users of semi-lunar knives likely found the tools more comfortable for repetitive slicing tasks than hafted elongated knives. The tradeoffs associated with leverage and extension were not missed when processing the soft parts of fish.

Two aspects of the Late Kachemak period suggest the development of localized social competition and, perhaps, low-grade alliance formation between villages. These are (1) elaboration of body ornaments (lip labrets and other jewelry) and other display objects such as stone lamps (Figure 3.8) and (2) distinctive mortuary treatments. (See discussions in Chapter 3.) Steffian and Saltonstall (2001) argue that labret styles show ethnic differentiation between regions of the archipelago as well as affiliations between groups living on adjacent portions of the Alaska and Kenai Peninsulas. Labrets are often made of high-grade coal (jet) from sources on the Alaska Peninsula. (At least one specialized production household workshop for coal labrets has been excavated on Kodiak at the Uyak site; Steffian, 1992a.) These developments suggest increased social tension and heightened attention to social status and position in the corporal and spiritual worlds. They may also indicate that people of the Late Kachemak witnessed the initial developments of institutional social asymmetry. This possibility is supported by the analyses presented in Chapter 9.

6.2.4. Act IV—Emergence of Social Asymmetry

Willing subordination of some members of a population is widely recognized as a key to the emergence of social inequality (Boone, 1992; Clark and Blake, 1994; Gilman, 1991). Understanding how such willingness emerges is a key issue of political evolution. Applying the theoretical perspective of behavioral ecology, subordination should emerge when it provides the best fitness outcome of available choices—when it is the best of a bad set of alternatives. Yet when could people find subordination a reasonable choice? Certain conditions appear to be critical; these include the structure of environmental variability (spatial and temporal), economic competition, and social organization.

Environmental structure involves a resource environment characterized by distinct spatial variability (or patchiness), not only in resource productivity but also in patch stability (or constancy through time). As witnessed in Act III, the situation becomes more complicated when people compete over resources. Competition in a spatially heterogeneous environment *can* lead to differential control over quality resources, if the resource ("store"), the extraction location ("patch"), and/or the technology ("means of production") are concentrated and defendable.

A variety of factors may account for the emergence of socio-economic inequality. These factors include storage, patch defense, and control over the means of production (Arnold, 1993; Boone, 1992, Hayden, 1994, 1995; Testart, 1982; Woodburn, 1980, 1982). In the case of incipient social inequality, though each factor may be involved, I place priority on the control of patches, from which other kinds of manipulations might follow. Stores are easily controlled by small household groups, because they can be watched from, even hidden within, domestic spaces. Social demands for sharing, however, will continue to impose strains on the accumulation and hoarding of stores. Initially, surpluses will not be generated in excess of what producers believe necessary for winter survival. (Unnecessary surplus would increase demands from "free-riders" or "poor cousins"—individuals who avoid the burden of producing their own stores when they can beg for support; Blurton Jones, 1987b.) We would expect surplus production (in excess of production as insurance against uncertainty) only when some value is gained from giving generously of surplus. As long as independent households produce their own stores (have access to patches), little motivation exists for producing excess surpluses for unequal redistribution.

A recent model (of the social and reproductive consequences of resource sharing) suggests one way that an overproduction of stores (and subsequent generous redistribution) provide households a payoff. The "show-off" hypothesis (Hawkes, 1991, 1993) suggests that male hunters often procure meat for redistribution beyond the domestic household to gain reproductive benefits (such as access to additional mates, or, in polygamous societies, attraction to additional spouses). This model provides insights regarding counter-intuitive observations of contemporary foraging societies, where hunting is socially valued even when it is economically suboptimal (Hawkes, 1993; but see Hill and Kaplan, 1993, for critique). This same model provides interpretive fodder for at least one prehistoric case, that of the apparent increase in big-game hunting in the California Late Archaic (Hildebrandt and McGuire, 2002). A similar "show off" factor might stimulate surplus production and generous redistribution in the North Pacific.

Because big-game hunting produces large resource packages in a single kill, foragers in a wide range of social and economic orientations practice big-game sharing (e.g., Lee, 1972; Kaplan and Hill, 1985; Wenzell, 1995; Winterhalder, 1986). North Pacific foragers' summer hunts frequently produced large sea mammal kills, and "show-off" males may have used their hunting success for differential access to a variety of benefits. But, if the "show-off" hypothesis is valid at all, it should apply universally and cannot be seen as the critical cause for the emergence of supra-household inequality. Why should we expect males to increase their "show-off" efforts at a particular time in prehistory? (This is a problem as well for Hildebrant and McGuire's (2002) "show-off" argument as applied to the Californian Late Archaic.) Overt showing-off is one of the activities that is most strenuously resisted in small-scale hunter-gatherer ideologies (Boehm, 1993). That does not mean, however, that good hunters did not receive social benefits in return for their activities (Hawkes, 1991, 2000; Hill and Kaplan, 1993).

Applying the "show-off" concept to redistribution of intensively-produced *stored* goods raises thought-provoking possibilities. Whereas individuals garner social and reproductive opportunities through generous redistribution of the resources they procure, sharing cumulatively-generated stored resources outside the household for personal gain conflicts with the interests of other household members who contribute to the store's production. In particular, a man or woman's affinal kin (spouses and their blood relatives), essential contributors in the capture and processing of time-limited resources, would become upset if any of the products of their labor were used by the man or woman or his/her blood kin for personal gain. Unmarried household members, however, might gain support for using surpluses for this purpose because a social, and potentially reproductive, gain may result for household kin.

Depending on the value placed on the surplus itself, "show off" sharing might be encouraged or discouraged among matrilocal vs. patrilocal, and monogamous vs. polygamous groups. Regardless of these permutations, as long as independent households maintain unobstructed access to resource patches and their means of production, accumulation of surplus *by itself* insufficiently encourages social inequalities.

Like control over stored goods, control over technology seems an unlikely source for economic benefits, since few extractive technologies used in hunter-gatherer production can be exclusively controlled (e.g., by control over critical materials or knowledge in technological manufacture or use). There are, however, circumstances when control over labor for specialty production can accentuate

Figure 6.3. Coin toss at Fourth of July festivities in Old Harbor. Photo by author.

economic and social asymmetries, as for example, when labor control facilitates the marketing of goods for long-distance trade that could not be traded economically without centralized control (e.g., Arnold, 1993, 1996c).

Before surplus production can be leveraged into asymmetrical power relations between households, a necessary precondition involves control over patches. Economists and behavioral ecologists discuss two kinds of competitions, scrambles and contests, that follow from distinct resource distribution patterns (Boone, 1992). *Scramble competition* occurs whenever each contestant has an equal chance to obtain goods. This condition is supported when resource distributions are equally unpredictable and equally accessible to all contestants (so-called "ideal-free" distributions). A good example of a scramble is a candy or coin toss for children at a birthday party (Figure 6.3). Up to this point in the model, all resource competitions are scrambles because individuals and households have more or less equal opportunities to capitalize on resource patches across the landscape. Sea mammals in open water, migratory birds, and wandering bears are distributed in an "ideal-free" manner.

As human population densities increase, rookeries, fish streams, halibut and cod patches, shellfish beds, and berry patches can be claimed and defended by factions drawn from local communities. *Contest competitions* arise when some individuals or groups monopolize goods because of their advantageous positions on the landscape and/or their enhanced ability to defend claims (due to high "resource holding potential"). Two factors support the emergence of contests over scrambles. First, concentrated, defendable resources comprise the highly structured landscape. As population densities increase, and it becomes more difficult to gain access to already-occupied, constricted patches (reduced "elbow-room"), groups arriving first have an advantage over late-comers. Second, encouraged by the advantages of exclusive control over limited quality patches, factions form for the purpose of resource control. Factionalization increases the defensibility of patches and makes the landscape more "despotic" in its resource distribution (Boone, 1992).

Because it involves cooperation of presumably self-interested individuals, group formation attracts attention in behavioral ecology (Smith, 1981; Kelly, 1995:205–221; Winterhalder, 1986). According to the fitness optimizing logic of behavioral ecology, cooperation only occurs when individuals recognize the personal benefits of economic and social cooperation. For tasks with economies of scale (more collaborators increase individual shares or reduce individual variance of returns), more participants are beneficial to all participants until the maximum scalar benefit is achieved (Smith, 1981). In egalitarian circumstances, collaborative work and resource/information sharing groups are limited by the nature of economic returns; and, as groups become larger, it becomes increasingly difficulty to detect and prevent free-riding (when people fail to contribute to the work of the group but continue to benefit from it). Accordingly, most egalitarian cooperative groups are rather small and/or work together for short duration. (For examples of group hunting see Lucier and Van Stone, 1995; Smith, 1991.)

In the context of competition for resource patches, factions or coalitions of individuals develop into resource-controlling units (sometimes called "corporate groups"). Following the logic of inclusive fitness theory (Trivers, 1971), coalitions based on kinship are most stable and effective. And as Vehrencamp (1983; Boone, 1992) has shown, cooperative groups of related individuals often exceed the size of groups of unrelated individuals. Close relations more willingly support, or overlook, free-riding in kin. Similarly, close relations tolerate unfair personal returns as long as the inclusive fitness benefits match or exceed loss in direct fitness to the individual. Over time, regularized alliances between close kin translate into greater household productivity (more cooperative producers), closer kin residence, and greater social identification with extended kin groups (lineage or clan affiliations).

Groups controlling the highest quality resource patches (particularly the ones with the greatest temporal stability and predictability) sustain production through good and bad years, while those households subsisting on more precarious patches periodically fall short in bad years. Economic asymmetry exists by definition under these circumstances. But how different social groups deal with the imbalance remains critical to the development of social inequalities as well as the emergence of greater social complexities. Three logical alternatives follow:

1) "rich" households refuse access to stores and patches while "poor" households are left to fend for themselves
2) "rich" households share, or permit "theft" of, excess production or allow the "poor" households limited access to patches (perhaps when each "rich" household achieves its target level of production)
3) "rich" households share stores or patch access in return for something of greater value than the units of production lost to the "poor" household.

Social costs associated with the first alternative would be high for closely related groups, but decrease as corporate household groups grow in size and self-sufficiency. Representing an accentuated "despotic" situation, this alternative seems unlikely, except perhaps when the majority of households are "rich" compared to a minority of "poor" households. Otherwise, "poor" groups could band together to depose the "rich" groups, continuing the leveling behavior expected of egalitarian groups (Boehm, 1993). Though isolation of well-off corporate groups might establish an unprecedented level of economic inequality or asymmetry, it would *not* stimulate significant social change or increased complexity; everyone would continue as before, but in greater inter-group isolation and with greater disparity between those who live in relative economic security and those who do not.

The second option, sharing without expectation of reciprocity, is based on the "tolerated theft" model of resource sharing (Blurton, 1987b). For production units of little value to the producer, sharing has minimal costs, and need not engender significant repayment. When the "tolerated theft" model is applied, significant changes in social structure and/or complexity are unlikely.

Applying the third alternative, however, stimulates new levels of production and competition. When relatively well-off factions recognize benefits through redistribution of surpluses, an incentive for producing excess surpluses (and allowing limited access to patches) presents itself. Benefits that "rich" corporate groups (or influential individuals within them) might receive include: recipients' good-will (as opposed to potential aggression)—breaking the "reverse dominance hierarchy," attraction as possible economic and/or political allies, and reputations as groups better able to support themselves and thus defend their territories. These circumstances also harbor an unintended consequence; advertising the productive benefits of corporate-controlled patches may attract the attention of aggressive competitors.

To summarize so far, in the early stages of developing inequality, we expect considerable competition between individuals and families intent on controlling resource patches and avoiding disenfranchisement. Over time, tensions emerge between the competitive and cooperative strategies of resource production, including the management of risk and uncertainty. Extended families, or even non-related individuals, recognize benefits from cooperating in the harvest and control of quality patches; within these groups, competition evolves concerning product dispensation rights. Competition for control over resources (and the status that follows) is facilitated by competition between families or villages. Cooperation between resources families and villages emerges as a result of political accommodation when defending from outsiders.

When competitors recognize the value of access to quality patches or of prestige through feats of bravery, competition may take the form of physical aggression. Endemic warfare emerges when individuals/groups of equal competitive ability skirmish to control resources or seek to attract subordinates and allies by advertising their prowess. War and negotiated peace expand in scale as long as neighboring elites find it advantageous to make political alliances with an expanding number of competitors. Such alliances are frequently unstable because of the dynamic tension between elites and their supporters and between potential allies/competitors.

As individuals and corporate groups seek a currency to advertise relative competitive ability, prestige economies may emerge. Evolutionary ecological models of *costly signaling* may account for this phenomenon (Boone, 2000, Neiman, 1997; Smith and Bleige-Bird, 2000; Veblen, 1953; Zahavi, 1987). Costly signaling refers to investment in expensive displays that advertise the producer's important qualities in a way that allows observers to estimate the signal's accuracy. Animals frequently practice this display competition (Fitch and Hauser, 1995; Manson, 1996). Rutting, for example, allows bucks to demonstrate their strength and agility to prospective mates who, in turn, use this information to select mates. Costly-signaling models in human behavioral ecology emphasize the development of symbolic currencies whose sole evolutionary benefit is the "honest" signaling of competitive advantage. When physical competition is expensive and hazardous, and when foregoing competition is even more hazardous, individuals and groups benefit when they can predict a contest's outcome prior to engagement. Accurate advertising of competitive ability (by combatants in positions of strength) is advantageous to both contestants, in that the physical engagement can be avoided.

With the development of contest competition and emerging differentials of access to productive resources, symbolic currencies can emerge through their role in costly signaling. Previously the dominant form of economy was subsistence-based. But the development of a vigorous symbolic economy pushes economic complexity to a new level. At this point, symbols become increasingly isolated from tangible consumables with indisputable fitness benefits. This development also dramatically changes the dynamic of local adaptation by increasing the complexity of the adaptive "landscape" (the environment that conditions individual and group opportunities and constraints).

Positive feedback, an "arms-race" of sorts, can arise in symbolic and physical competitions between competitors with similar productive abilities. Displays of wealth, elaborate feasts, give-aways, and the public destruction of property serve as mechanisms to advertise competitive abilities (Neiman, 1997). Individuals and groups unable to play the game may be forced to subsist on marginal resources and/or support the emerging elite as subordinates in return for access to food and defense in difficult times. At the same time, elites may recognize some advantage in supporting subordinates. Here we return to the benefits of giving away surplus resources.

Labor is one of the few services subordinates can exchange for food and defense that would be attractive to elites engaged in productive and costly-signaling competitions with other elites. Therein lies the emergence of patron-client relationships that solidify hierarchical social structures (O'Shea, 1981). Arnold (1993) forcefully argues that power over non-kin labor is a critical element of the emergence of institutional inequality and complex hunter-gatherer societies. This power can arise through various mechanisms, including debt encouragement and brute force (Hayden, 1995). But, without the emergence of a political economy based on the production of costly signals of competitive strength (in resource production or military facility), despotic corporate groups should be disinclined and unable to assert power over the labor of others.

With the creation of war-captive slavery, warfare can feed the political economy and its demand for productive labor. Non-captive subordinates recognize an advantage in allying with the most powerful elites to avoid becoming slaves. Because of increased productive labor and success in war, elites can offer better benefits to willing supporters. The elite's strong reputations discourage raids by all but the most audacious competitors (Hayden, 1995). Endemic warfare may then be seen as a product of physical competition for subsistence resources and labor, as well as a method for signaling competitive ability. At this point, rank and social stratification become solidified at the local level, and physical competition and defense expand to the regional level. Systems of craft specialization and trade in exotic goods become entrenched.

The extent of inequality supported by the aforementioned process depends on the degree of variation between the productivity and stability of different patches. Overexploited patches become less predictable and provide a poor foundation for competitive exclusion and ensuing inequalities. Conversely, if marginal environments become more productive and stable (due to climatic or technological change, for example), competition and inequality would decrease as elites lose economic leverage. Thus, inequality remains contingent on a particular kind of productive control: a control over stable, but defendable, patches surrounded by unstable and/or unproductive patches. The condition of unequal control over resources also requires patch controllers to value a patch's full productivity. Models of hunter-gatherer sharing behavior, already discussed, show that surplus production is worth defending only if it can be put to some desirable use (Blurton-Jones, 1987b).

Importantly, increased population density (in a spatially heterogeneous environment) drives this model for the emergence of social inequality. Even though some members of the population *must* experience ecological disadvantage to be willing to tolerate subordination, social inequality is explicitly *not* caused by population pressure or generalized resource stress (cf. Cohen, 1977; Keeley, 1988). Similarly, although some population members have exclusive access to relatively productive, reliable resources, inequality is *not* the result of simple ecological productivity or abundance (cf. Hayden, 1994, 1995). Inequality *is* fundamentally tied to *differential* access to stable, productive resources.

All of the behaviors enumerated in Act IV would be unthinkable in a society dominated by egocentric sharing networks (at local and regional scales), relatively equal access to productive patches, low population densities, or available free habitat. If people could move elsewhere to prevent exploitation and retain the same (or attain a better) condition of life and fitness, they would. For this reason, we do not expect supra-household social inequality to emerge in Act II, when people first experience limits of economic production. When resource unpredictability equally affects all individuals, no one has leverage to formulate long-term debts or dependencies. And, in the absence of significant levels of political competition between resource holders, there would be little reason to seek such leverage.

Box 6.4. Archaeological Predictions of Act IV: Emergence of Social Inequality

- greater infrastructure at defendable resource patches compared to earlier phases (reflecting patch-defense through physical occupation)
- enlarged houses or household compounds including facilities for exclusive interaction of faction members, entertainment of political rivals *qua* potential allies, and protection of corporately produced and owned products
- increased variation in house and/or compound size (To the extent that house-group status is marked in architectural variability [Coupland, 1988, 1996], rank or hierarchy would be apparent in a distribution skewed from many smaller dwellings to fewer large dwellings)
- increased frequency of "grammatically" distinct configurations of artifact design features such as family "crest" emblems and other social identifiers within clusters of households or villages (Political alliances or corporate groups with memberships from multiple villages might share similar symbols between sites.)
- distinctive body ornaments and garments to demonstrate group affiliation and rank, usually associated with larger houses as well as in a select few mortuary contexts (Peebles and Kus, 1978)
- increased frequencies of exotic, heavily-decorated "prestige" goods
- prestige goods distributed disproportionately in larger, or otherwise wealthier, houses
- increased construction of defensive facilities
- over time as local rivals become allies, and combatants come from greater distances, defensive sites enlarge and become strategically reoriented
- increased frequency of violent traumas to skeletons
- increased emphasis on military tools
- increased genetic diversity (especially mtDNA) in skeletons (reflecting trade and war-capture slavery between increasingly distant populations)

- diffusion of processing technologies over wider areas than styles and techniques of hunting tools (based on the ethnohistorically documented pattern for the North Pacific where women and children moved most often in marriage exchanges and as war-captives)
- greater diversity in mortuary treatment distinguishing slaves from non-slaves and high-ranking from low-ranking members of society (Archaeological evidence may not be available for all treatments. For example, on the northern Northwest Coast, dead slaves were tossed on the beach for scavengers rather than interred or otherwise preserved [de Laguna, 1990].)

The emergence of competitively-driven complexity need not result in overt rank or hierarchy. When productive, stable resource patches are densely packed across a landscape, or when the discrepancies between resources patches are narrowed, more competing factions can emerge and hold each other to relatively low levels of vertical differentiation. A competitive environment, with the ever-present threat of subordination or enslavement by other factions may sufficiently motivate unattached families into defensive political factions; a costly-signal driven political economic system may develop when identifiable "elites" are absent. This situation would rely on the continuing efficacy of Christopher Boehm's (1993) concept of "reverse dominance hierarchy," only in a more socially complex form than traditionally envisioned for egalitarian hunter-gatherer bands. This might provide a mechanism to explain the recently identified cases of relatively complex and yet minimally hierarchical (perhaps "heterarchical"; Ehrenreich et al., 1995) social configurations at Poverty Point (Gibson, 2001), Chaco Canyon (Feinman, 2000), and at times in Mesoamerica (Blanton et al., 1996).

6.3. CONCLUSION

Acts I–IV of this model described the evolution of a non-egalitarian rank society, where some individuals and family groups compete for prestige, power and wealth. Structural inequality results from a process of population growth and loss of equal opportunity under conditions of asymmetric resource distributions. Intensification, increased sedentism, storage, social aggregation, subordination, political competition (warfare/defense, trade, and rank), and formalized social obligation all contribute to this process. While Hayden (1994, 1995) and others are correct in noting that environmental productivity is necessary for cultural change to continue along this trajectory, they are incorrect in placing these circumstances in opposition to conditions of ecological stress. The critical factor in this model is an asymmetric resource environment, one with some individuals considerably disadvantaged by greater exposure to resource risk and uncertainty. Subordination results from this unequal distribution of ecological risk, amplified by socioeconomic competitiveness. With the possible exception of slavery, structural inequality can only develop when individuals are willing to sacrifice autonomy and equal returns; and, this only happens when subordination offers the best option from a limited set of alternatives.

This model avoids one flaw of some other models that seek to situate the motivation for political competition in innate drives for prestige or status (e.g., Clark and Blake, 1994; Hayden, 1995; Maschner, 1991; Maschner and Patton, 1996; see Clark, 2000). In this model, one pursues prestige if the expected reward is greater fitness or survival. Critical to the operation of this model, some involved individuals (perhaps a majority) find it in their best interest *not* to pursue positions of high status and, instead, support the aspirations of others. Thus, they choose to pursue positions of lesser status. The reliance on fitness as the ultimate currency of decision-making is a fundamental tenet in an evolutionary ecological approach. Innate or genetic differences across a population are unnecessary for explaining the evolution of new systems of social interaction. Instead, changes are predicted in dynamic relation to evolving socio-ecological contexts in which actors pursue fitness-related goals. Even though biological reductionism underlies the evolutionary ecological approach (a concept many find disturbing, especially when applied to cultural and ideologically-engaged systems), its explanatory framework can be applied cross-culturally because it relies on a set of behavioral motivations all humans share. Issues associated with symbolic interaction and ideological feedbacks emerge from, and to a certain extent are subordinate to, the fitness pursuits that mobilize them.

This model develops in a sequenced order of increasingly complex hunter-gatherer societies. Chronologically summarized, these societies reflect characteristics of: (1) colonization and expansion; (2) circumscription, reduced foraging ranges, and territoriality; (3) technological intensification to overcome seasonal variation, increased population density, and village aggregation; and (4) increased structuring of residential populations into corporate groups, localized competition, emergence of inequality and ranking, expansion of political alliances, trade, and warfare, and the emergence of a system of symbolic value capable of discriminating individuals on the basis of their access to resources, labor, and power networks. This model would play out differently in different environments. For example, colonization of small islands with intensive agricultural technologies and limited opportunities for emigration might lead more rapidly to despotic control and socio-political inequality. Polynesian expansion appears to have followed such a shorter trajectory (Kirch, 1984, 1990). Ancestors of the high-arctic Inuit, by comparison, lived in an environment in which sharing of information and resources was vital to survival; although there may have been periods of greater social competition than others (Savelle, 2002), the tendency for resources to be unpredictable in space and time rendered them unstable bases for the emergence of political and economic inequality.

This model constitutes an abstraction of a series of processes motivated by multiple factors. In any real situation, we must take into account the specific mechanics of the ecosystem, biogeography, and cultural history. Changes in the physical parameters of the environment (e.g., climate) influence the trajectory of change by making spatial or temporal resource variability more or less pronounced. And, at any stage of this developmental process, subsequent steps could slow or reverse the process. We might predict different trajectories in environments with low seasonal or spatial variability and/or less defendable resource patches (cf. Blanton et al., 1996; Hayden, 1995). Furthermore, because evolutionary trajectories of neighboring groups influence local developments, one needs to recognize that the aforementioned processes may be regionally integrated (Ames, 1995).

Chapter 7

The Sitkalidak Archaeological Survey Project

7.1. PROJECT GOALS

The Sitkalidak Archaeological Survey Project was conducted between 1993 and 1996. It was designed to identify and document the archaeological resources of the Sitkalidak Island region of Southeast Kodiak (Figure 7.1). The Old Harbor Native Corporation, the region's predominant land-owner, wanted a comprehensive inventory of cultural properties as a prelude to effective land management. The resulting archaeological data were anticipated as a source of regional data on changing settlement and land use patterns.

Priorities for the Sitkalidak Survey included basic chronological control, estimates of site function and significance in a larger settlement system, and information on scale and differentiation in economic and sociopolitical organization and territoriality. These goals were developed with the purpose of evaluating aspects of a preliminary version of the model outlined in Chapter 6.

7.2. METHODS

7.2.1. Survey Strategy

The research area designated for this survey was approximately 400 km^2. The region included Sitkalidak Island and the coastal section of Kodiak Island along the Sitkalidak Strait as well as the smaller islands and islets associated with this area (Figure 7.1). The region was thought sufficiently large to encompass the largest economically-integrated social unit at any time in the prehistoric past. In addition, it offered a geographically convenient unit of analysis (Sitkalidak Island

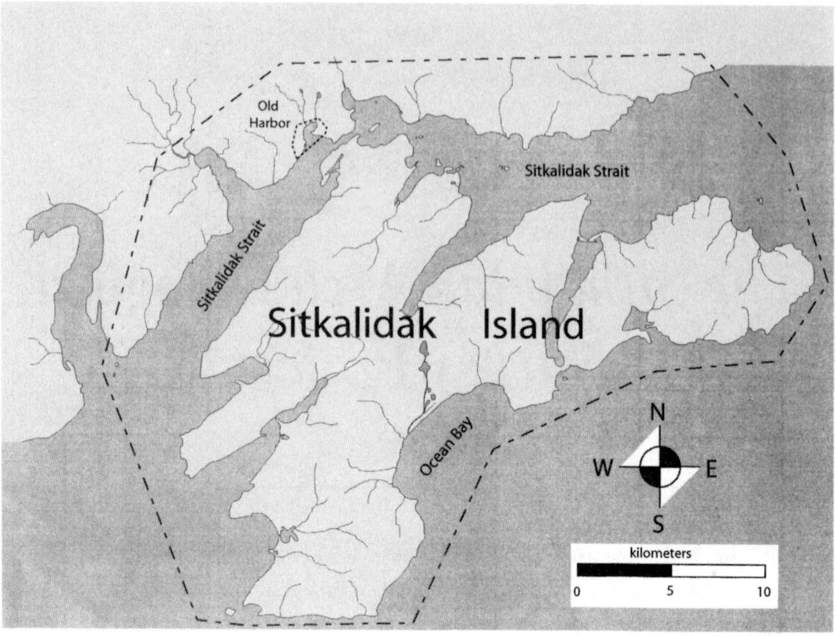

Figure 7.1. Sitkalidak Island and surrounding areas. Archaeological survey and excavations by the author in this region form the basis for the research reported in this book.

and Sitkalidak Strait), appeared manageable, and subsumed the goals of the Old Harbor Native Corporation.

The project crew surveyed all low-level shorelines accessible by boat, or within walking distance of boat landings (especially those with fresh-water and drained landforms), beaches and erosion profiles, low-level stream and lake shores, as well as defensible cliff-face islands, islets, and promontories (Figure 7.2). Other rocky and cliff-faced coasts received less attention. The interior areas within the survey, but away from low-grade streams, were not systematically examined. Steep slopes and mountain tops dominate those areas, and the dense summer vegetation and relatively low probability of finding sites render surveys there difficult and unproductive. Minimally, testing involved soil probing, shovel testing, and/or examining exposed erosion scarps. Sometimes more extensive test excavations were made.

Justification for this "stratified" survey scheme came from the established archaeological and ethnographic emphasis on coastal settlement and maritime economic and social orientation (e.g., Clark, 1984a). Only two classes of archaeological sites have been observed or rumored to occur on steep mountain slopes and ridges – cave sites and ridge-top stone cairns. Shamans used secret caves to hide ritual paraphernalia such as masks, and groups may have used them for the

Figure 7.2. Laada Bilaniuk and Libby Pontii surveying the Sitkalidak coast in 1993. Photo by author.

burial of important chiefs or shamans (Chapter 4). Mask and burial caves have been found occasionally throughout the Gulf of Alaska and Aleutians. A few Old Harbor elders may know where some of these are located. Others recall occasionally discovering them but have forgotten, or choose not to disclose, their locations. Although some members of the community would like to learn more about this aspect of their heritage, superstition surrounds the caves, and many in the community prefer that the caves remain hidden. We did not attempt to locate these sites.

Old Harbor elder Sven Haakanson, Sr. (personal communication, 1993) retold a story he had heard of large piles of beach rocks on ridge-tops around Kodiak. A few of these cairns have been located on a ridge in the Karluk area (Amy Steffian, personal communication 1996), and local pilots note seeing similar features on the Alaska Peninsula (Rick Berns, personal communication 1994). Unfortunately such features are not common. I have hiked many of the ridges on Sitkalidak Island over the years and have not encountered any features predating the 20th century.

A complex interaction of sea-level change and the cyclic nature of earthquake activity have shaped the history of land-level changes along the southern Alaska margin. Part of Sitkalidak lies within the region uplifted by the 1964 earthquake (see Chapter 2; Gilpin, 1995; Saltonstall and Carver, 2002; Plafker et al., 1992). Thus, in addition to prioritizing the survey of present coastal and river margins, the team also focused on regions that might have been closer to the coast prior to tectonic uplift. This procedure proved rewarding as it led to identifying several older sites from areas overlooked in previous surveys.

Conducted for three summers, the Sitkalidak Archaeological Survey documented 151 spatially discrete sites, including the 38 that had been previously recorded. Overall, the intensive coverage of land, which was examined with consistent subsurface testing measures, separates this survey from previous work in the area and accounts for the greater degree of success we had locating smaller or more deeply buried sites, especially older ones.

7.2.2. Site Identification

Sites were identified by a constellation of features including anthropogenic vegetation, surface house depressions, shell/bone middens, fire-cracked-rock, artifacts, and charcoal. Anthropogenic vegetation is indicated on Kodiak by an especially high density of cow parsnip (*Heraclium lanatum*), nettles (*Urtica lyalli*), and sometimes salmonberry (*Rubus spectabilis*) or elderberry (*Sambucus racemosa* L.) as compared to surrounding grass/tundra communities. Cow parsnip is one of the more successful plants on Kodiak, and, over the summer, it can grow to ten feet in height on nutrient-enriched or recently-disturbed soils. Today this plant often dominates the ground cover from June to October on archaeological sites dating to the last 2000 years. Nettles thrive on the calcium-rich soils of archaeological middens. Elderberry and salmonberry bushes sometimes grow more prolifically on very recent archaeological deposits (of the last 200–300 years).

In the survey, house pits were often detected from the surface. Surface indications were invariably subjected to intensive subsurface testing with soil probes and shovel pits. In most cases, subsurface charcoal deposits, artifacts, or dense faunal materials quickly confirmed depressions as cultural. Structural depressions at surface ranged from as little as one meter to over 10 meters in diameter and generally had slightly elevated walls (up to 15 cm) and deeper depressions (as much as 50 to 100 cm below surrounding ground surface). In addition to revealing an archaeological deposit, visible house depressions lent themselves to rapid measurement and mapping, providing valuable data on spatial organization (Figure 7.3). Semi-subterranean houses older than 3000 to 4000 years were rarely visible on the surface.

Erosion seriously compromises site preservation around Kodiak. Numerous sites have completely disappeared in the last several decades, exacerbated by tectonic submergence of many of the shores of Kodiak following the 1964 earthquake. While unfortunate, in some ways erosion facilitated the documentation process. Erosion tends to expose archaeological stratigraphy, including lenses of charcoal, house floors, hearths, and midden deposits. Beaches located in front of erosion faces were often littered with artifacts, sometimes yielding sizeable surface collections.

Figure 7.3. Koniag village site (KOD 110). Photo by author.

Profiles offered information about the intensity, duration, and reoccurrence of (or potential breaks in) occupation of the eroding segment of the site (e.g., thickness of deposits and number of discrete cultural strata). Stratigraphically-controlled samples of midden, charcoal, and other sediments were thus efficiently collected from cleaned erosion profiles.

The most reliable indicator of archaeological deposits was charcoal. In the absence of midden, artifacts, or clear house features, we used charcoal to identify the presence of former human occupation. Most researchers believe that Kodiak remained virtually unforested prior to 500–1000 years ago (Heusser et al., 1985; see Jordan and Knecht, 1988; Knecht, 1995; Saltonstall and Carver, 2002). Even today, only the northwestern half of the archipelago, many kilometers from the Sitkalidak area, is substantially forested. Groves of poplar constitute the only large tree colonies in the study area, and these occur only in limited locations. Shrub alder, willow, and dwarf birch are the only substantial wood species growing in the area. These bushes are frequently found on hillsides and stream banks, generally some distance from archaeological sites.

During the survey, I hypothesized that any wood charcoal found in subsurface deposits would be derived from focused human activity, not natural forest or brush fires. This hypothesis was overwhelmingly confirmed by the chipped stone debris (debitage), tools, or archaeofaunal material found in almost all cases where charcoal was recovered. This discovery meant that soil probes and shovel tests

Figure 7.4. Deborah Robinson and Laada Bilaniuk test for archaeological evidence. Photo by author.

could quickly, and with much certainty, determine the potential of a location for archaeological deposits.

7.2.3. Site Documentation

Once an archaeological deposit ("site") had been identified, team members established its location on a topographic map, took a Global Positioning System (GPS) reading of latitude and longitude, and recorded their observations on standardized survey forms. Various team members then set about mapping and testing the site itself (Figure 7.4). The primary goals of most test-pit excavations were the description of stratigraphy and the collection of dateable charcoal from each cultural stratum. Additionally, test pits were made in midden deposits to collect bulk samples of faunal remains.

Site maps were drawn using a compass and tape to establish map axes. House depressions were then plotted by tape or measured paces away from the given axes. Large sites were mapped in a series of measured transects. Every effort was made to accurately record the shape and proportion of each house as seen on the surface. In a few cases, where we spent longer periods at a site, more precise topographic maps were made using a surveyor's transit.

Artifacts (including debitage), faunal materials, charcoal, sediment samples, and wood samples, were catalogued and analyzed separately. Analyses directly contributing to the subject of this book included site component dating, typological and functional analyses of artifacts and their inter-site variability, as well as intra-site and inter-site spatial analyses of houses, sites, and regional settlement patterns. This chapter presents information on site chronology and material analysis, followed by brief summaries of the excavations conducted at the four most extensively investigated sites. The following two chapters (Chapter 8 and 9) apply these data to the analysis of spatial patterning in order to estimate changes in settlement and land use throughout the history of human occupation in Southeast Kodiak.

7.2.4. Analytical Units and Precision

For inter-site (regional) comparisons, the units of analysis are "components" rather than "sites." An archaeological *site* is defined as an area of past human activities where residues of these activities have accumulated, and where these residues are more-or-less continuously-distributed relative to adjacent space. A *component* is defined as a discrete chronological unit within a site, one that can be reasonably segregated from other units in time. This distinction is made to enable the assignment of chronological phases or periods to parts of extensively reoccupied sites. Given the imprecision of chronological dating methods employed (relatively few radiocarbon dates and more extensive coarse resolution typological dating), it is difficult to segregate site components into subphase units (i.e., smaller than the archaeological phases, such as Ocean Bay I or Late Kachemak). Additionally, for some measures, the larger period designations are more secure than the shorter phase attributions (e.g., Early Koniag).

A basic statistic from the Sitkalidak survey is a simple count of the numbers of site components from each time period. In Chapter 8, I start with this analysis as it provides an opportunity to introduce a number of issues relevant to subsequent analyses. Component frequencies, like all archaeological parameters, are influenced by the nature of component preservation and the archaeological sampling strategy employed. Any attempt to use these measures to infer actual prehistoric patterns of occupation demands that we consider sampling bias. As discussed above, the sampling strategy for the Sitkalidak Archaeological Survey was systematic and intensive, especially along the coast and lowland valleys. This strategy provided the best available approach for recovering a full range of preserved sites from all time periods.

We must also concern ourselves with the effectiveness of component recognition within sites. Here we are on less secure footing, because components could only be discerned in one of two ways. The least precise, but most applicable, method used was to identify components on the basis of the presence of diagnostic artifacts or features at a site. This might include a mixed combination of surface collections, house shapes seen on the surface, and diagnostic features in profile (e.g., red ocher lenses). A more precise method is to identify stratigraphically-distinct occupation layers with clear associations of datable artifacts or organic matter. This more precise method was used wherever possible; however, because site deposits rarely accumulate in analytically unambiguous "textbook" sequences, we can expect the stratigraphic identification of components to be complicated by depositional asymmetries (e.g., horizontal shifts in the location of primary deposit activities) and post-depositional processes (e.g., bioturbation, subsequent human occupation, etc.). This means that stratigraphic methods of component discrimination, at least in minimally-sampled sites such as those under discussion here, will often underestimate the number of discrete occupational events. Long-occupied sites could also yield continuously-deposited strata spanning multiple chronological intervals (analytical units). This problem is most acute in the later phases (especially Koniag and Alutiiq), when some sites such as winter villages appear to have been permanently occupied (Knecht, 1995). The implications of this bias are discussed in the course of data interpretation.

Finally, differential preservation of components must be considered. On balance, we can assume that the passage of time will play a significant role in the deterioration of cultural deposits. Physical erosion, chemical weathering, and biological agents are continuously operating on archaeological sediments (artifacts, organic remains, structures, and features). As a result, progressively older sites generally become less well preserved and more difficult to identify. In addition, older components on Kodiak tend to be smaller, and smaller sites are more susceptible to destruction and detection problems.

As mentioned earlier, one of the most common sources of site destruction on Kodiak is coastal erosion, when storm waves, gravity, and curious humans dig into cultural deposits. When we are lucky, the presence of fully-eroded sites remains evident by means of durable (usually stone) artifacts in the inter-tidal zone (Haggerty et al., 1991:213–214). Fortunately, Sitkalidak Island is located in an unusual area of tectonically-driven uplift (relative coastal emergence), with sites isolated above the shoreline, protected from the more devastating effects of coastal processes (Fitzhugh, 1996). This advantage is lost progressively as one moves northwestward from central Sitkalidak towards and across Kodiak. Today, several large sites erode into the Sitkalidak Strait, perhaps encouraged by local coastal submergence following the 1964 "Good Friday" earthquake (Chapter 2); these are much better preserved than sites on the northern and western sides of the archipelago where submergence was more substantial (Saltonstall and Carver, 2002). While we cannot ignore erosion as a biasing agent in the Sitkalidak sample, the dominance of uplift and coastal emergence across much of the research area has reduced the effect of erosion on the representativeness of our results. The large sample of early Ocean Bay period components testifies to this benefit and to the effectiveness of our survey strategy. The near absence of Early Kachemak sites indicates an unexplained anomaly.

7.3. SITE CHRONOLOGY

Attributing age to archaeological finds is a basic challenge in archaeological investigation. Indeed the problem of chronological attribution has been at the core of archaeological practice since it emerged from antiquarian roots late in the 19th century. Relative dating techniques, such as typological dating, seek to sort archaeological assemblages into relative temporal order. Absolute dating techniques, such as radiocarbon and other radiometric methods, seek to assign ages on calendric scales. Relative dating methods are usually less expensive and more inclusive than radiometric approaches, but less precise. Combining both techniques allows us to bring a larger number of assemblages under analysis while retaining some control over absolute chronology. Both typological and radiocarbon methods were used in this study.

7.3.1. Radiocarbon Dating

Radiocarbon analysis measures the time elapsed since the death of an organism (plant or animal) by measuring the proportion of unstable Carbon 14 remaining in its tissue. Wood charcoal is a common subject of this analysis as it occurs in abundance in many archaeological sites. Shell and bone are other common subjects for radiocarbon dating, but the dates derived from these materials are often less reliable, especially for marine shell and bone (Dumond and Griffin, 2002).

A total of 40 radiocarbon dates were obtained from 25 different Sitkalidak Survey (SAS) sites. Obtaining a sequence of representative dates for the larger collection of sites guided the selection of charcoal samples for dating. Most were derived from charcoal collected from the lowest archaeological component tested at each selected site. As such, the dates identify the presumed timeframe for site establishment, regardless of subsequent abandonment and reoccupation. In some cases, however, basal charcoal samples were too small for standard radiometric analyses, so samples from higher strata were dated instead. An attempt was made to date sites from each culture-historical phase, functional class (Chapter 8), and geographic region (bay, valley or coastal segment). Inevitably, sites with potential early dates were more intensively sampled, and allowances must be made for this when interpreting the data. Thirteen dates (the largest set for Kodiak, by far) are reported for the Tanginak Spring site based on research conducted as part of the Sitkalidak Archaeological Survey project and the subsequent excavations of the University of Washington Archaeological Field School in 1998, 1999, and 2001.

The radiocarbon dates reported in Table 7.1 are largely consistent with expectations derived from typological estimates (see next section and Figure 7.6). Care was taken to select charcoal that came from floor or hearth deposits in intact cultural strata. Thus the date samples should represent the general chronology

Table 7.1. Sitkalidak Survey radiocarbon dates[a]

AHRS Kod #	SAS Site #	Uncalibrated date (bp)[b]	Calibrated age range (BP)[c]	Lab # (Beta)	Stratigraphic context	Period[d]	Material[e]	Technique[f]
480	10	90 ± 70	285–0(1950)	78501	basal	AL	CH	C
473	2	120 ± 50	285–0(1950)	78508	single comp'	DKO/AL	CH	C
504	48	360 ± 60	515–290	78503	single comp'	DKO	CH	C
510	55	400 ± 50	525–310	78521	single comp'	DKO	CH	C
110	25	480 ± 60	555–330	78502	TP2 midden	DKO	CH	C
442	8	610 ± 60	670–520	71089	basal	DKO	CH	C
509	54	820 ± 90	930–640	78505	single comp'?	EKO	CH	C
106	126	930 ± 70	960–690	78518	basal	LKA/EKO	CH	C
558	124	960 ± 90	1055–685	78517	single comp'	EKO	CH	C
552	117	1090 ± 60*	1095–915	78514	upper floor	LKA	CH	AMS
86	116	1240 ± 60	1285–990	78513	basal comp'	LKA	BO	B
384	35	1490 ± 70	1180–900	78520	TP1-midden	LKA	MS	C
540	100	1500 ± 60	1520–1290	78511	single comp'	LKA	CH	C
541	101	1530 ± 60	1535–1300	78512	basal comp'	LKA	CH	C
566	86	1610 ± 60	1610–1350	78510	single comp'	LKA	CH	C
516	62	1720 ± 60	1740–1510	78506	TP2-2nd floor	LKA	CH	C
564	49	1750 ± 60	1815–1525	71092	Fea. 1-top comp'	LKA	CH	C
564	49	1890 ± 70	1970–1630	91316	Floor 1	LKA	CH	E
578	143	2110 ± 50	2295–1945	91318	single comp'	EKA/LKA	CH	C
554	120	4330 ± 90	5260–4630	78516	basal	OB II	CH	C
497	36	4390 ± 70	5280–4840	71091	basal	OB II	CH	C
89	127	4450 ± 100	5320–4835	83510	TP7-floor	OB II	CH	C

530	82	4560 ± 50	5435–5040	91317	TP3-mid-depth	OB I-II	CH	C
89	127	4610 ± 100	5585–4980	83511	TP7-floor	OB II	CH	C
522	68	4850 ± 120	5890–5310	78507	single comp'	OB II	CH	C
564	49	5300 ± 110	6300–5770	78509	110cm-base?	OB I	CH	C
555	121	6340 ± 100	7395–7005	78519	basal	OB I	CH	C
481	12	6380 ± 50	7380–7190	71714	upper strata	OB I	CH	AMS
481	12	6600 ± 230*	7890–7005	76738	basal stratum	OB I	CH	C
481	12	5370 ± 60	6291–5952†	134790	above top ocher floor	OB I	CH	AMS
481	12	6070 ± 60	7158–6746†	161616	top upper midden	OB I	CH	AMS
481	12	5630 ± 70	6619–6289†	161617	lower upper midden	OB I	CH	AMS
481	12	5710 ± 70	6718–6312†	134789	above white tephra	OB I	CH	AMS
481	12	5880 ± 80	6887–6494†	134791	below white tephra	OB I	CH	AMS
481	12	6250 ± 60	7303–6994†	161618	Stratum H	OB I	CH	AMS
481	12	6430 ± 60	7432–7250†	161619	Stratum L floor	OB I	CH	AMS
481	12	6579 ± 60	7574–7339†	161612	Level 12	OB I	CH	AMS
481	12	6280 ± 60	7317–7011†	161613	Level 13 -above sterile	OB I	CH	AMS
481	12	6490 ± 60	7555–7271†	161614	Level 13 -above sterile	OB I	CH	AMS
481	12	6480 ± 70	7558–7266†	161615	Level 12 -above sterile	OB I	CH	AMS

[a] All dates processed by Beta Analytic Inc. using the Libby ^{14}C half life (5568 years) and referenced to 1950.

[b] Conventional dates are corrected (normalized to the modern PDB-1 international standard) using the C13/C12 ratio for all but two dates based on values typical of the material sampled. Two dates, marked with (*) are uncorrected.

[c] Calibrated dates presented as maximum ranges at 2 sigma. Unmarked date ranges derived by Beta Analytic, Inc. using the Pretoria/Beta Analytical Dendro-calibration (Stuiver and Long 1993; Talma et al.1993; Vogel et al. 1993). Calibrated ranges marked by (†) were derived by the author using Calib 4.3 (Stuiver et al. 1998a, 1998b; Stuiver and Braziunas 1993).

[d] Period: AL = Alutiiq; DKO = Developed Koniag; EKO = Early Koniag; LKA = Late Kachemak; EKA = Early Kachemak; OB II = Ocean Bay II; OB I = Ocean Bay I.

[e] Material: BO = Bulk Organic Sediment; CH = Charcoal; MS = Marine Shell

[f] Technique: AMS = AMS 14C date; B = bulk low carbon; C = standard 14C date; E = extended count.

of the Sitkalidak area. Accordingly, these dates suggest several conclusions about Sitkalidak culture history and the SAS project. Figure 7.5 plots the conventional and calibrated dates.

The first point is that a relatively high percentage of the dates for this survey fall early in the archaeological sequence (Ocean Bay I and II). This is an extraordinarily large number for such a small area, compared to the small number of previously identified Ocean Bay sites throughout the Kodiak Archipelago and on the Alaska Peninsula.[1] Two explanations can account for this high density. First, our intensive survey strategy clearly increased our rate of success in locating deeply-buried older sites as compared to strategies often used in Kodiak archaeological survey. Second, the unique geological conditions of southeast Kodiak possibly account for the success in locating early sites. As discussed in Chapter 2, the southeast side of the Kodiak Archipelago has experienced tectonic uplift overall throughout the Holocene compared to other parts of the region. This process appears to have helped preserve archaeological sites in the Sitkalidak region, where they could be found with a survey strategy that included somewhat inland and elevated landforms.

The second, and most striking, pattern in the Figure 7.5 plot is the complete absence of dates between 4000 and 1800 bp. Several explanations are possible for this pattern: sampling bias, regional abandonment, and/or geological processes of selective site destruction. Selection sampling bias seems unlikely, given the number of sites sampled and the intensity of the survey. Regional abandonment seems more plausible, given the appearance of Ocean Bay II tradition sites in Kachemak Bay on the Kenai Peninsula around the beginning of this interval. The apparent increase in occupation density on Kodiak's larger rivers between late Ocean Bay and Late Kachemak time periods may also play a role (Jordan and Knecht, 1988; Knecht, 1995). Nevertheless, the mechanisms for regional abandonment are far from apparent (see discussion in Chapter 10). Geological processes that led to the destruction or submergence of sites in the 4000 to 2000 bp interval seem possible. At least two mechanisms could have caused site loss from this interval. While tectonic-induced tilting or faulting of the North American Plate has normally led to coastal emergence in much of the study area (Gilpin, 1995), it is possible that the geometry of tilting planes differed in the past. A shift of the hinge-line (the axis of earthquake-induced coastal displacement) of only 10 km towards the Aleutian trench would dramatically increase coastal erosion in the Sitkalidak area. Or possibly an encroaching tideline deteriorated or obliterated sites during a regional sea-level high stand. The Early Kachemack on Kodiak is known for its characteristic oily black matrix. Stein (1984a, 1984b, 2000) has argued that local sea-level rise can lead to diagenesis of shell midden into such a matrix, and submergence is a reasonable hypothesis for the Early Kachemak deposits at Old Kiavak and elsewhere (Clark, 1996, 1997; but see Yesner, 1998).

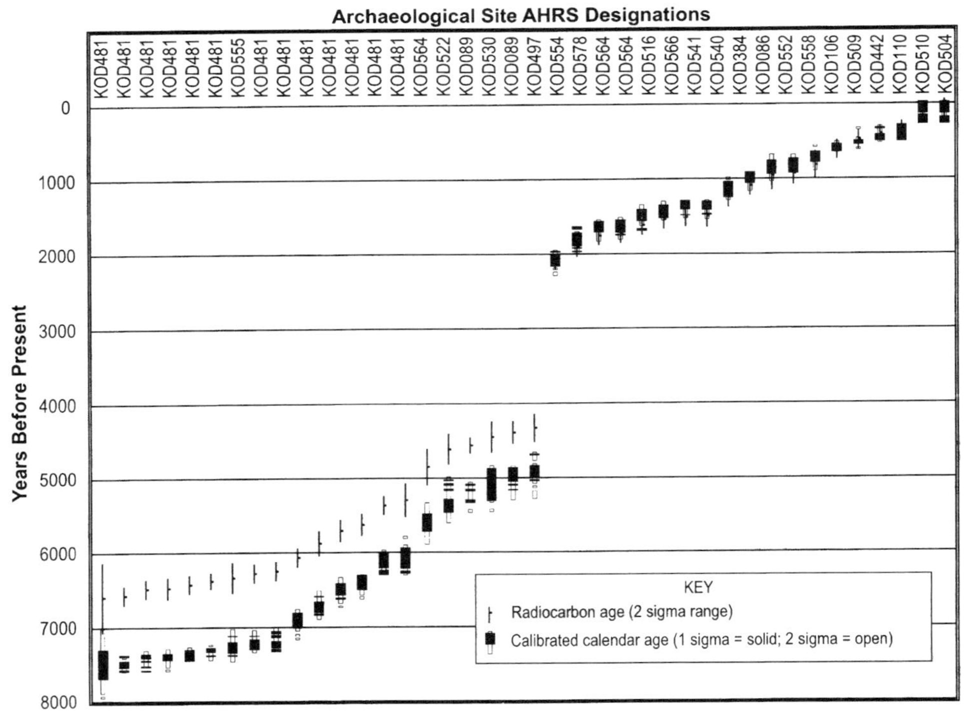

Figure 7.5. Radiocarbon dates from the Sitkalidak Archaeological Survey showing both uncalibrated (14C) dates and calibrated age ranges at 2 standard deviations. Note the large gap in dates between 4000 and 2200 bp. Explaining this gap remains a challenge for continuing research.

The third notable aspect of Figure 7.5 is the constant rate of site formation indicated over the past 1800 years. This linear array of dates covers the Late Kachemak, Koniag, and Alutiiq phases. Since these dates, for the most part, represent the basal components (i.e. the oldest occupation) identified at each dated site, this smooth array suggests continual and gradual establishment of new sites during the last two prehistoric phases without any noticeable change in the rate of settlement establishment (one dated site about every 100–200 years). These data tentatively argue against any significant change in settlement mechanics as expected if Kodiak was abandoned by Kachemak people and then reoccupied by the Koniag, or if the Koniag were a militaristic population overcoming the large Kachemak groups already occupying the area (see Chapter 3).

7.3.2. Typological Dating

With only 25 radiometrically-dated sites, it is necessary to extend chronological control over the larger sample of 151 sites using established typological indicators. Sustained archaeological research in the archipelago has identified a series of reasonably reliable chronological diagnostics, as enumerated in Figure 7.6. Three typological categories have been used to date Kodiak sites. Structural information includes red ocher "floors," single-room pit-houses, and multiple-room pit-houses. Technological classifications derive from the dominant lithic manufacturing methods, with chipped stone dominating early stone tools and various forms of ground stone taking precedence later. Chronologically diagnostic artifacts include microblades, plummet sinkers, notched flat-stone sinkers, ulus, coal labrets, incised pebbles, pottery, trade beads, and iron.

As this survey frequently depends on typological-dating indicators, we must test the accuracy of typological associations. For this purpose, Figure 7.6 includes correlations of the typological traits associated with radiometrically dated site components. Though all dated sites of each phase did not include every identified trait, the overall pattern is consistent with the ideal typology. This pattern reinforces our confidence in the accuracy of typological dating for the area, in general, and suggests that few, if any, of the dated samples were seriously contaminated or out of context.

The only unexpected results of the comparison in Figure 7.3 came from two fragments of ground-slate found in an upper layer of a stratified site with an Ocean Bay I lower layer. Although ground-slate is found occasionally in Ocean Bay I contexts, especially slate rods (the author's unpublished field notes; Clark, 1979; Knecht, 1995:105; Steffian, personal communication, 1996), the undated upper layer may have been occupied closer to, or within, the Ocean Bay II phase. Ground-slate is now reported in Ocean Bay I sites after 6000 bp (Steffian et al., n.d.). Also, a notched stone sinker was found in a shallow Developed Koniag deposit. While only one occupation zone was noted in the test excavations, an older occupation was

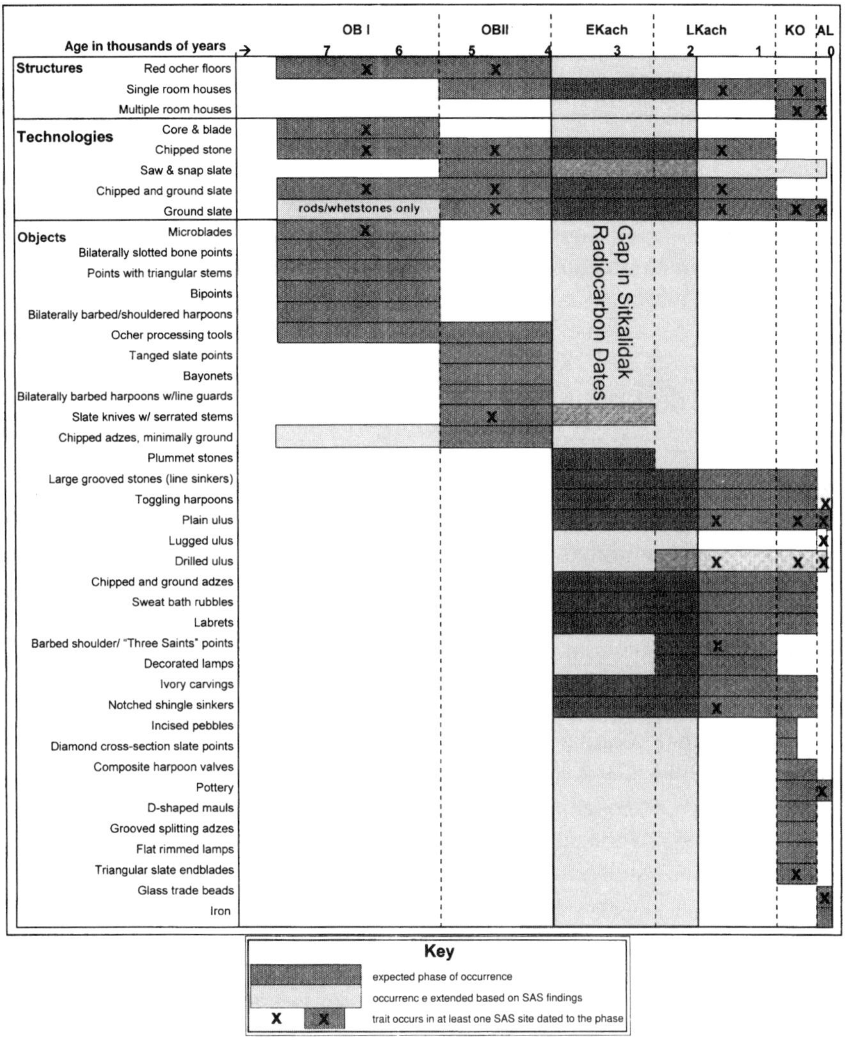

Figure 7.6. Typological correlates of archaeological phases. Dark bars show accepted temporal distributions of traits. Light bars suggest extensions of temporal trait distributions based on observations from the Sitkalidak Archaeological Survey. Marks of "X" or other text indicate traits occurring in an SAS site dated to a particular phase. Generally, the radiocarbon date correspondence justifies the use of material traits as temporal diagnostics for other sites that were not radiocarbon dated. Adapted from Fitzhugh 1996 and Steffian et al. 1998.

found just beyond the limits of the Koniag village. This second locus was less well-preserved, but it could have been a Kachemak occupation. Otherwise, this is the only case of a notched stone sinker found in the Sitkalidak survey to be associated with Koniag deposits, despite their occurrence at the "New Karluk" (KAR-001) site (Knecht, 1995). Accordingly, the typological indicators can be used to attribute relative age to site components that have not been radiometrically dated.

This dependence on typological attributes forces me to present the substantive analyses of this book by typologically-defined periods and phases. Our ability to consider variability and incremental changes out of synch from periods and phases is thus inhibited, but this is a common problem of chronological scale in archaeology.

7.4. MATERIAL ANALYSIS

7.4.1. Artifacts

The artifact sample generated by the Sitkalidak survey is small and eclectic because of the recovery methods used (surface collection and limited test excavation). Erosion profiles, when present, often provide opportunities for recovering artifacts, but the resulting collections are generally displaced and can be linked only indirectly to different site occupation layers. Test excavations rarely yield large numbers of artifacts. While this collection is appropriate for typological dating of many sites, artifact assemblages from most sites are too small for meaningful comparison with published collections from other sites. Nevertheless, our collections allow us to draw a few conclusions when aggregated into groups of similar age and functional classification. These analyses are presented in summary in Chapter 8 and in detail in Fitzhugh (2002a).

7.4.2. Faunal Materials

I hoped to collect sufficient faunal samples to address questions of economic change through the prehistoric sequence as well as to examine social inequality of access to high quality resources within larger village sites, especially of the later time periods. Regrettably, the faunal preservation in the Sitkalidak area was insufficient for the first goal, and time constraints in the field and lab limited the second. Over 40 bulk midden samples, ranging from one to several gallons per stratum, were collected from Koniag period sites throughout the survey area. These samples provided fine-grained economic and seasonal information for their respective sites. No faunal material associated with the Ocean Bay through Kachemak periods was found. Time and shipping costs prevented systematic sampling of more than one or two faunal columns from larger Koniag sites. While these samples are useful for examining aspects of local economy and site seasonality, it was impossible to use these data for examining intra-site spatial variation in faunal composition as originally intended.

Preliminary analyses indicate that a variety of shellfish, fish, birds, and sea and land mammals were harvested in the Koniag period. Shellfish dominated most deposits and included butter clams, blue mussels, cockles, whelk, periwinkle, chitons, limpets, and a variety of small snails. Sea urchin shell fragments were often found in dense pockets in midden deposits, suggesting that huge meals of sea urchins were consumed and their shells discarded in discrete pockets. Cod outranks salmon as the most common fish species reported in a preliminary analysis of eight midden samples from five Koniag sites. Halibut and red snapper are also identified, in lower percentages than either cod or salmon. Seal, sea lion, and porpoise are represented in several sites, although they occur in such low density that few bulk samples had many representatives of these taxa. Larger taxa, including sea mammals, fox, and birds dominate discretionary (non-bulk) faunal samples. (These discretionary collections were made to insure that the presence of larger taxa was recorded even if bulk samples missed them.) Finally, whale bones were observed and documented at sites where they were found, but left uncollected. Whale remains were limited to Koniag period sites (usually villages) around the outer coast regions of Sitkalidak and the outer extremities of the Sitkalidak Strait.

The faunal samples from the Sitkalidak Survey remain useful for seasonal dating of site occupation and identifying the degree of economic specialization at sites of different function and location in the Sitkalidak region. Working towards these goals, students at the University of Washington have been sorting these samples, but the analyses remain to be completed.

7.5. EXCAVATIONS

In addition to the survey and small test excavations, we carried out more extensive test excavations at three sites to help redress some of the limitations of a surface-oriented survey. These were the sites at Tanginak Spring (SAS 12/KOD 481), which we have been excavating for several summers; and Bear Island (SAS 49/KOD 564) and Partition Cliff (SAS 2/KOD 473), which we excavated for a few weeks each. Each was discovered by the Sitkalidak survey and represented a class of sites poorly documented in Kodiak excavations. A fourth site, Lighthouse (SAS 127/KOD 089), was partially excavated in the springs of 1995 and 1996 by a team of Old Harbor junior high and high school students as part of a school archaeology project under my supervision. As these four sites span the archaeological history of Kodiak from over 7000 years ago until the terminal Russian Alutiiq phase (ca. 1865 AD), they provide site-specific archaeological detail that complements the regional-scale spatial data presented in the following chapters.

7.5.1. Tanginak Spring Site (KOD 481)

The Tanginak Spring site is located on a small knoll eight meters above sea-level at the northeastern end of Sitkalidak Strait on Sitkalidak Island (Figure 7.7). The site overlooks a marsh that likely was an open bay or lagoon at the time of occupation. Today the site is located about 300 meters from the active beach at

Figure 7.7. Tanginak Spring site excavations in 2001. Laska Fitzhugh in foreground. Photo by author.

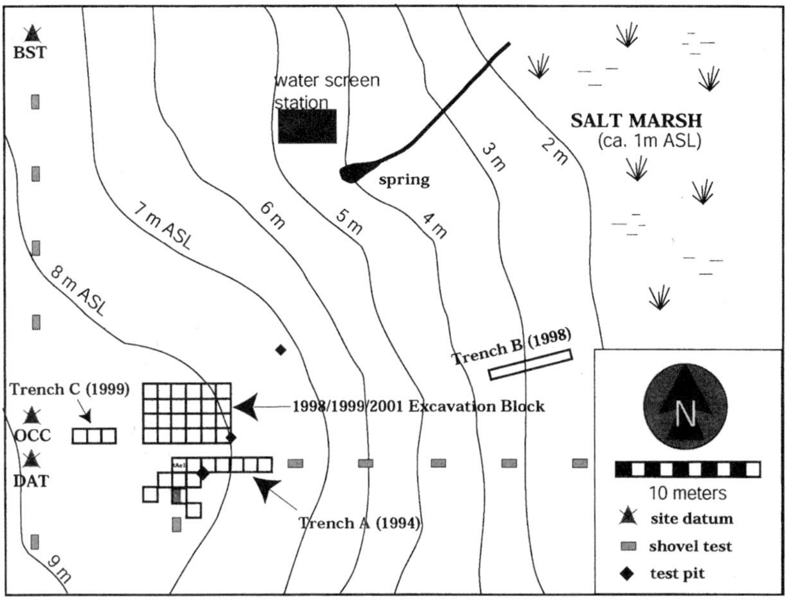

Figure 7.8. Plan of the Tanginak Spring site (KOD 481), showing topographic contours, reference benchmarks (datums) and areas of excavation through the 2001 excavation season.

Tanginak Anchorage. A small spring that may have provided the drinking water for occupants flows out of the knoll about four meters below the site.

The site was discovered, tested, and radiocarbon-dated in 1993, with results suggesting a very early occupation date. Further investigation began the following summer, when we excavated a 6 × 1 meter trench in the site's core area (Figure 7.8). This excavation revealed a much deeper, artifact-rich, and stratigraphically complex deposit than we had expected from 1993 tests (Figure 7.9). We returned in 1998, 1999, and 2001 when students from the University of Washington and Old Harbor expanded these tests into a full-scale excavation.

Figure 7.8 illustrates the topographic configuration of the site and the location of primary excavations conducted through the 2001 field season. A single radiocarbon date of a carbonized fragment of willow collected from a small test pit in 1993 returned a date of 6380 ± 50 bp (AMS, Beta 71714). The 1993 test pits had only been excavated to about 45 cm below surface and had revealed abundant red ocher staining, an asymmetric, stemmed, chert biface (knife), and a microblade. Our 1994 trench excavation (Trench A) extended 150 cm below surface before striking sterile clay and glacial till (Figure 7.9). We exposed at least five occupation surfaces, thousands of fragments of lithic debitage, and hundreds of worked stone tools. In 1998, 1999, and 2001, we turned our attention to the excavation of a 6 × 4 meter block just north of Trench A (Figure 7.8).

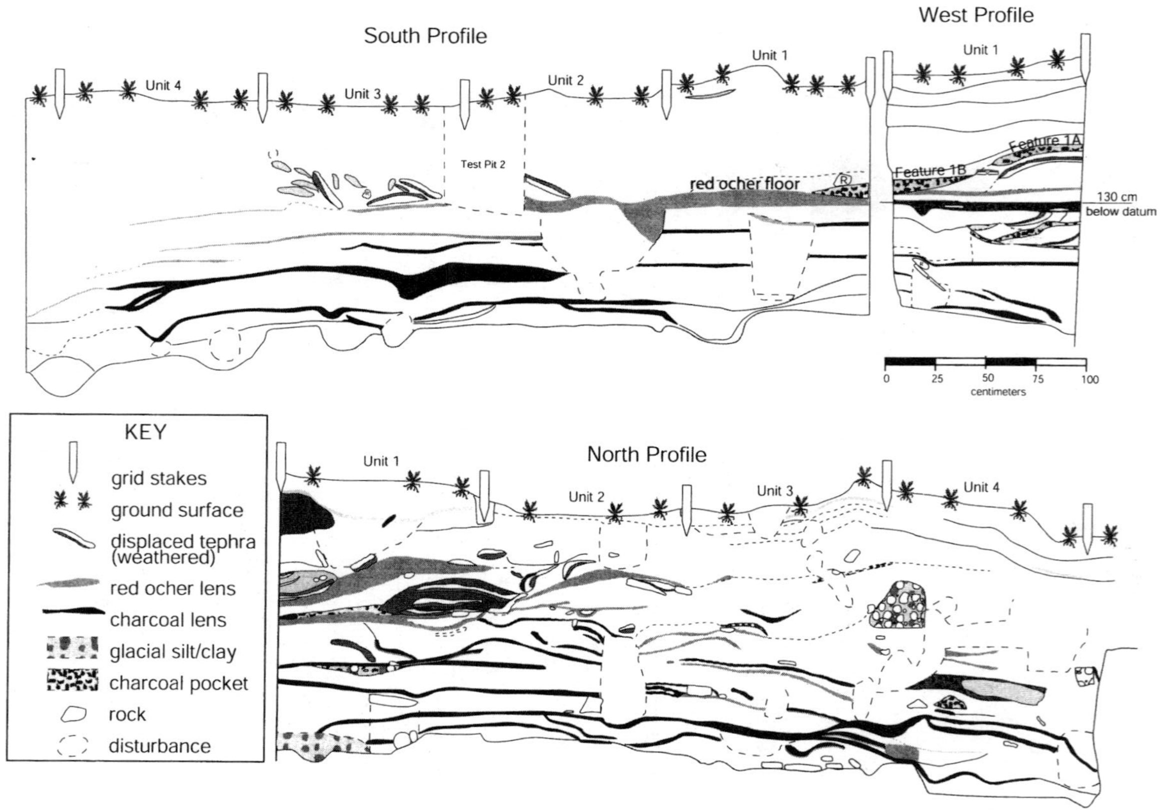

Figure 7.9. Stratigraphic profile of the 1994 test trench excavated in the Tanginak Spring site (KOD 481). Shown are compacted occupation lenses of red ocher and charcoal separated by silt-clay cultural fill. Short sloping packets of multi-color volcanic ash are interpreted as sod packets, piled on the edges of the two top-most (red ocher) floors, perhaps to anchor the base of skin tent walls. Similar ash packets are found at lower levels in other areas of the site.

Pending a more detailed report, the occupation sequence is briefly summarized here. The main site deposit (centered on the eight meter-high knoll) is capped by an organic soil about 25 cm thick. Near the top, this zone revealed a discontinuous dusting of 1912 AD Katmai tephra (volcanic ash), and within the next 5 cm down, we begin to find cultural debris, where newly-formed soil had developed into the cultural deposit. No obvious occupation surfaces were apparent in the top 50 cm of the site (Zone 1), but the zone contained artifacts of all sizes, from flakes to ocher-grinding palettes. Zone 2 is marked in Trench A by two dense, superimposed red ocher lenses, each no more than 7 mm thick and 5 cm apart. Charcoal and artifacts were found laying in horizontal positions on these surfaces. A hearth, along with other features found on the top ocher layer in a horizontal exposure excavated in 2001, suggests the ocher functioned as an interior floor. This floor dates to 5370 ± 60 bp (Beta 134790), or between 4335 to 4044 BC (at 2 sigma). Around its margins, especially in the Trench A profiles (Figure 7.9), we found overturned blocks of stratified tephra. The ocher floors extended flat under the tephra chunks, unlike semi-subterranean features that are commonly dish-shaped in cross section. We interpret this as evidence of tent occupation, with tent walls held down by sod blocks. Outside of the ocher floor area, in the block excavation to its north, Zone 2 slopes down hill and merges with a thick, black gravel midden 30–50 cm thick. Although we found many stone artifacts in the midden, our research revealed few faunal remains due to poor organic preservation. We did, however, recover small amounts of burned porpoise, bird, and fish bones; but these are too small and deteriorated to help address early Ocean Bay subsistence variability.

Below Zone 2 in Trench A are a series of thin, compacted, occupation surfaces, composed of charcoal and ocher (ca. 3–5 cm thick) and separated by silt-clay "fill" mixed with charcoal and artifacts. This is termed Zone 3 in Trench A. In the ongoing block excavation (see Figure 7.8), this zone is more complex and cannot yet be correlated with layers in Trench A. In the eastern end of the block, towards the front of the knoll, occupational debris is stratified into dense, horizontal layers with little fill separating them. In the western end of this block, occupational layers are thinner and have considerably more fill and sterile sediments between them (especially in the deeper layers). Near the top of Zone 3, a tephra horizon is found in most squares of the block excavation, but not in Trench A. The ash is more intact in the eastern-most squares and is less pristine in the center and western squares closest to Trench A. Radiocarbon dates closely bracket this ash between 5710 ± 70 bp (Beta 134789; above tephra) and 5880 ± 80 (Beta 134791; below tephra). The unweathered nature of this lens suggests that it had little exposure to weathering before being buried. This indicates that the site was reoccupied within years, not centuries. An arc of post-holes was found below a thin occupation surface just under the ash lens in one 2 meter square. We were, however, unable to follow the post-holes in other squares, where they should have shown up if forming a ring or other structure. A large slab hearth was found above the tephra in the same square. Otherwise, the excavation has turned up very few features. The lowest occupation lens, reached in the western end of the excavation, has been dated to 6570 ± 60 bp (Beta 161612) or 5623–5382 BC (2 sigma, outside range). A date from the Trench A profile is comparable, but with a wider margin of error (6600 +/– 230 bc, Beta 76738, or 5940 to 5,055 BC).

The primary goals of this excavation were to determine the antiquity of this site, the nature of habitation construction, and the composition of artifacts belonging to such an early Kodiak occupation. Abundant charcoal was collected from each natural/cultural and artificial excavation layer in each excavation square of the site. As previously mentioned, the dates returned so far suggest that Zone 3 of this site is comparable to the dates of 6,620 ± 60 bp from KOD 324 in Khizhuyak Bay, 6,220 ± 70 bp from KOD 224 on Uganik Island (dates cited in Jordan, 1992:130), a basal date of 6,080 ± 90 bp from the Rice Ridge site (KOD 363; Knecht, 1995:33), and a date of 6390 ± 70 bp (Beta 130189) from the Zaimka Mounds site (Amy Steffian, personal communication, 2002). The date of 7,790 ± 650 bp from the basal component of Crag Point (Jordan, 1992) is considered unreliable (Mills, 1994). This makes Tanginak Spring one of the earliest known archaeological sites in the archipelago.

How does this site match expectations from the model in Chapter 6? As already discussed, the site appears to lack semi-subterranean house features. Using sods to hold down tent edges may not be significantly more labor-intensive than hauling rocks from the beach, and sod blocks were probably used because large rocks were not immediately available. The post molds identified so far average around 10–15 cm in diameter, appropriate for poles around 5–10 cm with packing around them. These are considerably smaller than the posts found in the Koniag New Karluk sod houses, where at least one post preserved in place was over 25 cm in diameter (personal observation 1987). Taken together with the considerable association of red ocher with occupation surfaces, and the common occurrence of ocher grinding implements (see below), it seems likely that Tanginak houses were hide-covered tent structures, with small pole frames and ocher-coated skins. The close correspondence between the cessation of red ocher "floors" and inception of definitively semi-subterranean houses throughout Kodiak suggests that ocher pigment was functionally linked to the earlier habitation structures. Possibly ocher was used to process hides that served as tent covers; in addition, ocher may have been used as a water-proofing agent (Jewitt, 1987; Philibert, 1994).

The material remains recovered from the Tanginak Spring Trench A excavation are extensive in comparison to the rest of the Sitkalidak Archaeological Survey collection. (Results from the 1998 to 2001 excavations, beyond Trench A, are not sufficiently analyzed to report in this summary.) This trench produced 150 bags of flakes, each with several hundred pieces of debitage, in addition to 43 microblades, 69 microblade and flake cores, 23 bifacially flaked projectile points of various forms

Figure 7.10. Artifacts from the Tanginak Spring site (KOD 481): a.–b., e.–f., i.–j. bifacially flaked points; c. polished bit scraper; d., m., p. bifaces; g. microcore platform rejuvenation flake; h., n., r. macroblades; k.–l. pendants; o., q. chipped and polished adzes; s., x, microcores.

from stemmed points to bipoints, bifacially flaked chert knives, hammerstones, ocher grinders and a grinding palette, abraders, adzes, chert scrapers and cobble spall scrapers, and a serpentine pendant. The basal floor had a unique assemblage of full-sized blades, two unifacially flaked projectile points made from blades and two pecked stone lamps (Figure 7.10).

As discussed in Chapter 5, the series of blades, microblades, and projectile points appears to represent an evolution of technology through the occupation levels of this site. Large blades and core/blade manufacture of blanks for projectile points are isolated on the basal floor. Microblades co-occur in this level and continue through subsequent levels. Projectile points in the middle occupation floors are dominated by stemmed bifacially flaked points with pinched shoulders. Bipoints co-occur with these pinched-shoulder points and continue above them into the upper layers. One pinched-shoulder point was recovered from the Ocean Bay I layers of the Sitkalidak Roadcut site in 1963 (Clark, 1979:335, plate 1, item b), but the type is uncommon in Ocean Bay assemblages and probably is an early horizon marker. Bipoints are more common in Ocean Bay I assemblages and continue in low frequency alongside ground-slate tools through later Kodiak prehistory.

In sum, Tanginak Spring is a fascinating stratified site with great potential for clarifying many mysteries of early occupation in the Kodiak archipelago. Excavations are continuing to uncover a larger area and to further address questions of site formation, habitation construction, and technological development.

7.5.2. Bear Island Site (KOD 564)

The Bear Island site is located on the southwest side of Sitkalidak Island in a deep fjord-like bay. It is situated on top of a small, seven-meter-tall islet in an intertidal lake, closed off from the ocean today by a cobble beach. The islet is about 70 meters long and 30 meters wide with gently sloping sides, unlike any defensive contexts identified in the survey (Chapter 6). During our visits in 1993 and 1994, the lake levels were rarely over 15–30 cm and tide water only entered the lake at highest tides when sea water backed up the short outlet stream.

In our first visit to this location in 1993, we encountered a pair of adolescent brown bears wrestling along the lakeshore a few meters from the islet. We moved on, unaware of the archaeological significance of the landform. When we returned a month later, we were able to climb the islet and look around. We found a series of small house depressions along the crest of the island. A test pit near the top revealed a stratified, multi-component series of occupation floors, red ocher layers, volcanic ash lenses, and a significant collection of artifacts from both a ground-slate and flaked chert tradition. A radiocarbon analysis of charcoal from one of the lower components about a meter below surface returned a date of 5300 ± 110 bp (Beta 71092), which calibrates to 4350 to 3940 BC and 3860 to 3820 BC. The second interval is fairly far out on the tail of the probability distribution (all between 1 and 2 sigma), and it is more likely that the actual date falls within a few hundred years of 4000 BC. The presence of house pits on the top of this site, with a series of occupations reaching back into the Ocean Bay I phase, made this a hopeful location for examining the transition to semi-subterranean habitations.

In 1994, excavations at Bear Island began by opening two 2 × 2 meter squares in one of the more distinct house depressions at the top of the islet (Figure 7.11). These squares were oriented in checkerboard fashion so that the walls of the units collectively bisected the house in two directions, leaving a record of the house profile and stratigraphy (Figure 7.6). The excavation proceeded by natural stratigraphic layers. Dry screening was attempted in a quarter-inch mesh rocker screen, but was ultimately abandoned when it proved impossible to get the wet silt-clay substrate through the screen. Artifacts were piece-plotted on level forms, and charcoal, calcined bone, and debitage were collected by 1 × 1 meter quadrants per level. In ten days, a crew of six managed to excavate the larger units to only

Figure 7.11. Nicole Misarti and Laura Starro excavate a unit at the Bear Island Site (KOD 564).

about 60 cm below surface, through two components. We had to discontinue the excavation at a thin layer of volcanic ash that more or less marked a sterile transition between upper and lower occupations. A single 1 × 1 meter quadrant was excavated down to the sterile substrate to get a controlled record of the full stratigraphic profile, and an additional trench (250 × 50 cm) was excavated off of the north side to extend the house pit profile, and to capture the transition from the zone of house construction to the natural stratigraphy outside the house. Under the pressure of time, as our chartered fishing boat waited to move us to the next field site, we drew profiles, back-filled the excavation, and closed up camp.

The house targeted for excavation was roughly circular and measured approximately 4 meters in diameter. Figure 7.12 illustrates the two bisecting stratigraphic profiles of this excavation. For our purposes, this excavation can be divided into three stratigraphic zones. The top zone, Zone 1, is the actively weathering, (O/A) soil horizon. A cultural deposit (Feature 1) is located at the base of this zone. A radiocarbon date of 1720 ± 60 bp (1815–1525 BP: Beta 78506) from this deposit marks the terminal occupation. This date falls in the middle of the Late Kachemak comparable to a date from the floor of the semi-subterranean house (see below), suggesting that the Feature 1 charcoal was originally deposited on the

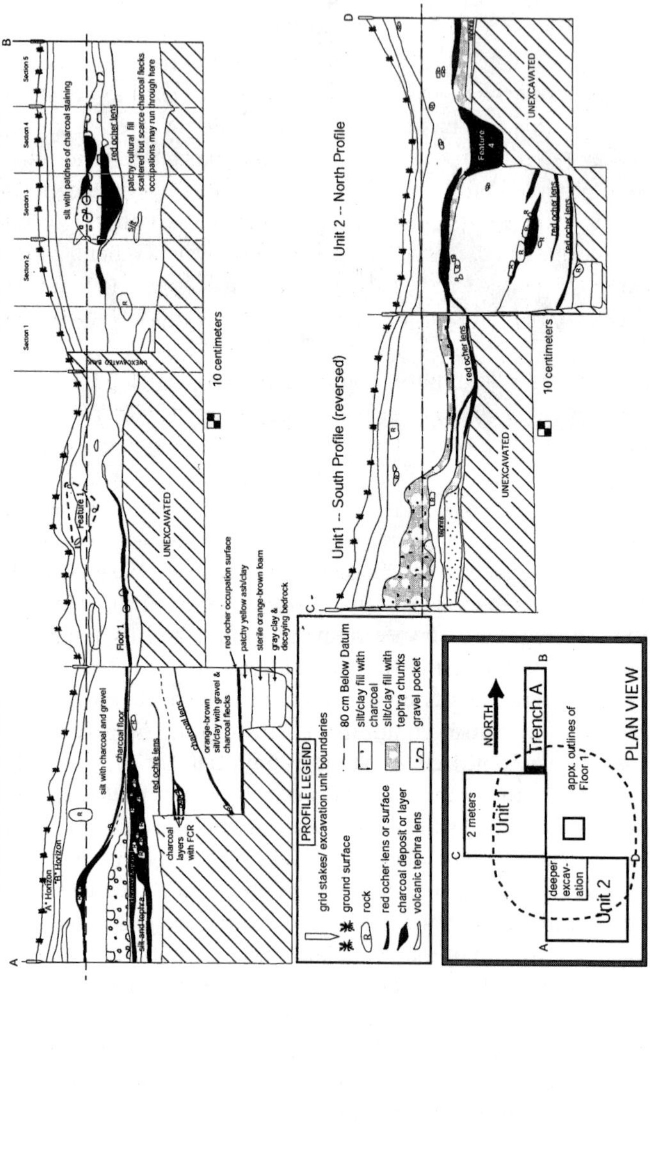

Figure 7.12. Stratigraphic profiles and reference plan view of the 1994 excavations in the Bear Island site (KOD 564). See text for description.

roof sods of the pit-house or possibly in a midden (now decomposed) that rested against the house walls.

Zone 2 begins at the base of Feature 1 and the O/A-horizon. About 30 centimeters of silt (with scattered charcoal and gravel) separate Feature 1 from the first distinct house floor. This fill contains various ground-slate artifacts similar to those associated with the occupation layers below it, including ground-slate knives with serrated stems, diagnostic of the late Ocean Bay II and Early Kachemak phases (Figure 7.13). Given the Late Kachemak date from Floor 1 (below), the tools found in the fill (roof sods) may relate to an older deposit, disturbed in house construction. At the base of this fill, Floor 1 is a thin but dense layer of charcoal and compacted earth. Its profile preserves a semi-subterranean contour and fits the house definition observed on the surface. Found on this floor were ground-slate points, knives, lance fragments, and other implements. Two ulu fragments found in this and the lower strata of this second zone support the Kachemak rather than Ocean Bay affinity. A radiocarbon analysis on charcoal from the Floor 1 deposit produced a conventional radiocarbon date of 1890 ± 70 bp (1970–1630 BP, Beta 91316).

The southern end of the north-south profile (A–B in Figure 7.12) shows the complex stratigraphy underlying the wall associated with Floor 1. A large gravel deposit is pinched off by the Floor 1 deposit. Below that is a dense layer of charcoal with fire-cracked rock and a few patches of silt. This appears to have been an occupation surface that was post-dated by an event of gravel deposit (possibly a tsunami-surge deposit or construction to solidify muddy surface) and then, after an interval of time, the construction of the semi-subterranean house associated with Floor 1. This scenario is complicated by the convergence of the two occupation deposits in the center of Floor 1; nevertheless, Floor 1 and its associated house appear to have cut into this next lower cultural stratum.

Zone 2 ends below the second dense charcoal deposit resting on silt clay sediments with cultural refuse intermixed. A lens of pure beige volcanic ash is preserved in several areas of the excavation, below Zone 2. The ash is disturbed near the center of the excavation area, presumably by the subsequent activities of the site's occupants. This distinctive feature is useful as a stratigraphic marker, since it is continuous in many areas; and, where it has been cut through, the sediment contains pockets of the ash mixed into the matrix, providing a means to easily distinguish the nature and extent of disturbance. This ash appears very similar to an ash deposit found in excavations at the Old Harbor Lighthouse site (SAS 127/KOD 089), which has been tentatively dated to about 4500–5000 BP by association with a dated Ocean Bay II floor. If the two deposits originated in the same tephra fall, the ash in the Bear Island site strata supports an Ocean Bay II age for the cultural deposit immediately above and below it. In all but a single 1 × 1 meter quadrant, the 1994 excavation was discontinued at the level of this ash.

Figure 7.13. Artifacts from the Bear Island site (KOD 564). A mix of Ocean Bay and Kachemak materials.

Zone 3 is known only from the 1 × 1 meter quadrant excavation (and the lower sections of the 1993 test pit). This zone is distinctly different from the dense cultural deposits above. Thin lenses of red ocher and charcoal slope down to the south. Artifacts include flaked chert points, an asymmetric biface, a hammerstone, and a spherical ball of soapstone or clay of unknown use (about the size of a cue-ball, with a thin axial perforation) (Figure 7.13).

In summary, the test excavation at the Bear Island site revealed three components. The earliest occupants maintained a chert flaking technology, occupied the site sometime between 5000 and 6500 years ago, and presumably left the excavated part of the site unoccupied for long intervals between short visits. Somewhat later, people began to use the site more intensively about the time of a major ash fall (or just after). This second occupation left abundant charcoal deposits and ground-slate tools known to be associated with both Ocean Bay II and Early Kachemak traditions. Finally, a Late Kachemak deposit clearly included a semi-subterranean habitation, while older deposits at the site do not preserve distinctive structural features in the sediments excavated to date. Future research at this site may yield important insights regarding the development of semi-subterranean house structures, the transition from Ocean Bay to Kachemak, and/or the poor representation of Early Kachemak sites in the Sitkalidak region.

7.5.3. Partition Cliff Site (KOD 473)

The third extended test excavation was conducted at the Partition Cliff site on the Pacific side of Sitkalidak Island. This site is located on top of a cliff 15 meters above a boulder-covered, surf-pummeled beach, about 2 km from the *Awa'uq* Refuge site. On the first day of our survey in 1993, we visited this location on a whim to get a different view of the refuge and the outer coast of Sitkalidak. The site area is limited to about 7 × 7 meters and appears to be eroding over the cliff. No house features were identified, but several large whale bones were found projecting out of the sods.

Tests in 1993 revealed a pavement of flat slabs and a fishing line sinker beneath a whale cranium. In 1994, a crew of nine returned for a day to expose a 4 × 4 meter area of the site. Figure 7.14 is a plan view of the cleared portion of the site. Three whale vertebrae, a single cranium, a mastoid process, and several fragments of ribs and other bones suggest that the assemblage includes the remains of a single small whale (around 10 meters long) of unknown species (Figure 7.15). The pavement included cobbles, flat slabs, and angular rocks, probably brought up from the beach below. The orientation of flat rocks and slabs suggests intentional placement to create a flat surface, near the western side of the excavation. A hearth area with charcoal and calcined bones (primarily fish) was discovered in the southwest corner. Charcoal from this feature generated a date of 120 ± 50 bp (AD 1665–1950: Beta 78508). Based on the pre-contact nature of the deposit, I favor the pre-Russian contact end of this range (AD 1665–1784). The shallow nature of the cultural deposit argues against any antiquity, so an occupation within the 18th century seems likely.

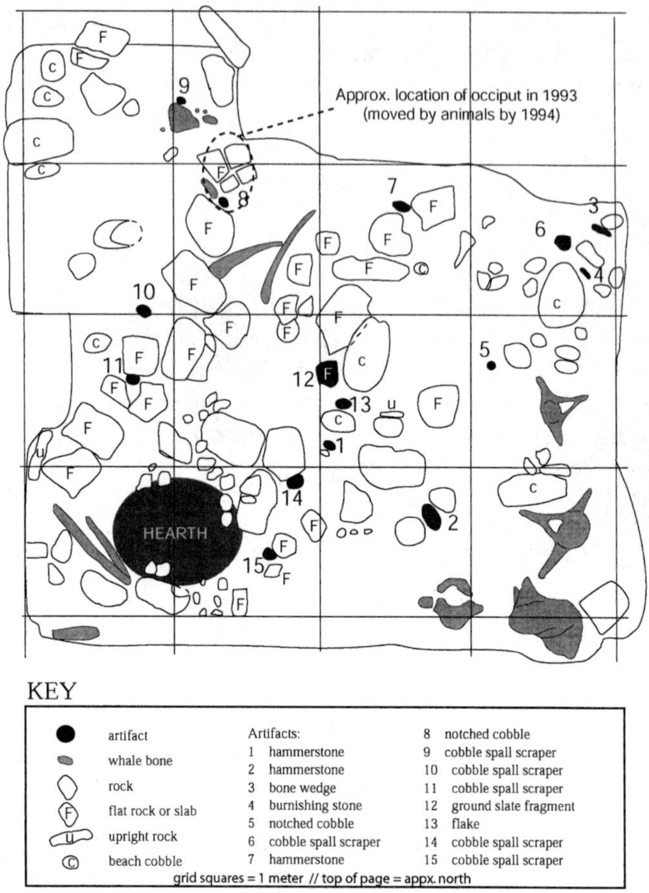

Figure 7.14. Plan of 1994 excavation at the Partition Cliff site (KOD 473).

Recovered artifacts include five large, crudely-formed cobble choppers/ scrapers, three hammerstones, a whale-bone wedge, three grooved-cobble fish-line sinkers, and a burnishing stone. In addition, we found a decomposed fragment of something that may have been pottery, and a large beach rock with a hollowed out "lamp-like" (but apparently unmodified and uncharred) depression.

Two activities represented in this artifact assemblage are supported by the site layout. Most notably, the dominance of large and crude implements (choppers, hammerstones, and a bone wedge) implies a processing function for the site. The presence of partial remains from a single whale suggests that the site was used for a short period to butcher a whale. Whaling practices in the late prehistoric and early historic period often relied on luck, vigilance, and knowledge of local currents to

Figure 7.15. Close up of whale cranium from 1993 testing at the Partition Cliff site (KOD 473). Photo by author.

land whales in appropriate places. One might imagine that this particular whale washed ashore in the small cove beneath this cliff, forcing hunters to deal with it there.

The second activity recorded in this assemblage is fishing. Three fishing weights and the presence of burned fish bones in the fireplace suggest that at least one meal of fish was caught near the site and cooked and consumed during the short occupation. Lack of evidence of structures suggests that, if this location had been occupied for more than a single day, habitations were hastily built and perishable. Russell (1991:21–22) records use of temporary shelters made traditionally by Alutiiq in the bush. These included branches of mountain alder tied together or cut and formed into a teepee, with grass and moss used to cover and insulate the structure. An alder shelter would not leave many archaeological traces after a few decades in such a moist environment; but, if it was standing during the delineation of activity areas, patterns of stones, artifacts, and faunal materials might reveal its location. To date, the spatial analysis is inconclusive.

The Partition Cliff site is unique in the Sitkalidak assemblage as a processing location with clear functional correlates, and it is peculiar in its context on top of a cliff. More than any other site in the Sitkalidak area, this provides a window into late prehistoric subsistence activity and non-habitational land use. Its character is consistent with ethnographically documented hunting practices (Chapter 4).

7.5.4. Lighthouse Site (KOD 089)

The Old Harbor Archaeological Field School project was organized to introduce village students to the archaeology of their region and give them access to the rich historical and cultural heritage buried beneath their feet. The Lighthouse site is situated at one end of the village of Old Harbor (Figure 7.16). Semi-subterranean house pits (barabaras) can still be seen there. In the late 1980s, a road cut through the area exposed cultural stratigraphy and artifacts characteristic of the Russian-American phase. Bulldozers, however, obscured many of the house depressions. In April 1995, we tested areas of the site and selected two adjacent deposits for excavation by the Old Harbor junior high and high school students. One of the areas was a slight depression with a few large stones forming an apparent wall on one edge. The other area was a more clearly delineated house pit without side rooms. We established a datum point at the highest point of the site (100N/100W) and gridded both excavation areas from this point. Excavation proceeded in a checkerboard grid, leaving every other 1 × 1 meter square unexcavated. This preserved north-south and east-west profiles at meter intervals and saved 50% of the excavation area for future examination. This strategy was considered most appropriate for student excavators. Local datums were established for each pair of excavation squares. Excavation was conducted by arbitrary 10 cm levels, each broken by distinct natural stratigraphic breaks when noted. Notes were maintained by level within each unit, and included three-point plotting of artifacts, large rocks, and features. All sediment was sifted through quarter-inch mesh shaker screens.

The 1995 excavations lasted over three weeks, and exposed two Russian-American floors. In April and May 1996, we returned for a second season. Time constraints limited the dig to a single week, but we were fortunate to have the high school students for five full school days. The students completed excavation of the squares opened in 1995, from approximately 30 cm below surface, down to sterile silt and glacial till underlying the site at an average depth of about 100 cm below surface.

Both floors exposed in 1995, as well as most of the historic debris found in surface collection, date to the late Russian-American phase in the interval between AD 1855 and 1865 (as recorded in distinctive ceramics traded into the region only in this interval), and both structures appear to have been semi-subterranean. The House 1 excavation revealed a floor covered in fire-cracked shale (FCR) and charcoal. Birch bark rolls and salmonberry seeds were also common on the floor, as were various fragments of iron barrel hoops, green wool, burlap fabric, buttons, glass, and ceramic fragments. A small number of ground-slate tools and worked-slate pieces were found above the floor, and may be either contemporaneous or earlier in date, having been raised above the floor deposit in roof sods.

The floor in the southern squares was different in character. Charcoal and FCR were absent. The floor was only noted by a heavily-compacted earth layer

Figure 7.16. The Three Saints Orthodox Church in Old Harbor. Sven Haakanson Jr. and the author in foreground. Photo by L. Bilaniuk.

Figure 7.17. High school students Alexandria Christiansen and Gordon Naumoff, Jr. during the 1996 excavation of the Old Harbor Lighthouse site (KOD 089).

with some cloth and other historic debris. This floor was connected to the charcoal floor but was separated by a step, or drop, of about 5 cm from the burned floor to the clean earth floor. The nature of these two areas is unclear without further excavation. The FCR-covered surface may have been a cooking or storage area, based on the extent of FCR and charcoal and evidence of storage vessels (iron hoops from wooden vessels), fragments of ceramic plates and cups, and a thick hearth-like ash and burned-earth feature near its center. Alternatively, this might have been a privy or garbage deposit area, as suggested by the (indigestible) salmonberry seeds and overall clutter of the floor in this area; in this case, it was probably a side room that was converted into a dump area when the room was no longer needed for other purposes.

House 2, partially excavated in 1995 by the junior high school students, revealed a charcoal-stained floor and possible hearth area. The floor was associated with similarly aged Russian-American ceramics, a few fragments of iron, and ground-slate tools. A small area was exposed and few artifacts were recovered, making it difficult to say much more about this house on current evidence.

The 1996 excavations below the Russian-American period floor of House 1 exposed several cultural strata dating to the late Ocean Bay II interval and possibly an even earlier occupation layer. Figure 7.18 shows a north-south and an east-west profile with the House 1 floor shown near the surface along with several

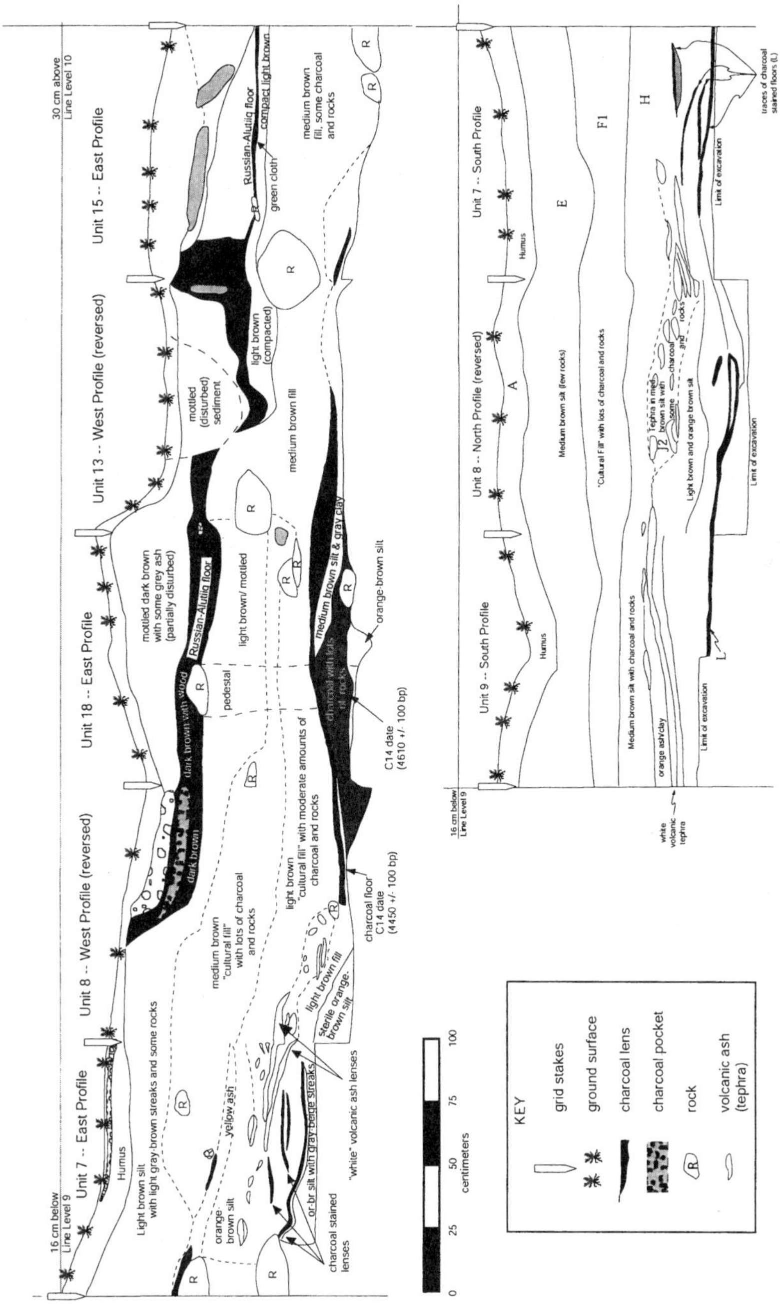

Figure 7.18. Stratigraphic profile of the 1995–1996 Old Harbor Field School excavations at the Old Harbor Lighthouse site (KOD 089) showing a late Russian era floor overlying Ocean Bay II fill, charcoal and volcanic ash lenses.

underlying strata. Below the Russian-American floor (C), a zone of sterile silt (E) covers a thick layer of cultural fill (F1), with lots of charcoal and gravel mixed with silt. The artifacts in this layer suggest an Ocean Bay II to Early Kachemak association, similar to the intermediate component of the Bear Island excavation. The dominant diagnostic artifact in this assemblage is the same notched-stemmed ground-slate knife as recovered from Bear Island site.

Below the cultural fill layer, two distinct lenses are present in different parts of the site. At the northern end, away from the Figure 7.18 profile, the cultural fill graduates into more homogenous brown silt (H) and rests on a more or less continuous lens of white volcanic ash (J2), which slopes down to the south. This ash becomes increasingly discontinuous, moving south along the Figure 7.18 profile, until it disappears and is replaced by a dense black charcoal-rich feature (I and K). This feature pinches off into a thin charcoal-stained occupation surface at its edges and was more difficult to trace away from the center of the feature. This feature also contains an upper (I) and lower (K) lens of charcoal separated by a pocket of white tephra or fire ash away from the profile, and by light brown silt (J3) in the profile. Charcoal from the upper and lower lenses of this feature was collected from a test pit in 1995. Radiocarbon dates on the two lenses were 4450 ± 100 bp (5320–4835 BP, Beta 83510) for the upper lens and 4610 ± 100 bp (5585–4980 BP, Beta 83511) for the lower. The large overlap in these two dates renders them virtually indistinguishable from one another in probable age, suggesting little more than a resurfacing of the feature. When encountered in the test-pit excavation, the lenses were thought to represent distinct floors; but results from the larger excavation suggest a single hearth feature based on the small circumference and thick nature of the deposit. The occurrence of what appeared to be volcanic ash (rather than fire ash) between the two layers of charcoal would suggest a post-tephra occupation for the upper lens. The lower lens could have been deposited prior to, or after, the tephra fall, depending on how the ash got into the feature. Alternatively, this ash may, in fact, be fire ash with a coincidentally similar appearance to the volcanic ash less than a meter away. This would be more consistent with a hearth interpretation for the feature. Geochemical analysis will ultimately resolve this issue.

This hearth feature clearly dates to the Ocean Bay II phase and may have been inside a semi-subterranean house or tent. If the structure were a pit-house, the overlying cultural fill could be contemporaneous midden residue that was on, or against, the outside wall and fell into the house after its collapse. Alternatively, it might be decomposed midden from habitations located outside of the limits of the excavation. The stratigraphic position of the tephra deposit and the Ocean Bay II hearth and floor surface suggest that the ash fall did not precede the dated occupation by any long period of time. As at the Bear Island site, the continuous nature of the ash away from the immediate zone of occupation suggests that people did not remain at the site after the ash fall, giving it time to settle and become sealed. However, occupation must have followed the tephra deposit reasonably closely, covering areas of ash before it could weather and disturbing the tephra in the center areas of the excavation where the hearth feature and floor are found.

In addition to the notched-stemmed ground-slate knives, noted above, the cultural fill and underlying feature and floor strata contained several ground-slate lance points as well as worked slate. By far the dominant tool, the point or lance constituted 50% (N = 16) of the artifacts from these levels. Several basalt blades and a chert microblade were also found in the cultural fill and charcoal feature layers. It is possible that these artifacts were dredged up into the fill of the Ocean Bay II occupation from an older Ocean Bay I phase occupation at the site.

* * *

With the detailed presentation of surveys and excavations conducted, in this chapter we established the parameters of the data that are available for analysis and model testing. In the following chapters we proceed to examine the hypotheses developed in Chapter 6 with focused analyses at different scales, from sites and settlement comparisons in Chapter 8, to houses and population estimates in Chapter 9.

Chapter 8

Site Scale Analyses

8.1. INTRODUCTION

In this and the next chapter, we examine evidence from the Sitkalidak Archaeological Survey project that addresses predictions of the "four act" model in Chapter 6. To do this, we need to reorganize the predictions into data-specific groupings. The lines of evidence available to test these predictions, controlled for time, addressed in this chapter include: component frequencies as a measure of changing settlement density; site size measures of population aggregation (site area and number of houses per site); measures of site function variability (spatial and geographical parameters and artifact assemblage character); and settlement patterns. In Chapter 9, we examine social inequality and trends in demography by means of: house attributes as a measure of social variability; and two methods of population estimation (the house/site area method, and the carbon-date frequency method). Predictions specific to each data category will be presented at the beginning of each analytical section and the findings summarized at the end of each section. Chapter 10 assembles all predictions and their results by time period, enabling a comprehensive view of the relative success of the model.

8.2. COMPONENT FREQUENCIES AS A MEASURE OF CHANGING SETTLEMENT DENSITY

The model in Chapter 6 predicts an increase in settlement densities throughout the archaeological history of the archipelago. Stabilization may occur when population packing first causes resource depression, but then population should increase again when labor and technological intensification "solve" the initial population bottleneck. Because most sites documented in the Sitkalidak Survey are residential in character, component frequency can be used as a measure of changing settlement density through time in this region (with the possible exception of the Early Kachemak). As a measure of occupation intensity, we expect that actual frequencies are underestimated, that earlier sites are underreported due to deterioration, and that later sites are underreported due to continuous occupation through phase transitions. For reasons to be discussed later in the chapter, the Developed Koniag is probably better represented than preceding phases.

Table 8.1 and Figure 8.1 document changes in component frequencies by period and phase in the Sitkalidak study region. Frequencies per period and phase are included in Figure 8.1 (dark bars) to illustrate the importance of correcting

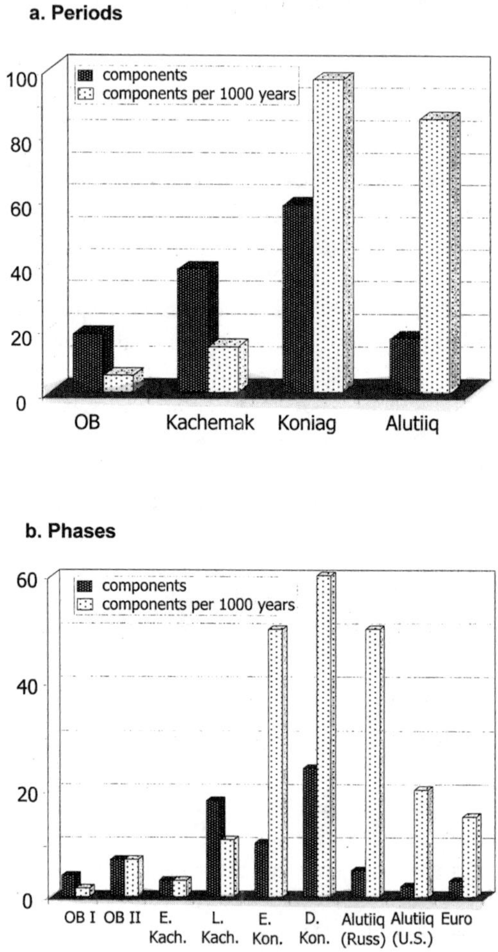

Figure 8.1. Frequency of archaeological components identified in the Sitkalidak Archaeological Survey. Components are chronologically distinct occupations, in this case defined by culture historical (a) periods or (b) phases. Multi-component sites are counted once for each component represented.

Table 8.1. Number of archaeological components from the Sitkalidak Archaeological Survey identified to (a) period, (b) phase.

(a)

	Ocean Bay	Kachemak	Koniag	Alutiiq
components	18	38	58	17
components per 1000 yrs	5.14	14.07	96.67	85.00
duration of period	3500	2700	600	200

(b)

	OB I	OB II	E. Kach.	L. Kach.	E. Koniag	D. Koniag	Alutiiq (Russ)	Alutiiq (U.S.)	European
components	4	7	3	18.00	10	24	5	2	3
components per 1000yrs	1.60	7.00	3.00	10.59	50.00	60.00	50.00	20.00	15.00
duration of period	2500	1000	1000	1700	200	400	100	100	200

for interval length. The time-corrected patterns (light bars) are considered more representative of changing site densities. Figure 8.1a includes components that could only be classified to period but not phase (increasing the statistical reliability of the pattern, but reducing temporal resolution). These data support a general increase in site occupation from Ocean Bay through Koniag periods, with a slight decrease in the Alutiiq period. As we will see, grouping at this coarse scale, while statistically robust, hides important chronological variability.

Figure 8.1b shows the site frequency data by phase (excluding components that were only attributed to period). The higher resolution of the corrected data better shows variation in settlement and land use throughout the prehistoric epoch. The Early Kachemak decline is expected from the gap in radiocarbon dated sites between 4000 and 1800 bp, discussed in the last chapter. Thereafter the data confirms a steady increase in settlement intensity from the Late Kachemak to Developed Koniag. This is consistent with expectations for increasing populations and land use diversification through time, although neither issue can be directly addressed with these data. It is important to note that these figures ignore variation in site character and function, and they do not accurately reflect changes in relative population density through time. These parameters will be examined in more detail later in this and the following chapter.

8.3. SITE SIZE MEASURES OF POPULATION AGGREGATION

Site size is a useful measure for a number of reasons. Under appropriate conditions, it can help reveal regional-scale settlement organization and population size (e.g., Brumfiel, 1976; Stepponaitis, 1978, 1981; see Chapter 9). Site size can be measured in different ways. Here we consider two: site area and number of houses.

According to the model, we predict a continuous expansion of site size, from the uniformly small residential sites of the Ocean Bay I and II phases (with room for fewer than 5 small tents or houses) to the largest sites in the Koniag period (a few with 50 or more houses). We also expect increased variation in site area through time, as modest outpost sites continue to be used seasonally by small groups in the Kachemak and Koniag periods.

8.3.1. Site Area

Site areas are determined by field measurements of site dimensions, where discernable. Site perimeters are characterized by a rapid decline in the density of cultural material and features towards a depositional environment predominantly generated by non-cultural processes. In many cases, soil probes, test pits, erosion profiles, visible landform changes, and anthropogenic vegetation provided sufficient criteria for approximate measurement. In other cases, deeply-buried components could not be measured directly. No attempt is made here to compensate for site loss where erosion had occurred, and we assume, for reasons discussed earlier in this chapter, that site destruction has not significantly affected general patterns in this area. While this assumption may be questionable (Donald Clark, personal communication, 1996), we have few alternatives. In general, site area (not component area) is measured because of the almost insurmountable difficulties associated with isolating separate component boundaries within a multi-component site on the basis of minimal excavation. This method reflects an exaggeration of areas for components in multi-component sites, a bias that must be acknowledged and minimized through careful analysis.

In the primary analysis of Sitkalidak site area measurements, only period data are used because phase sample sizes are too small to generate robust comparisons. This raises a problem for one of the model predictions, that of expected changes between Ocean Bay I and II phases (Acts I and II). According to the model, there should be no significant change in site size between these two phases. Despite the small samples, phase averages provide tentative support for the prediction. The average component area for Ocean Bay I is 1146 m^2 and for Ocean Bay II is 1216 m^2, and the samples are statistically indistinguishable ($t = -0.838$ at 2 df). The Ocean Bay II sample has a much greater range of variation, including the smallest and largest Ocean Bay sites. This is expected for a larger sample drawn from identical populations. Given these observations, we can combine Ocean Bay I and II samples together with other Ocean Bay sites that could not be specified to phase for comparison with other periods.

Figure 8.2 summarizes the metric area data from the Sitkalidak sites by period. As expected, these graphs show a general trend of increasing component size through time, until the Alutiiq period. Ocean Bay sites are generally small, from 25 m^2 to just under 3500 m^2 (mean = 804 m^2). As with the other histograms in Figure 8.2b, the Ocean Bay distribution is skewed to the right, with the greatest frequency in the low range. While a few sites appear to be larger, contemporaneous occupation areas probably did not exceed 750 to 1000 m^2.

Despite the trend of increasing size through time, both plots show that Ocean Bay and Kachemak samples are quite comparable in median and quartile ranges, and the t-test of means also shows no significant differences. Analysis of the Kachemak sample suggests a second mode of site size developing in the 2000–4000 m^2 size range; these larger sites suggest the development of a "village" class of sites in the Kachemak period (see below). One site stands out as unusually large (10,800 m^2); it is an unexplained outlier in this analysis. This site is one of few Early Kachemak sites documented in the Sitkalidak Survey and is unique in many respects other than size: deep and oddly shaped pit depressions and deep greasy black sediments. Future research could help discover reasons for these anomalies.

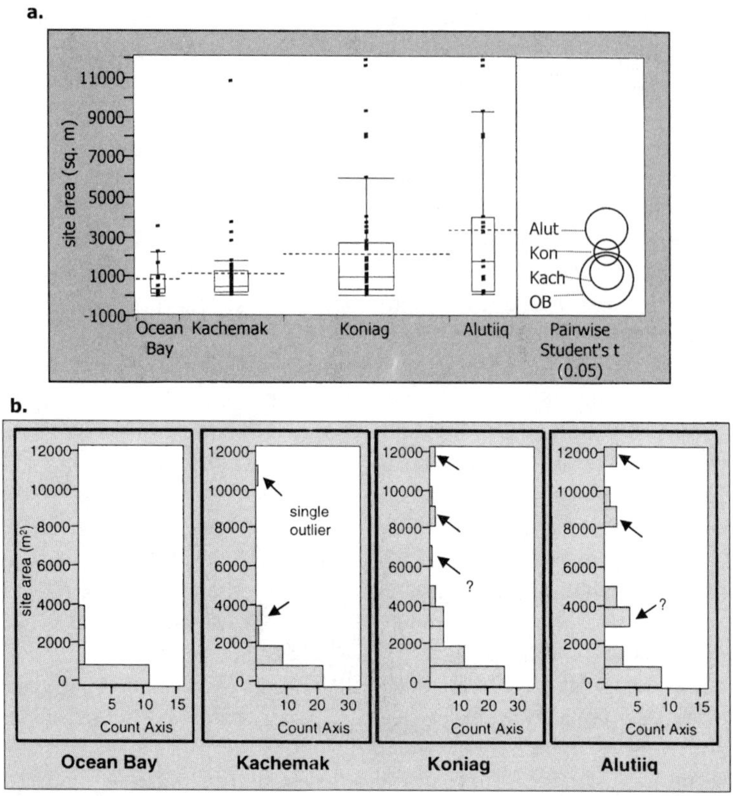

Figure 8.2. Comparisons of site area data by period. (a) Quantile box and dot plots and arithmetic means (dashed lines). Pairwise Student's t comparison circles show statistical similarities based on arithmetic means and standard deviations. Increasing circle overlap indicates greater similarity. Where the outside angle of interception between two circles is less than 90 degrees (including complete separation) the samples are statistically different at the 95% confidence level. Pairwise Students t assumes normality, which is violated in these cases. The circles are shown for their ability to illustrate degrees of statistical difference between samples. (b) Histograms of site area by period showing distribution shapes and subsidiary modes (arrows). Two largest modes in Koniag and Alutiiq sites are from the same multicomponent sites.

While these data suggest many similarities between Ocean Bay and Kachemak sites, there are also significant differences. The small number of large sites in the Kachemak sample suggests a significant reorganization of settlement. As well, Ocean Bay deposits lack the surface pit-houses which dominate Kachemak components. One might propose that some large Ocean Bay sites are exaggerated in size by overlay (palimpsest) deposits that have grown out of proportion to any single

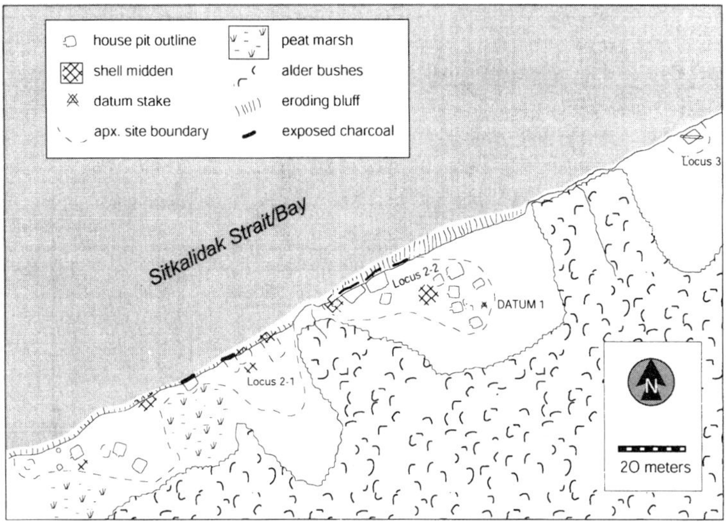

Figure 8.3. Plan of the Tanginak Bluff site (KOD 384). There are four areas or loci of occupation, which may or may not be contemporaneous. An unknown amount of the site has eroded into the bay. While the primary occupation is Late Kachemak in age, the location was occupied in Ocean Bay and Koniag periods. An Early Kachemak village is only a few hundred meters to the east.

occupation by the unconstrained nature of tent use, while larger Kachemak sites better represent the area of use due to the recycling of existing semi-subterranean features.

We expected to see an increase in larger site sizes for the Kachemak period as a result of village formation at this time (Haggarty et al., 1991:123; Jordan and Knecht, 1988:231–232). Villages involve the establishment of semi-permanent residences in aggregations larger than the economic unit of production (family, camp, or band). Superficially this documented change is not markedly apparent in site areas from the Sitkalidak region. In addition to the anomalous outlier on Figure 8.2b, one other Late Kachemak site may be considerably larger than the norm (Figure 8.3). The Tanginak Bluff site (KOD 384) is composed of three loci. (Because of their discrete spatial orientation, each locus is included as a different component in Figure 8.2.) The site's aggregate size is estimated at 3950 m^2, but it is one of the few in the Sitkalidak area that show signs of extreme, and ongoing, erosion. The original site was probably considerably larger. Nonetheless, the increased variation in site sizes, and the inference of villages, suggest that the most frequent small mode of sites in the Kachemak period may relate to more specialized purposes than comparable sites in the previous Ocean Bay period. We consider this further in additional analyses below.

In the Koniag period we see much greater variation in component size. Four modes can be distinguished in Figure 8.2b (at less than 500 m^2, 4000 m^2, 8000 m^2, and 10,000 m^2). These modes fit expectations for a logistically-fixed settlement and land use strategy with large, aggregated winter villages, seasonal settlements, and camps or locations for targeted extraction of resources. Most of the variation in this sample appears in the better represented Developed Koniag phase, but the pattern probably relates to the Early Koniag as well.

The Alutiiq distribution is very similar to Koniag and appears to retain the same multi-modal structure (the larger modes are dominated by sites attributed to both periods). This observation supports the notion that much of Alutiiq traditional settlement and land use practice continued into the Russian and American phases, despite drastic social, political, and demographic effects of Russian, and then United States', colonization and dominance. It is important to point out that most of the larger sites in the Alutiiq distribution date to the first 55 years of the Russian phase, before diseases devastated the populations and before Russian officials mandated resettlement into relatively few Alutiiq villages (Crowell and Lührmann, 2001:60; see also Clark, 1984a:187).

Despite the abandonment of a large number of settlements around the Sitkalidak area, traditional modes of subsistence continued to dominate into the Alutiiq-American phase, when we still see a few small sites that were probably seasonal subsistence camps. One of the large settled villages of the Alutiiq-Russian phase was near the early Russian outpost of Three Saints Bay (just beyond the limits of this survey). This village appears to have been moved first to Barling Bay, where it was later mapped by Alphonse Pinart (Clark, 1974a:7–8), and finally to the modern location of Old Harbor, sometime in the late 1800s. The largest Alutiiq-American phase site reported in these data is found in Barling Bay and could have been the same site that Pinart identified. Measurements for the modern village of Old Harbor are not included, although photographs from the 1880s and 1890s show a large village (Figure 8.4). The site is under the modern town and little of its original definition can be discerned.

The data in Figure 8.2 generally support the expectation that sites increased in area and variability from Ocean Bay and Kachemak to Koniag periods. As predicted, Ocean Bay I and II sites are not significantly different in size. However, we also see less site enlargement between Ocean Bay and Kachemak periods than expected. While the larger sites of this time period include the first recorded villages, the site area measurements do not show very much difference in site size before the Koniag period, which began only 850 years ago. This could mean that the expectation of expanded site size with the emergence of intensive storage-based subsistence economy is incorrect, that most Kachemak winter sites were relatively small, or that we have a bias against recording the larger Kachemak sites. How likely is it that larger Kachemak or Ocean Bay sites are buried beneath large Koniag villages?

Figure 8.4. Residents of Old Harbor and visitors pose in front of a sod house in 1888–1889. Courtesy of the U.S. National Archives (22-FFA-266).

Attempting to answer this question, we can look at the number of pre-Koniag components that were identified in, or under, Koniag components. It is assumed that only the larger Koniag sites could mask larger earlier sites, so the focus is on those 18 Koniag deposits that are listed as having site areas greater than 2000 m^2, the upper limit of most earlier distributions. Of these, only three sites had evidence of underlying Kachemak components and none revealed Ocean Bay strata in test excavations or erosion profiles. (One large Alutiiq village site, KOD 093, did have a deeply buried Ocean Bay II component). While site testing was limited, a greater number of older components should have been identified if large Koniag sites commonly covered older deposits. Note that several multi-component sites were identified in the survey sample, but most were smaller than 2000 m^2 and would not hide a larger pre-Koniag component.

Diagnostic house forms might also provide a clue to site overlap, assuming that areas within Koniag or Alutiiq sites preserved Kachemak house-pits observable on the surface. House area statistics reported in Chapter 9 suggest a limited amount of overlap (as might be expected). However, using these same 18 Koniag sites, only two contain obvious Kachemak houses, and three others are possibilities. This analysis suggests that large Koniag components have not swamped or obscured many earlier components, even though occasional unidentified buried components probably do exist. One possible explanation might be the tendency for older sites around Sitkalidak to be elevated and landlocked by ongoing tectonic processes, thus creating a "horizontal stratigraphic" effect (Fitzhugh, 1996; Gilpin, 1995). Another explanation may involve village locations shifting to outer-coastal areas in Koniag times, discussed later in this chapter. Both processes are likely.

In summary, general trends in site size indicate minimal changes in settlement sizes prior to the Koniag period, and they specifically point to significant reorganization of settlement and land use in the Koniag period. These data support the model predictions, although a more dramatic increase in site sizes during the Kachemak period was anticipated. This evidence suggests that most Kachemak villages were relatively small by comparison to the larger Koniag sites. An alternative measure of site size (numbers of house structures per site) follows.

8.3.2. Number of Houses Per Site

Measurements of the number of houses per site offer the best available method for examining trends in settlement size in the later prehistoric periods. Figures 8.5 presents the house frequency data for this comparison. Because the Early Kachemak, Early Koniag, and post-contact Alutiiq phase samples are too small to be meaningful, this analysis focuses on the period comparisons. Throughout these comparisons, sites marked with "0" houses had no semi-subterranean pit-houses; they may, however, have had other forms of above-ground habitations.

Figure 8.5 shows the Sitkalidak data for number of houses per site, including a statistical summary and histograms. Beginning in the Kachemak period, data reflect a nearly uni-modal distribution of sites with variable numbers of houses. Within the main mode (0–10 houses), the most common numbers of houses are 2, 5, and 10 (not captured in the graph). A single site with 19 houses forms the second mode and stands out as truly large. The large site is the same anomalous Early Kachemak site discussed in the previous section. If data from the different loci of the Tanginak Bluff site (KOD 384; Figure 8.3) were combined, the total would be 22 houses. Other than these two big sites, the Kachemak period is characterized by small sites, with few exceeding 10 houses. The sub-modality in this sample could indicate a seasonal shift between aggregated and disaggregated sites; but, because few very large aggregation sites were used, the data suggest a pattern of fairly small villages or settlements dispersed over the landscape.

The Koniag period builds on the Kachemak pattern and adds a previously unknown class of sites, very large ones. Three modes are identifiable in Figure 8.5b—small (1–15 houses), large (23 houses), and very large sites (35 houses). Within the small mode, sites form two groups (not captured in the graph): very small (1–4 houses) and small (5–15 houses). The modality in the Koniag data fits the ethnohistorically-reported seasonal strategy of winter aggregation in a few large

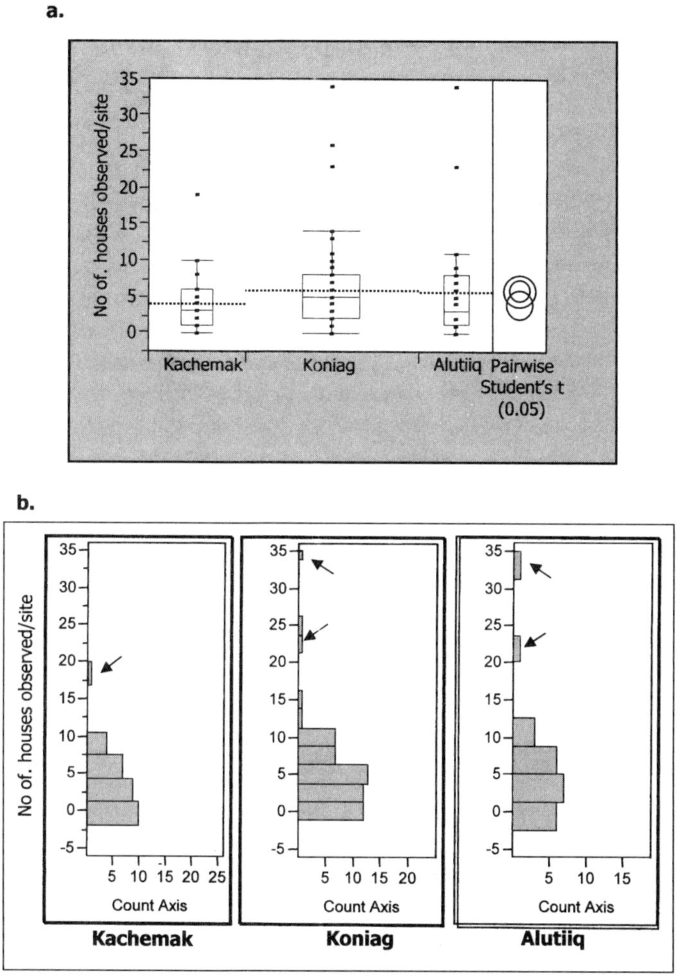

Figure 8.5. Numbers of surface house features recorded per site by period. (a) Quantile box plots and average numbers of houses per site. Pairwise student's t circles as in Figure 8.2. (b) Histogram distributions. Arrows show secondary modalities.

villages followed by dispersal from spring to fall into smaller settlements and short-term camps. The sub-mode from 5 to 15 houses could represent seasonal settlements of extended lineage groups, or small winter villages without the means to attract the aggregated populations of the few much larger villages. These issues are considered further under the discussion of site functions and settlement patterns.

The Alutiiq histogram preserves the Koniag pattern, primarily due to the continued occupation of Koniag sites through the early part of the Russian occupation.

The behavioral patterns generating the two lowest modes in the Koniag distribution continue into the Alutiiq-American phase (data not shown).

In summary, the data on site size and house frequencies from the Kachemak and Koniag period indicate the gradual addition of larger sites through time, and the development of increasingly distinct settlement size classes (with an increase in variation). This pattern matches expectations for the elaboration of more formalized, repetitive, and complex patterns of seasonal land use through time. Koniag patterns appear qualitatively and quantitatively different from preceding periods, both in scale and organization. While these patterns may indicate a systematic sampling bias against large pre-Koniag sites, an attempt to detect such a bias finds little support. The more likely explanation is that settlement patterns did, in fact, change significantly at the Koniag transition (see Settlement Pattern analysis below). This follows expectations of a major change in the organization of political economy and warfare. Post-contact patterns exhibit similarities to the preceding Koniag, but also reflect the gradual reduction in numbers of large villages with depopulation and resettlement.

8.4. SITE FUNCTION VARIABILITY

Site function characterizes an aspect of variability in archaeological sites. Following a derivative of Binford's (1980, 1982) hunter-gatherer functional classification that is consistent with ethnographic variability in maritime hunter-gatherer settlement and land use (Mitchell, 1983), the site-function classification used here includes *locations*, *camps*, *settlements*, *villages* (*large* and *small*), and *refuges* (defensive sites). (Site function labels are italicized to indicate their classificatory status.)

Locations potentially include lookouts, kill sites, butchery stations, signal fires, or any other archaeological manifestation of a short interval non-residential land use. The protohistoric Partition Cliff site, discussed in Chapter 7, is interpreted as a butchery station. Other *locations* include a lookout/flaking station of Ocean Bay affiliation, two or three flaking stations of unknown age, and a couple of charcoal stains in locations that suggest signal fires. Signal fires, as reported by local oral histories, were used as part of a well-developed communication network around the island in the late pre-Russian periods (Sven Haakanson, Sr., personal communication, 1993).

Camps, *settlements*, and *villages* are types of residential sites, differing in their size, intensity of occupation, and nature of construction (with or without pit-houses). *Camps* are minimally constructed, without semi-subterranean houses (or, at most, one), and generally smaller than 1000 m^2. *Villages*, by contrast, are large residential sites with numerous semi-sedentary structures (visible at ground surface), deep middens, and otherwise thick deposits. From ethnographic analogy and ecological theory, we expect villages to have been occupied primarily in winter, when resource extraction and travel would have been limited by inclement weather and the absence of the economic opportunities that promote dispersal in the summer months. As outlined in the model, villages of this sort should post-date the development of intensified resource procurement and storage. We can also assume that some individuals may have remained at the villages year-round, as was the case ethnographically, so the term "winter village" becomes a cultural construct that underestimates the seasonal occupation of these aggregation sites.

Settlements are an intermediate category including larger or more permanent versions of camps (sites larger than 1000 m^2 without pit-houses, and pit-house sites with fewer than 5 structures). *Refuges* include any site with residential features situated on defendable landforms, such as cliff-faced islets and promontories. Generally these defensive sites lack a convenient source of fresh water and would be difficult bases for foraging, given the problem of launching boats from their exposed steep and rocky shores.

Several predictions anticipate changes in the variability of site function through time. From the model, we expect little variation or change in residential sites through the Ocean Bay Period. People should move serially from one camp to another. Logistical foraging could generate *locations* (kill sites, butchery sites, look-outs, etc). We expect to find few *location* sites, however, because, unlike terrestrial hunter-gatherers, maritime hunter-fisher-gatherers would accomplish many logistical tasks on the water and could bring home most game (with the exception of very large sea mammals) for butchery because of the greater transport capability of boats (Ames, 2002). For the Kachemak period, we anticipate increased diversity of residential site use, with intensive extraction of storable resources from camps or settlements located adjacent to seasonally productive patches, especially salmon streams. From the Koniag Period, we expect further elaboration of site variability, with aggregated villages, seasonal settlements at corporately-owned patches, and smaller camps at more limited resource locations. We also anticipate the use and growth of defensive sites corresponding to the advent and expansion of endemic warfare.

8.4.1. Site Functions from Spatial and Geographical Parameters

Site components are classified according to a multivariate procedure that accounts for modality in the data on site area and house frequencies (discussed above), as well as the character of occupation, and the geographic setting. Ideally, this classification seeks to 'cleave nature at its joints' (Spaulding, 1982), in an effort to match observable archaeological units to functionally distinct site classes. On a more practical level, this classification is designed to create site types for meaningful comparison with a variety of other parameters (Dunnell, 1986). The utility of the classification will be born out in its ability to reveal systematic changes in site use, through time and across space.

While designating *locations* and *refuges* is straightforward for the sites in the Sitkalidak sample, distinguishing between residential classes is more difficult. For this study, *camps* are distinguished from *settlements* as follows. *Camps* are sites lacking evidence of house depressions and less than 1000 m^2 in area. *Settlements* are sites smaller than 1000 m^2 with house-pits, as well as some sites larger than 1000 m^2, but with very few house-pits. Finding a convenient break between small house-pit sites (termed *settlements*) and large house-pit sites (termed *villages*) is more difficult. Figure 8.6a shows a bivariate scatterplot of site area by number of houses. Density contours have been added to show the structure of the data as the result of the interaction of both factors. The contours can be used to isolate modalities in the data distributions. Using this plot, I chose to divide sites into *settlements* and *small villages* by means of a subtle break in modality between the densest concentration of points in the lower left corner, and a somewhat less dense "ridge" of points extending up to the right. Seven points are completely separated from the cluster and have been termed *large villages*, either for their much larger size and/or their greater number of house-pits. In general, *settlements* tend to have fewer than six house-pits. Figures 8.6b and 8.6c present a third dimension, total house area, against the dimensions of site area and number of houses, respectively. These plots confirm the consistency of the attributions developed from Figure 8.6a in that each class primarily clusters with its own group, with very little overlap.

It is important to note that this classification is heuristic. While the functional attributions should be broadly correct, the breaks between units most certainly misattribute some sites. For example, many of the smaller sites classified as *villages* may have really functioned as seasonal resource extraction settlements by larger-than-average subgroups of the local village, while some of the larger *settlements* may have actually been villages of smaller social groups. This classification can be evaluated by seeing how well it performs when measured against other variables. The main purpose here is to evaluate how well the classification identifies pattern changes in site use through time. First we will explore how this classification breaks down by chronological period.

Table 8.2 and Figure 8.7 summarize the Sitkalidak component data by function class. The raw data are included in Appendix A. Table 8.2a provides a cautious tabulation, including only the 89 components for which I felt confident about the period attribution. Table 8.2b includes all 133 components with at least tentatively-assigned chronological period. Despite the resulting discrepancy in size, the proportional contributions of each class to period are broadly similar, as shown in Figures 8.7.

These data can be summarized as follows. First, the Ocean Bay period is dominated by *camps* and *settlements*. The *settlement* category in this period reflects the discovery that a few Ocean Bay sites are horizontally extensive and frequently reoccupied. I believe these sites were regularly reoccupied choice camping grounds for several foraging families, a hypothesis driving ongoing excavation at the Tanginak Spring site (KOD 481). As previously suggested, tents erected in slightly different areas of an unconstrained surface would result in a site of greater horizontal extent than any single occupation. Interestingly, the problem of horizontal spread is

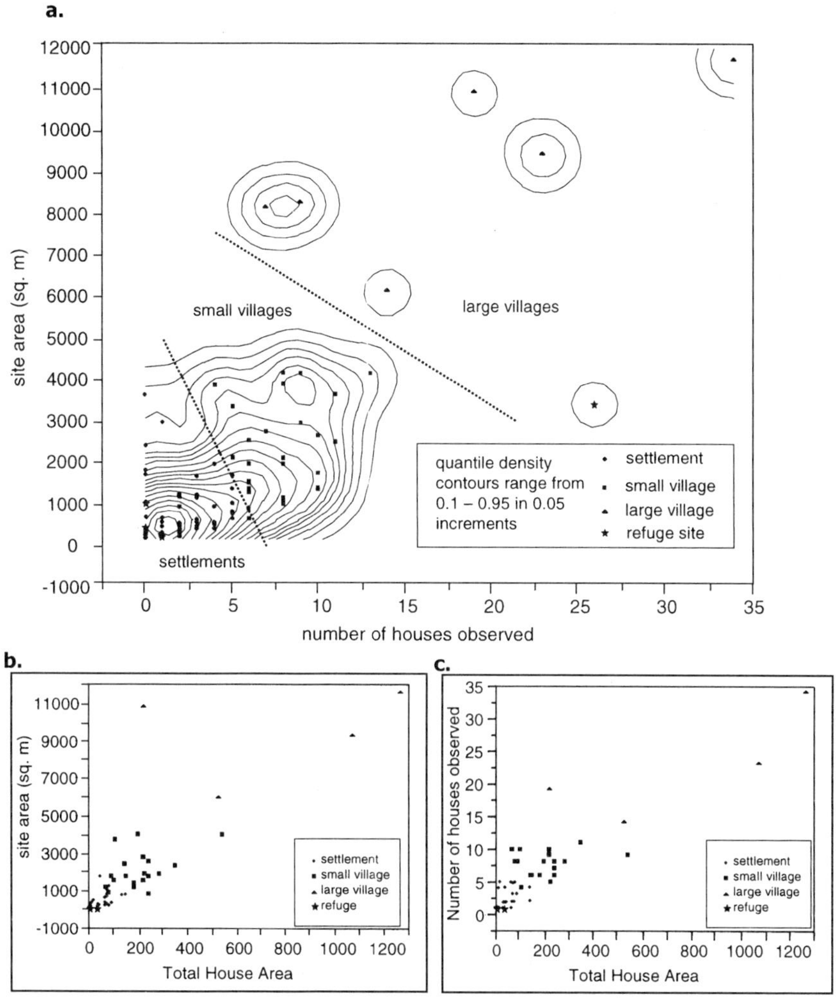

Figure 8.6. (a) Bivariate plot of site area against number of houses with nonparametric density contours. Dashed lines indicate cleavage points for discriminating between *settlements*, *small villages*, and *large villages*. *Refuge* sites are also included. (b, c) Bivariate plots of total house area per site against site area and number of houses, respectively, showing consistency of classification in all plot combinations.

less significant in later periods, when semi-subterranean features provide attractive foundations for subsequent occupation, as is well documented in the New Karluk and Uyak Site excavations (Jordan and Knecht, 1988; Knecht, 1995; Steffian, 1992b). While Ocean Bay II sites have produced pit-houses in excavation (and part of a depression has been reported from the excavations at the Ocean Bay I levels of the Rice Ridge site: Donald Clark, personal communication, 1996),

Table 8.2. Tabulation of functional site classes by period. (a) conservative estimate (excluding sites of uncertain period attribution), (b) liberal estimate (including sites of suggested but uncertain period attribution).

(a)

	Ocean Bay	Kachemak	Koniag	Historic	Total
location	1	0	0	2	3
camp	6	4	1	0	11
settlement	5	10	15	3	33
small village	0	3	16	7	26
large village	0	1	3	4	8
refuge	0	2	5	1	8
Total	12	20	40	17	89

(b)

	Ocean Bay	Kachemak	Koniag	Historic	Total
location	2	0	0	3	5
camp	9	4	1	0	14
settlement	5	23	24	6	58
small village	0	8	22	8	38
large village	0	1	5	4	10
refuge	0	2	5	1	8
Total	16	38	57	22	133

the Sitkalidak Survey and associated excavations identified no Ocean Bay period surface depressions.

Settlement and *small village* classes dominate the Kachemak sample. Few camps are identified, although many of the undated components in Appendix A are *camps* (and *settlements*). Many of the undated *camps* likely fit in the Kachemak and Koniag samples, suggesting that the true variability in Kachemak and Koniag ranges from *camp* to *settlement* to *village*. Only one *large village* is identified in the Kachemak sample—the anomalously large site KOD 553, already noted in the discussions above. The first defensive *refuge* sites appear late in the Late Kachemak phase, both dated after 800 AD.

The Koniag sample shows an increase in the proportion of *villages* to *settlements* and includes three to five *large villages*. Five *refuge* sites were in use at this time. Only one *camp* is reported, but we can expect more to exist in the unattributed sample. Finally, in the Alutiiq period we see a few of each site category, except *camps*. Unfortunately, we do not yet have the means to subject the

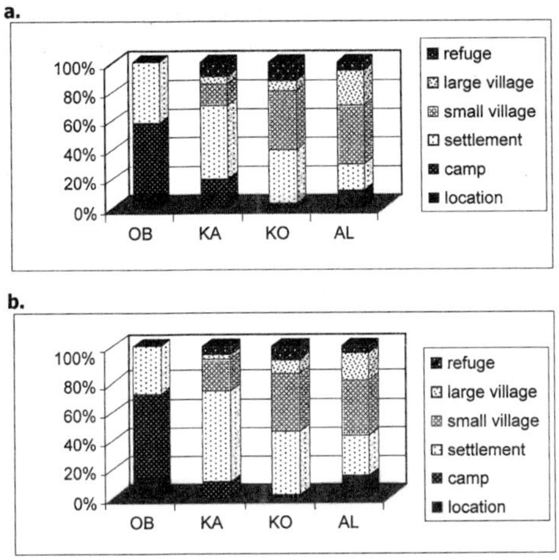

Figure 8.7. Proportional change in site function classes by period. Data from Tables 8.2a,b and Appendix A. (a) only including components of secure period attribution; (b) including all components with possible period attributions. See text for discussion.

Alutiiq *settlement* and *camp* profiles to the scale of chronological resolution needed for examining the land use trends in this short interval of dramatic reorganization.

8.4.2. Artifacts Assemblage Character

Artifact assemblages are useful for characterizing the kinds of activities conducted at different sites. Here we examine a modified version of an analysis first developed by Fitzhugh (2002). This analysis identifies patterns of land use by looking at artifacts deposited on sites of different functional classes and through time. Using artifacts as an independent measure of land use patterns, we can evaluate the site function classification as well as extend it in interesting ways.

Artifacts from the Sitkalidak survey are grouped into seven functional groups (Table 8.3). These classes discriminate between subsistence procurement (hunting, fishing), resource processing, site maintenance, tool manufacture and rejuvenation, and ornamentation. With the exception of very small hammerstones (included in "tool production"), the category of hammerstones was eliminated from most analyses because of the generalized utility of this tool in processing, site maintenance, and tool production (and other activities). Because the grouping variables subsume a wide number of artifact classes, analyses are only valid in the comparison of proportional changes in that category to others across some other axis (e.g., temporal change or site function).

Table 8.4 tabulates the average proportional abundance of each artifact class by site function class and by period. Figure 8.8 (a and b) presents these data in

Table 8.3. **Artifact Classification**

Functional grouping	Artifact class
Ornamental (N = 50)	Beads Painted pebbles Labretts (lip plugs) Other ornaments
Hunting (N = 60)	Projectile points Lances Harpoons
Fishing (N = 51)	Grooved cobbles Notched shingles
Resource Processing (N = 301)	Bifaces (not proj point) Scrapers Macroblades/ microblades Semilunar knives (*ulus*) Slate knives (non *ulu*). Utilized flakes/flake tools Pottery
Site maintenance (N = 16)	Ocher grinders Axes Nails (historic) Post remnants Bone wedges
Tool production (N = 675)	Abraders Adzes Debitage and other production debris Possible burins Burnishing stones Chisels Cores Gravers Grindstones Unsuccessful blades/microblades Tool preforms Worked slate Worked bone Small hammerstones
Hammerstones (N = 37)	Large hammerstones

graphic form. *Locations* are excluded from Figure 8.8a because the two components are too small to provide a meaningful measure for comparison with other site categories. The number of components for *camps*, *large villages*, and *refuges* is also small. These problems, and the generally small numbers of artifacts found per component, render the site function trends only suggestive. Larger samples are needed to establish statistical confidence. The period trends are more robust given

Table 8.4. Artifact proportional abundance by site function (a) and period (b).

	Number of components	Ornaments	Hunting	Fishing	Processing	Site Maintenance	Tool production
a. **Site function**							
Locations	2	0.00	0.17	0.10	0.30	0.00	0.43
Camps	8	0.01	0.08	0.00	0.16	0.01	0.75
Settlements	39	0.05	0.04	0.05	0.34	0.05	0.47
Small villages	30	0.08	0.02	0.07	0.34	0.04	0.46
Large villages	8	0.11	0.00	0.02	0.42	0.02	0.43
Refuges	6	0.17	0.17	0.00	0.33	0.00	0.33
b. **Period**							
Alutiiq	16	0.30	0.05	0.02	0.34	0.05	0.24
Koniag	38	0.03	0.03	0.04	0.41	0.04	0.44
Kachemak	28	0.01	0.04	0.10	0.25	0.03	0.56
Ocean Bay	12	0.01	0.08	0.00	0.24	0.01	0.66

the larger number of sites in each period category. The following patterns are observed:[1]

(1) The tool production and resource processing (cutting) categories dominate the primarily stone artifact record of the Sitkalidak area in all site types and time periods. Tool production tends to impose a strong influence on the other categories, especially resource processing. Because of this, in both

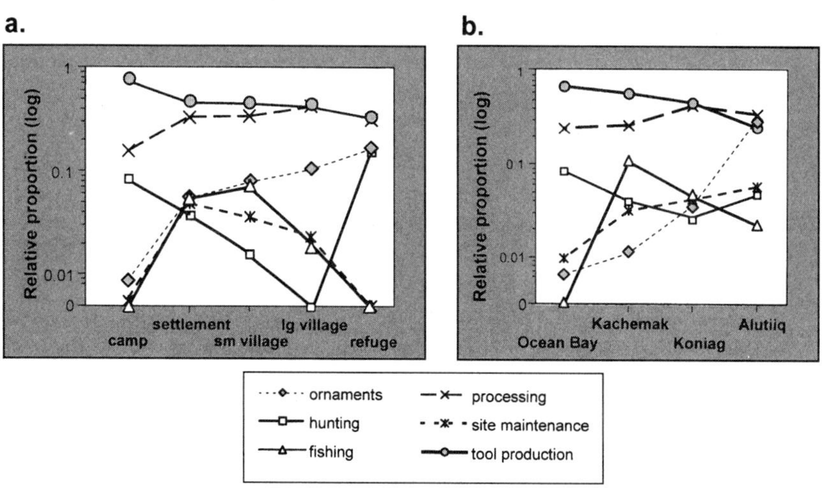

Figure 8.8. Proportional abundance of artifacts by (a) site function and (b) period. See text for discussion.

plots in Figure 8.8, the variable (Y) axes are presented using a logarithmic scale. This provides better representation of changes in all artifact classes.

(2) Averaging 75%, tool production artifacts are found in greater proportion in *camps* than in any other site classes. In other site classes, tool production artifacts average below 50%, declining slightly from *settlements* (47%) to *villages* (small = 46%; large = 43%). Tool production appears least in *refuges* (33%). As shown in Figure 8.8b, this is clearly a product of change through time, probably due to the decrease in importance of flaked tools following the development of ground-slate technology that creates much less waste material. Plotted according to cultural period, tool production declines steadily from Ocean Bay to Alutiiq.

(3) Resource processing tools are the second most numerous artifact class. Their proportional distribution increases from *camps* (16%) to *large villages* (42%), in reverse of the tool production pattern. In the period plot (Figure 8.8b), resource processing tools also increase in proportion to the decrease of tool production from Ocean Bay to Koniag. Resource processing appears to decline slightly in Alutiiq sites. The largest proportional increase in processing activity appears to occur with the shift from Kachemak to Koniag, *not*, as we would have thought, from Ocean Bay to Kachemak.

(4) Hunting artifacts decline from *camps* to *settlements* to *small villages*. None were recovered from large villages. As expected for sites where defensive weapon use would be emphasized, hunting tools increase once again in the refuge sample. In the period plot, hunting tools dominate Ocean Bay sites and decline relative to other tool classes both in the Kachemak and Koniag periods. An increase is seen again in the Alutiiq period. This pattern is consistent with a decline in the relative importance of hunting to more intensive harvesting practices beginning in the Kachemak period (reflected in the data on processing and fishing). The increase in status display and other artifacts of social importance through time (as reflected in the increase in ornamentation through time) also contributes to this decline in relative importance of hunting objects. Hunting tools are most prevalent in *locations* (not shown) and *refuges*. This pattern is particularly significant since *refuge* sites are otherwise most comparable to *villages* (which have the lowest representation of hunting tools). Hunting tools are presumed to have served also as weapons in warfare.

(5) Fishing tools are most prevalent in *small villages* and in the Kachemak period. No fishing implements were identified in *camp* or Ocean Bay assemblages. This category is dominated by notched stone weights that are most common in Kachemak sites around Kodiak. Organic fish hooks, leisters, and fish harpoons are reported from sites with better preservation elsewhere in the Kodiak Archipelago. Had they occurred in our sample, we can predict they would slightly elevate the fishing category in the Ocean Bay and Koniag samples, but not sufficiently to change the overall

patterns seen here. These patterns suggest that fishing changed form from Kachemak to Koniag times.

(6) Site maintenance is a small category dominated by Ocean Bay I ocher grindstones from the Tanginak Spring site (KOD 481), scored as a *settlement*. The decrease in importance of this class is not considered reflective of the change in importance of sites as much as a change in the nature of site construction. (Ocher tent hide and pole preparation required less distinctive, durable tools than sod and wood construction.) If this interpretation is correct, we would expect wood-working tools to increase even as site maintenance tools decline overall. We do not have the data to evaluate this proposition.

(7) Ornamental items serve as a reasonable index for the relative importance of social competition and status display (whether of a segmentary or stratified nature). As expected, this class increases steadily from Ocean Bay to Alutiiq and from camps to large villages. Because trade beads are counted individually in this analysis, they tend to elevate the proportional importance of ornamental artifacts in Alutiiq sites. The increase in ornamentation from Ocean Bay to Koniag is consistent with the emergence of heightened social competition and ritual in the (Late) Kachemak period and of institutionalized ranking, slavery, and a prestige economy by Koniag times (Fitzhugh, 2003).

Adding artifact classes to the analysis of site function clearly supports, and extends, the site functional analysis. It is, however, the temporal dimension of artifact variation that contributes most directly to model evaluation. As predicted, the Ocean Bay to Kachemak transition initiates a major change in subsistence focus (hunting to fishing) and increases the importance of resource processing. In the Kachemak to Koniag transition, we see less dramatic change in relative emphasis on subsistence technologies with the exception of processing, which intensifies most dramatically across this transition. From the Kachemak to Koniag samples, we see more dramatic developments in aspects of political competition through ornamentation and warfare. Sustained increases in processing tools through time (along with their great importance in *settlement* and *small* and *large village* categories) remain consistent with increases in surplus production, first for winter storage, and eventually for political purposes and to feed a burgeoning population (see Chapter 9). Contest competitions and the emerging prestige economy would have demanded ever increasing processing activities by larger cohorts of laborers.

8.5. SETTLEMENT PATTERNS

Having discussed the spatio-temporal variation of site sizes, numbers of houses, occupational intensity, and function, it is now appropriate to examine

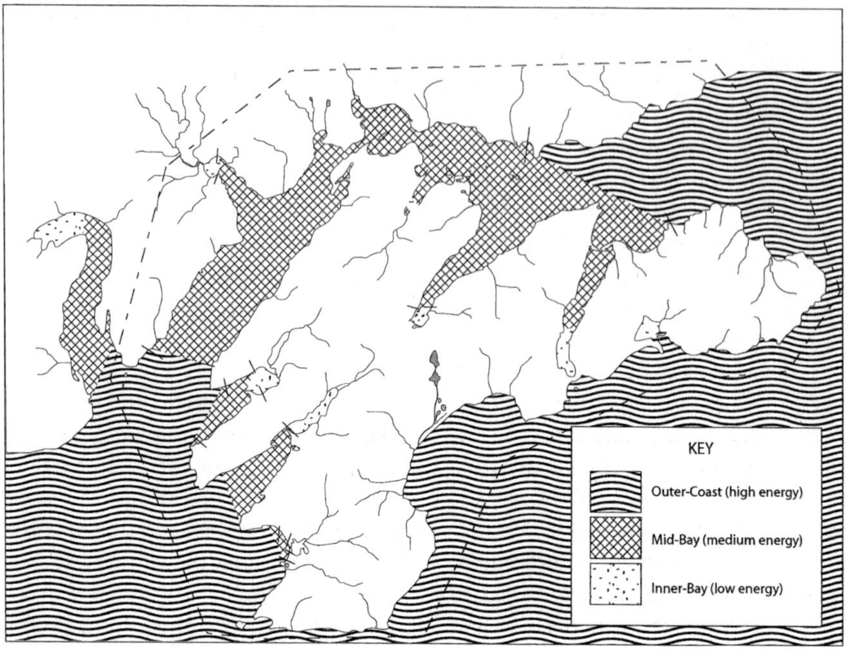

Figure 8.9. The Sitkalidak study area divided into outer coast, mid-bay, and inner bay zones. Settlement pattern changes are analyzed with reference to these zones. See text for explanation.

settlement patterns as distributions of sites across a variable landscape. To protect the integrity of archaeological sites, detailed settlement distribution maps are not provided. Instead, settlement patterns are developed using tabulated data on site location.

Figure 8.9 shows the study area divided into broad ecological zones: outer-coast, mid-bay, and inner-bay/riverine areas. These divisions correspond to variation in the degree of exposure to inclement weather and proximity to different categories of subsistence resources. These ecological zones are adopted from Haggarty and colleagues (1991).

Table 8.5 tabulates these sites by geographical position and function. The following discussion relates primarily to the relationship between time period and location, shown in Figure 8.10. Ocean Bay sites were found throughout the survey area. Note that Ocean Bay sites tend to be located in areas classified as mid-bay. From these positions, Ocean Bay people could maintain efficient access to diverse habitats, consistent with the findings in other settlement analyses for the Gulf of Alaska (cf. Erlandson et al., 1992:53; Haggarty et al., 1991:209–213). The outer-coast zone remains unsettled in Ocean Bay times, but presumably was

Table 8.5. **Sitkalidak sites of different periods, subdivided by geographic position and functional classification.**

	Ocean Bay			Kachemak				Koniag				Alutiiq			Total
	inner	mid	other	inner	mid	outer	other	inner	mid	outer	other	inner	mid	outer	
location	0	2	0	0	0	0	0	0	0	0	0	1	0	1	4
camp	0	2	4	0	1	1	2	1	0	0	0	0	0	1	12
settlement	0	5	0	5	13	2	3	4	14	6	0	0	5	1	58
sm village	0	0	0	2	4	2	0	5	7	7	3	0	5	3	38
lg village	0	0	0	0	1	0	0	0	2	3	0	0	2	2	10
refuge	0	0	0	0	1	1	0	0	1	4	0	0	0	1	8
unknown	1	1	0	0	0	0	0	0	1	0	0	0	0	0	3
Total	1	10	4	7	20	6	5	10	25	20	3	1	12	9	133

logistically exploited from home bases less exposed to the open ocean. No sites were identified on salmon streams or at their outlets (inner-bay or riverine locations), supporting the view that intensified salmon harvesting and processing were not important subsistence activities at this time (at least not in this region). If representative, this distribution suggests salmon were not strategically significant in the settlement of this part of Kodiak for nearly half of the prehistoric period. In other areas of Kodiak, Ocean Bay sites are sometimes located where important salmon streams are found today. These include the Karluk Lagoon, Karluk Lake (Jordan and Knecht, 1988:230–231), Afognak River (Clark, 1979, 1997), and Buskin River (Patrick Saltonstall and Robert Kopperl, personal communications, 2000, 2002). Ocean Bay people, undoubtedly, targeted river resources during periods of peak

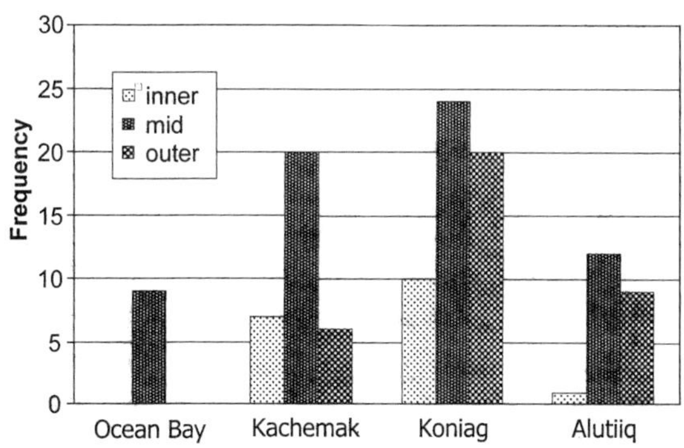

Figure 8.10. This graph shows a change in the distribution of sites across different coastal zones from Ocean Bay to Alutiiq time periods.

availability. However, the absence of evidence for riverine sites around Sitkalidak, along with the lack of intensive technologies for prolonging this resource into winter, argue against a prominent role for riverine resources prior to Kachemak times in this region.

In Kachemak site distributions, inner-bay sites become established adjacent to seasonally-productive salmon streams. This pattern supports the emergence of intensive salmon stream fishing, reflected in the increase of fishing and processing artifacts previously described. Interestingly, in this study area, Kachemak sites in the inner-bay locations, adjacent to salmon producing streams, produce few, or no, netsinkers compared with sites in mid-bay/mid-strait regions. We may infer that the movement into river areas reflects the development of weir fishing and other methods for capturing anadromous fish, such as spearing.[2] Based on the association of Kachemak netsinkers with sites near wide, shallow beaches, I suspect that they were used for net fishing of migratory salmon and herring, where these fish schooled close to shore. Absence of netsinkers found around the Sitkalidak region in the Koniag period may reflect the abandonment of off-shore net fishing, in favor of more exclusive focus on river fishing in the later period. Prime fishing locations at river mouths, or along river courses, were more easily controlled than sections of open coast. And, increased importance of the whale hunt during the spring in Koniag times could have conflicted with netfishing for herring.

Six Kachemak sites are recorded from outer-coast areas, but these are primarily camps, small villages, and refuges of the terminal Kachemak. The low proportion of Kachemak sites in the outer-coast zone suggests a continued commitment to protected regions for settlement through most of this period (or at least the Late Kachemak, for which we have data). At the same time, the larger sites (presumably villages), found in intermediate or semi-protected zones, reflect a continued interest in access to ecologically diverse resources for those parts of the year when the salmon streams were not productive. Another factor limiting the establishment of winter villages in shallow inner-bay zones may have been the tendency for the shallow water in these areas to freeze during cold winters, unlike medium and high-energy coastal waters. Small refuge sites were established for the first time late in this period (see below).

In the Koniag period, the patterns of previous distributions are maintained, but the outer-coast zone takes on a significantly different role than in previous periods. Not only do Koniag habitation sites move into the high-energy, outer-coast zone, but the zone appears to become the primary focus of village occupation. Two phenomena likely account for this dramatic shift in settlement pattern: whale hunting and the increasing need for defense against attack. With the development of whale hunting as a major economic activity, it was advantageous to locate populations in the regions closest to whale sightings and kills. Given the nature of Koniag poison-lance whaling reported ethnohistorically (Crowell, 1994), retrieval of killed whales would be low unless people were in the vicinity of the kill several days later, when the whale eventually floated to the surface and began to drift. Moving villages out to the hunting areas increased the opportunities for sighting whales, made it possible to launch larger whale hunting parties, improved the chances of a successful strike, and enhanced the odds of recovering killed whales. In addition, whales, or their products, would be difficult to transport to distant villages, and when kills were successful, it would be best to have populations situated close to the landed whale. Support for this explanation can be found in the distribution of whale bones on archaeological sites. Koniag sites located adjacent to known whaling areas in, or close to, the outer-coast zone, were the only locations to yield significant whale remains. All of these sites are classified as villages and most constitute the largest villages in the sample. The economic interpretation of these findings suggest whale hunting was fairly successful, supported large populations, and reorganized local economy and settlement strategy in dramatic ways.

In conjunction with its presumed importance to the subsistence economy, whale hunting interacted with changes in social organization and political competition during the Late Kachemak and Koniag time periods. Whale hunting was dangerous, difficult, and highly unpredictable; like warfare, it provided a mechanism for status competition, as hunters could leverage their success in the hunt to political success in their community. In a social context, where some people gained benefit through status competition, whale hunting took on exaggerated importance. As reported in Chapter 4, ethnohistoric accounts suggest whale hunting was pursued by a feared, and revered class, of hunters. Knowledge of secret rituals and magic arts contributed to hunting success.

Defense reasons likely also prompted village relocation to outer-coast zones. The three largest Koniag villages are situated in strategic locations near highly defensible *refuge* islets in the outer-coast zone. The associated *refuges* are minimally big enough to hold the populations of the largest village in times of attack. Ethnohistoric reports of warfare suggest that villages would relocate on such islands during the raiding season, while young warriors were off attempting to sack the distant villages of enemies (Chapter 4; Townsend, 1980). Raiding was probably carried out in early summer, after the worst winter weather subsided and before intensive labor efforts were needed to procure stores (late summer to fall).

Despite the shift in settlement focus, more Koniag than Kachemak components were found in the inner-bay/riverine areas, where salmon fishing was a seasonal focus. Most streams with any kind of a salmon run support settlement by the end of this period. The pattern suggests that Koniag groups continued to depend on salmon resources, and families were forced to rely even on limited production streams. This finding suggests that outer-coast sea mammal hunting (whale, seal, sea otter, and sea lion) was not sufficient to support all groups throughout the year (or that some were excluded from it). Given seasonality and low preservation value of sea mammals, the continued use of inner-bay resource patches is not unexpected. But, the heavy settlement of even minor stream systems calls for

Table 8.6. Sitkalidak region defensive sites

KOD (SAS) Site No.	Periods	Base ^{14}C Age (yrs bp)	Size (m^2)	No. houses	Location
86 (116)	Lt Kach/ Kon	1240+/−60	875	0	Sitk' Strait
552 (117)	Lt Kach	1090+/−60	16	1	Sitk' Strait
558 (124)	Lt Kach	960+/−90	50	1	Sitk' Strait
442 (8)	Late Koniag	610+/−60	3300	26	outer coast
450 (85)	Late Koniag	no date	(> 3300)	Nr	outer coast
576 (141)	Late Koniag	no date	288	0	outer coast

explanation. Based on the increased size of population inferred from village sites (see Chapter 9) it seems reasonable that subsistence needs required exploiting even the smaller streams. This pattern coincides with findings associated with greater political competition for prime resource locations and the exclusion of less powerful groups from the prime patches. Less fortunate families were likely then forced to exploit less productive or unpredictable streams and other patches (see Chapter 6).

The size, location, and occupation dates of defensive sites are critical for understanding regional patterns of political organization, alliance, and warfare. Defensive sites in the Sitkalidak region occur in the Kachemak and Koniag periods (Table 8.6). Three small refuge sites have been identified along the inner shores of the Sitkalidak Strait. The basal components of these sites all date to the last few hundred years of the Late Kachemak Period. All are small, with room for no more than one or two houses, and only one has overlying Koniag deposits (near the southwest entrance of the Strait). This Late Kachemak pattern suggests localized, small-scale conflicts between households or independent settlements.

In contrast, select Koniag defensive sites are large (26 or more houses) and constructed on the tops of cliff-flanked islets in locations fully exposed to ocean swells and strong winds. The three largest are strategically located at both entrances of the Sitkalidak Strait and at points of island vulnerability (Figure 8.11).[3] One of these, KOD 450, is the *Awa'uq* refuge site used in AD 1784 by the aggregated populations of several large villages in an unsuccessful attempt to escape confrontation and subjugation by Russian invaders (Figure 4.2). This pattern of defensive sites, along with ethnohistoric evidence of large-scale military cooperation at contact, suggests that by the late Koniag period, Sitkalidak populations were organized into a regional alliance (with two to four large aggregate communities sharing the strait and bays), cooperating in matters of defense and war. By the late Koniag period, most warfare was waged against enemies from places as distant as the Eastern Aleutians, Alaska Peninsula, Kenai Peninsula, and Alexander Archipelago, several hundred kilometers away.

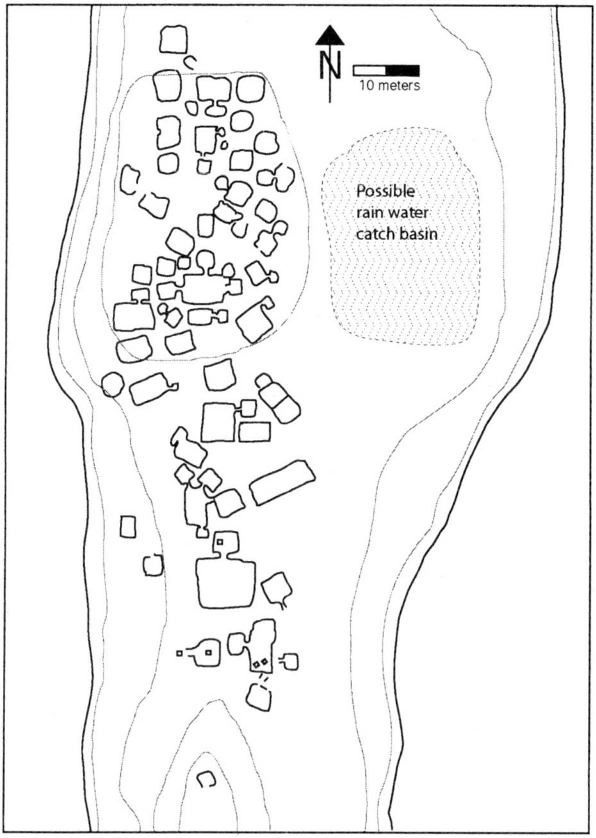

Figure 8.11. Plan of a Developed Koniag defensive site (KOD 478). Some of the smaller pit depressions may be "borrow" pits for extraction of construction sods.

Returning to the settlement analysis, the Alutiiq sites in Table 8.5 include those occupied through the decades of contact as well as those occupied by the Alutiiq people in the Russian and early American periods. Contact period villages were occupied for several decades following the military campaign at *Awa'uq* rock. At least some of these villages survived until the smallpox epidemic of the 1830s and the imposed resettlement of the 1840s (Crowell and Lührmann, 2001:60)

Following the abandonment of contact-period villages, the Alutiiq pattern reflects a dramatic shift of settlement, back towards the Sitkalidak Strait and away from the outer-coast zone. Only one post-contact Alutiiq village site is located on the outer coast of Sitkalidak; likely it was also a contact period site, but Lisianskii failed to document it in 1804 (his travels did not take him to the outer coast of Sitkalidak; Clark, 1987). This site is large with considerable historic trade materials (beads, glass, copper ring). I suspect it did not endure long in the face of the demographic, social, and political pressures of the Russian phase.

Changes in settlement and land use suggest that Alutiiq subsistence strategies and politics were no longer focused on competitive whaling and warfare. Two features of the Russian period facilitated this change. First, disease and conscription diminished the pool of hunters available for both activities. Second, Russian authorities prohibited warfare and disrupted social systems that rewarded those activities. Aspiring young men (who might have hunted whales and led raids) were organized into sea otter hunting brigades; they were encouraged to supply large numbers of furs in exchange for trade goods, favored treatment, or punishment avoidance. Some whale hunting, however, was continued during the Russian phase to provision colonial food reserves. Father Gideon (in Black, 1977:89) refers to facilities for blubber rendering and whale meat storage at the Three Saints Bay fort. The Russian fur company administrators commonly used Alutiiq labor, employing traditional subsistence techniques to provision their colony.

Once diseases had run their course, the depleted population may have satisfied basic subsistence needs from the Sitkalidak Strait area without resorting to the outer zones. Interestingly, economic dependence on store-bought staples did not significantly impinge on traditional subsistence practices (aside from whaling), until after the establishment of the commercial salmon industry in the AD 1880s. In excavations at the Old Harbor Lighthouse site (KOD 089), we found numerous traditional stone tools (lances, *ulus*, ground-slate scraps) along-side trade ceramics dating to the last decade of Russian rule (Chapter 7). Lead and other imported materials were sometimes modified into traditional artifact forms such as fishhooks. At the turn of the 20th century, Alutiiq people still lived in sod houses and paddled skin kayaks.

Had the native whaling tradition continued into the late 19th century, it would have competed with a commercial whaling station established on the shores of the Sitkalidak Strait at Port Hobron. This station was commissioned to process whale blubber from large whales commercially hunted with harpoon guns south of Sitkalidak Island (Reeves et al., 1985). Catches were numerous until hunters gradually depleted stocks and the station was abandoned in the mid 20th century.

The settlement data presented here, like other examined categories of data, support the model predictions. Ocean Bay period *camps* were usually situated along the sides of bays and the Sitkalidak Strait, with logistic access to both the outer-coast and inner-bay/river patches. These positions insured nearby access to winter stocks of shellfish, fish, and other foods in low densities. Presently we are unable to assess the degree of mobility from camp to camp. Better dating of individual components (at Tanginak Spring and other sites) will improve our understanding of residential mobility. The addition of river and river-mouth sites during the Kachemak period supports other evidence of intensified harvesting/processing of more r-selected resources at this time. Small defensive sites after 1200 bp suggest increases in political tension between local families. Coupled with evidence of considerable labor invested in ornamentation, ritual activity, and elaborate mortuary treatment in the Late Kachemak, localized violence appears to reflect a period of incipient social competition, rivalry, and claim-staking. By the mid- to late-Koniag period, this tension dissipated at the local level, and defense became a cooperative affair for one, or even several, large village(s). The movement of villages to the outer coast in Koniag times is consistent with a more politically motivated settlement pattern. Possibly whale hunting served as a form of subsistence intensification that paralleled the development of warfare, social inequality, and prestige trade. Additionally, whale hunting may have represented a form of political competition among individuals within villages and corporate groups.

8.6. SUMMARY

This chapter presented a series of analyses concerning the spatio-temporal variation of archaeological settlement and land use (at the scale of archaeological sites or components), along with their changing relationships to the surrounding environment. These data provide strong support for many of the model predictions developed in Chapter 6. We see little change in site density or site size variation from Ocean Bay I to II. More or less functionally homogenous *camps* and small *settlements* dominate the Ocean Bay site sample. Located primarily in geographically intermediate locations, these residential sites afforded logistical access to a range of resource zones, from rivers and inner bays to the rugged outer coast.

The Kachemak period shows less change in site characteristics than anticipated. Most sites remain small, similar in size to the larger Ocean Bay sites. The nature of site construction, however, changes substantially as semi-subterranean houses become the standard form of habitation. This change, reinforced by increasing importance of site-maintenance tools, signals relative occupational permanence. Thus, *settlements* replace *camps* as the dominant functional site class and *small villages* appear for the first time, illustrating an expected outcome of increased settlement density and constriction of foraging range (due to population packing). The small number of considerably larger sites (*small* and *large villages*) fit the expectation for village aggregation in the Kachemak Period. Most village sites remain small compared to the much larger Koniag villages that follow. Technological change towards a focus on mass-harvesting and storage of processing intensive r-selected resources (especially fish) is clearly demonstrated by the development of inner-coast/river zone settlements, increased importance of fishing, and processing tools. Political changes are apparent in the Late Kachemak with the initial use of small defensive sites supporting one to a few households (in what must have been fairly localized skirmishes). Increases in the relative proportion of ornamental artifacts suggest localized status competitions. Combined with previously known aspects of Late Kachemak mortuary ritual and artifact characteristics, these data fit a pattern of considerable social unrest.

The Koniag data suggest an explosion in the scale of sites. Site density reaches a peak, primarily driven by the expansion of seasonal resource extraction *settlements* occupying almost every fixed resource extraction location. Structurally, the settlement system merely expands the Kachemak system to include larger site classes; but, the shift of the largest sites into the outer-coast zone signals a major change in the scale, and orientation, of socio-political competition. This resettlement allows further resource intensification (whale hunting) and provides more effective defense from attack by distant rivals. Technological changes in the Koniag period follow expectations for a politically-driven economy. Artifacts of hunting, fishing, and tool production decline in relative importance. Artifacts related to processing, site maintenance, and ornamentation continue to gain importance, as expected in a political economy dominated by surplus production and prestige competition. Defensive sites expand to encompass one or more large village populations.

In the next chapter we consider social inequality in relation to an analysis of house variability and develop estimates for demographic change. These analyses address several of the remaining predictions from the Chapter 6 model.

Chapter 9

Social Inequality and Demography

In this chapter, we consider the evolution of social inequality and other aspects of social organization with an analysis of house size variability in the Kachemak and Koniag periods of the Sitkalidak region. Data of this analysis are then considered along with data discussed in the previous chapter to model change in population levels through the archaeological history of the Sitkalidak region. Population change is also tracked using radiocarbon date frequencies for the Kodiak region.

9.1. HOUSE ATTRIBUTES AS A MEASURE OF SOCIAL VARIABILITY

In the Sitkalidak survey, houses of the Kachemak, Koniag and Alutiiq periods were mapped and measured at a variety of sites, allowing several kinds of comparisons. In this section, we examine house size variation between periods as well as between sites of the same period. This allows evaluation of predictions for the emergence of social asymmetry in the development of Kodiak complex hunter-gatherers. The central variable in this analysis, house size, is measured both by total house area and main-room area. As a second variable, number of side-rooms is briefly considered as well. This analysis assumes that house size measures are suitable proxies for household population size. Other assumptions include the following: that enduring political power in middle-range societies is predominantly based on the size of political factions or corporate groups, and that kinship and co-residency are sources and manifestations, respectively, of corporate organization.

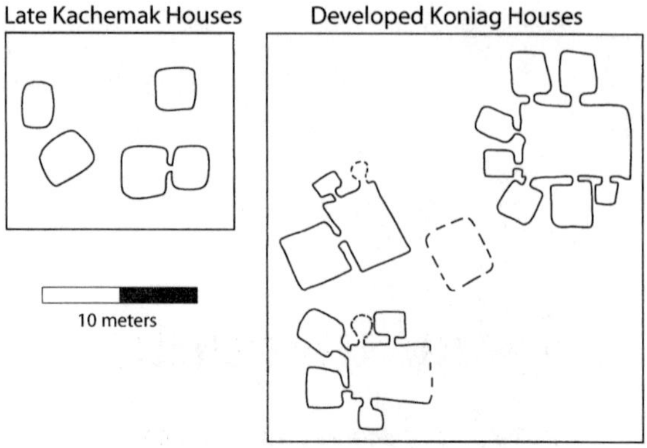

Figure 9.1. Comparison of Late Kachemak and Developed Koniag house form and size. Kachemak houses tend to be small, averaging 15 m^2 while Koniag houses vary more in size (avg. 47 m^2) and often have large central rooms and multiple side rooms.

During the Sitkalidak Archaeological Survey, we mapped all features visible on site surfaces. These maps were then used to calculate the attributes, such as house area and side-room frequencies, used here.

Figure 9.1 illustrates the difference between typical Kachemak and Koniag houses. Kachemak houses in the Sitkalidak sample are rectangular and small, averaging 15 m^2 in surface area. In excavations at the Uyak site, Steffian (1992b) found Late Kachemak houses to be rectangular with raised benches around the periphery, clay-lined storage pits, small central hearths, and somewhat sunken 'cold-trap' entrances (designed to keep cold air out of houses). Sitkalidak houses from the Developed Koniag phase were much larger (avg. 47 m^2) and typically had between one and several side-rooms extending off a central main-room. According to ethnohistoric accounts, the central main-room was used as a kitchen area and a general living room (Davydov, 1976:5). Special celebrations and feasts were often held in the large central spaces. While sometimes reserving one room for a sweat-bath chamber, side-rooms (*zhupan* in Russian) were used primarily as sleeping quarters. Archaeological evidence of Koniag houses from the New Karluk site (and elsewhere) confirms ethnohistoric accounts. Large central hearths are found in the main-rooms, where adequate space facilitates numerous daily activities for many inhabitants and visitors (Figure 9.2). Large storage boxes are placed against the walls of the main-room. Side chambers make appropriate bedrooms. Some side-rooms, found filled with fire-cracked rocks, likely served as steam bath areas (Jordan and Knecht, 1988; Knecht, 1995).

In general, the transition from Kachemak to Koniag house forms and sizes reflects a shift from nuclear to extended family living (Jordan and Knecht, 1988:236). According to Lisianskii (1814; cited in Clark, 1987), in 1804 Alutiiq households

Figure 9.2. 1987 excavations of the New Karluk (KOD 001) showing two superimposed floor levels, each with posts and plank covered drainage trenches. The higher floor was designated House Floor 1. Photo by author.

averaged 18–20 adults. Holmberg's (1985:43) account provides a comparable estimate. The shift in residence patterns likely paralleled a reorganization of community social relationships as well as the emergence of lineage solidarity and intra-village political competition. If true, house architecture would represent the emergence of unequal factions as an increase in the *variability* of house size, not just a shift in its average. If this development paralleled that of social ranking (by subordinating individual families with restricted access to the resources of subsistence and political power), a skewed distribution of house sizes (with many small houses and few larger ones) would result. Modality of a rank/skewed distribution may be a manifestation of household stratification (Fitzhugh, 1996; cf. Brumfiel, 1976). Also, uneven or multi-modal distributions of house sizes are expected when co-residential faction sizes increase with the addition of nuclear-family units, rather than the addition of individuals.

Figure 9.3 shows frequency distributions of total house size and main-room size from the Kachemak, Koniag, and Alutiiq periods.[1] Because all Kachemak people lived, cooked, ate, and slept in their small single-roomed houses, main-room area distributions are considered more comparable across the time periods than total structure area distributions. Kachemak social visits were limited by the number of people who could be accommodated. With the addition of separate

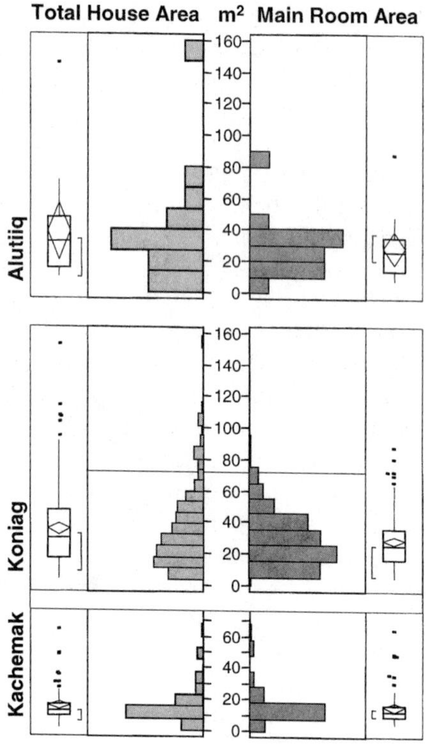

Figure 9.3. Frequency histograms of total house size and main-room size by phase. The development of multiroom houses in Koniag times complicates comparison between Kachemak and Koniag house areas. Main room comparisons are used to render the structures more comparable.

side-rooms for sleeping, bathing, and storage in the Koniag period, activity areas became more specialized, and house area increased out of proportion to any change in household population.

These data confirm a significant shift in overall house size between Kachemak and Koniag periods. Both sets of distributions in Figure 9.3 show mostly small Kachemak houses and larger Koniag and Alutiiq houses. Surface walls of Kachemak houses are between 3 and 5 meters per side. These dimensions are consistent with Steffian's (1992b) data for excavated Late Kachemak house floors at the Uyak Site (excluding the area of entrance passages). Most Sitkalidak houses in the Kachemak sample range in area from 6 to 40 m^2, slightly skewed to larger houses. Nevertheless, at least two modes are apparent: one at 12 m^2 and one at around 50 m^2. A house at 64 m^2 is a very unusual structure for the Kachemak and is considered an outlier in this distribution. Steffian (personal communication, 1996) notes a similar range of variation in Late Kachemak houses at the Uyak site and at KAR 039 in the Karluk Lagoon (see also Jordan and Knecht, 1988: figure 3). The 50 m^2 mode, including house sizes three times larger than most, suggests a small degree of structure size differentiation among the Kachemak.

In the Koniag sample, robust distributions show a dramatic extension of house size (by either measure) in the upward direction. In both cases, the primary mode is at the small end of the distribution (20 m^2 for total area; 16 m^2 for main-room area). Possible secondary modes are seen at 85 m^2 and 107 m^2 in the total area distribution. Because of its exceptional number of side-rooms (7), an outlier in the total area sample stands alone at 153 m^2. This house has a sizeable, but not unusually large, main-room (53 m^2). In the main-room data, only one secondary mode is apparent at about 75 m^2. The Alutiiq sample is statistically inseparable from the Koniag sample and represents a continuation of the Koniag pattern during the early contact period, before depopulation and settlement reorganization.

These distributions suggest three conclusions. First, though incipient household status differentiation may have been present in the Kachemak period (at least by the end of the period), this differentiation was limited as represented in the house size data. Second, the right-skewed distribution in the Koniag pattern matches expectations for ranked or stratified household differences. Third, the skewness in both sets of distributions suggests Koniag social ranking at the residential scale was pronounced.

The number of side-rooms per house provides complementary evidence. As shown in Figure 9.4, Kachemak (and Early Koniag) houses are primarily single-room structures with no side-rooms. But, contrary to normative descriptions in the literature, Koniag sites also appear to be dominated by single-room houses, a tradition that carried through into the Alutiiq period (Figure 9.5). The gradual decrease in frequency of houses with increasing numbers of side-rooms supports

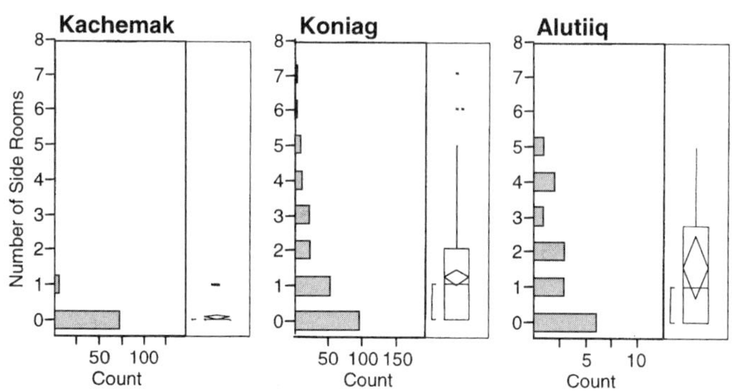

Figure 9.4. Frequency histogram of number of side-rooms per house by phase. This plot illustrates the range of variation in the number of ancillary chambers (sleeping rooms, storage rooms, and bathing chambers) in Koniag houses.

Figure 9.5. Alutiiq sod house ("ciq'lluak" or "barabara") in Karluk in 1889. Photo credited to "T.H.B.," courtesy of U.S. National Archives (22-FFA-1164).

the house-area analysis, and again suggests the presence of ranked co-residence patterns in Koniag times; the lack of secondary modes in this distribution suggests an even gradation from large to small houses. This pattern strongly reinforces a view of ranked household variability in the Koniag period, whether side-rooms indicate household population, storage needs, or both. The Alutiiq sample, discontinuous and clustering cases at the lower end, is statistically inseparable from the Koniag distribution.

Together these data strongly suggest social differentiation was weakly developed in the Kachemak sample and well established by the mid-Koniag. A relatively small proportion of Koniag houses contain a disproportionate amount of living area. Before we can settle on a socio-political interpretation for house variation patterns, two complications need to be considered.

First, the modest differentiation in Kachemak, along with the extended differentiation in Koniag, may result from pooling house measures from sites of different function into single distributions. Seasonally-distinct residential sites may have had different house-size needs. Winter villages contained aggregate populations that spread out to smaller settlements during other seasons of the year. According to the ethnohistory, winter was a time of socializing and larger structures were needed to host guests. Smaller structures sufficed when people were spread throughout the region in smaller settlements, busily collecting and processing food.

As a second complication, these data may reflect the addition of non-residential structure types in Koniag times. Noted in the ethnohistoric literature, the most likely candidate for large non-residential structures would be the ceremonial house (*kazhim/kazhu*, similar to the *karigi* of the Yupik Eskimo and other northern Alaskan groups). Ethnohistoric reports (Davydov, 1976:20; Gideon 1989:44; Merck 1980:100) document large ceremonial structures in Kodiak villages. Other reports claim that dancing, feasting and other activities occurred in the central room of residential houses (Black, 1977:93; Huggins, 1947:124; cited in Russell, 1991:32). Thus, the large structures observed in the Koniag sample possibly were *kazhims*, owned collectively or individually (Black, 1977:93; Davydov, 1976:18; Holmberg, 1985:54).

The first complication (that house sizes differ by functional site types) can be examined by comparing house area measures from the Koniag against their associations with functional site categories (*settlements*, *small* and *large villages*, and *refuges*). Figure 9.6 shows box plot comparisons for Kachemak and Koniag samples. In the Kachemak sample, the functional classes are not statistically different on the basis of house size (Wilcoxon/ Kruskal-Wallis test, $p = 0.48$). Interestingly,

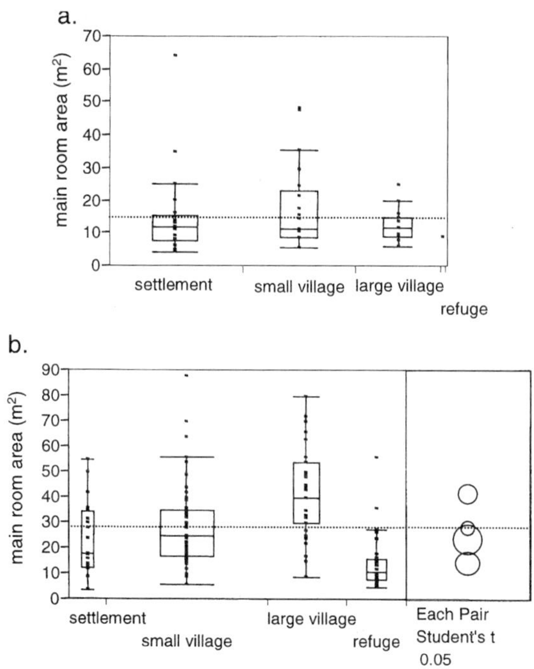

Figure 9.6. Box plot comparison of house area by site function categories used to examine the degree of house size variation within and between site types: (a.) Kachemak houses; (b.) Koniag houses.

the largest houses are found in the *small village* class. Our only site in the *large village* category (KOD 553) is the atypical Early Kachemak site discussed in the last chapter; if we exclude it, the Kachemak sample shows the expected trend of smaller and more uniform (normally distributed) houses in *settlements*, while *small villages* have a positively skewed distribution (more small houses and fewer large houses) befitting a ranked set.

Greater variability in the Koniag sample signals increased settlement complexity. All samples are statistically distinct (pairwise student's t test, $\alpha = 0.05$), with the exception of *settlements* and *small villages*. If house size is an indication, most Koniag sites classified as *small villages* might be better classified as large *settlements*. These data show the greatest variation in house size is found in the Koniag Period (when this variability increases markedly), and in villages (where we would commonly expect co-residential and social/ceremonial aggregations). The smallest Koniag houses are found on refuge islets, where many people congregated in cramped quarters to better defend themselves from attacks. House variation is, thus, greatest in *villages*. The combination of residential and non-residential structures remains a potential explanation for house-size variation, and we must consider other evidence before we can draw conclusions of social differentiation from these data.

We turn next to interrogate the second complication (that house size variation, in *villages*, is a product of different uses of structures). If large structures were non-residential (ceremonial halls or otherwise), they should lack typical residential features. Side-rooms of Koniag houses were used for sleeping, storage, and bathing (Knecht and Jordan, 1986). Communal/ceremonial structures should be few, large and lacking side-rooms (compared with smaller residential structures). If large structures are non-residential, all houses with side-rooms (regardless of their number) should be small. If, however, all structures are residential, and larger structures belong to powerful co-residential kin groups, a correlation between main-room size and side-room frequency is expected. Figure 9.7 shows the relationship between main-room area measures and side-room frequency. This plot includes data from 63 houses in three Koniag villages (SAS 11, 25, and 140), selected because they were the only villages in the sample with ten or more intact, mapped house features.

Figure 9.7 illustrates that some large structures have side-rooms, and thus these are interpreted as residential. Because some of the largest structures lack side-rooms (or have only one), this sample appears to include a few non-residential structures as well (perhaps ceremonial *kazhims*). Even though the sample size is too small for statistical treatment, the data weakly support expectations associated with the second hypothesis (that the largest numbers of side-rooms correlate with the largest residential structures). Houses with the biggest main-room areas often contain multiple side-rooms, and the houses with the largest number of side-rooms tend to be larger structures. In general, more side-rooms correspond with larger structures, suggesting that larger kin groups resided in them.

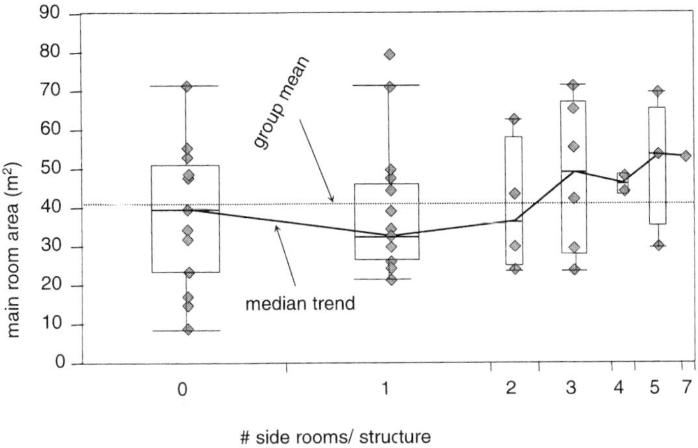

Figure 9.7. Box plot comparison of Koniag large village main-room areas by numbers of side-rooms per house. This plot suggests that house size and complexity is largely a function of differential domestic residential populations, leading to the conclusion that some families were larger and presumably more politically powerful than others.

The presence of several large structures with relatively few side-rooms, and the presence of a few small structures with large numbers of side-rooms suggest partial independence between main-room area and side-room frequency. Given archaeological support for ceremonial structures, along with ethnohistoric evidence that these features were often owned by individuals and their families (Black, 1977:93), the Koniag data imply a complex form of residency and political power (somewhat different from the patterns evident among the Tlingit and other Northwest Coast groups). Where corporate-kin group membership and political power on the Northwest Coast can be conveniently measured archaeologically using residential features (such as longhouses; Ames and Maschner, 1999; Coupland, 1985, 1988, 1996), politically powerful Koniag families could have controlled both a residential structure and a ceremonial structure. Furthermore, corporate group alliances may have included multiple residential households.

These cases being in the minority, I conclude that a larger portion of the Koniag house-size variation (in main-room area, total house area, and numbers of side-rooms) relates to the differences in co-residential group size. Many larger structures housed larger groups (probably kin), with members contributing to a common household economy and sharing in social relations at enlarged scales compared to the previous periods. The multi-modality seen in the house-size data should relate to differences of household leaders' abilities to solidify close kin into stable co-residential units, to host social and ceremonial activities in their homes, and/or to monopolize the acquisition of slaves. Smaller dwellings indicate either unattached, or lower status, family groups affiliated with larger households.

Future investigations might apply methods of fine-grained household archaeology to examine when, and how, large houses formed, and whether different house "life-histories" can prompt inferences about the dynamic rise and fall of fortunes of individual household groups. Excavations of Early Koniag houses already completed at Settlement Point might provide data for such an analysis (Saltonstall, 1995, 1996, 1997).

The Sitkalidak house data suggest that social differentiation began modestly in the Kachemak period, but did not become marked until the Koniag period, and perhaps only during the Developed Koniag phase. This development was predicted to occur well after economic intensification, and in concert with several other developments such as the emergence of a prestige economy and endemic warfare. Discussed in Chapter 8, the analyses show how prestige objects and defensive activities both increased, beginning in the Late Kachemak and continuing into the Koniag periods. The slight variation observed in Kachemak house sizes most likely corresponds to the end of this period, when other indicators of economic and political competition are considered.

9.2. TRENDS IN POPULATION CHANGE

Given the importance of population change in the evolutionary model presented in Chapter 6 (as cause and consequence of other dimensions), we need a reliable estimate of relative pre-contact population change for Kodiak. In this analysis we consider two separate estimates. One estimate uses Sitkalidak house and site area to infer population changes for the southeast region of Kodiak. The second estimate uses radiocarbon dates from the Kodiak Archipelago, accumulated in 500-year intervals as indicators of population trends at the archipelago level.

Errors in any archaeologically-based population estimates can result if the archaeological sample is biased, or if the assumptions are invalid. Biased samples result from selective site formation and transformation processes (Schiffer, 1987), and from non-representative data collection. Erosion and decay, as well as changes in cultural practices, bias comparisons. For example, the shift from single-roomed to multiple-roomed houses indicates a change in domestic space use. What this change involved, and how it altered per capita allotment of house area, is less clear. For these reasons, as with any population estimate based on archaeological information, the results must be considered broad probabilistic generalizations.

9.2.1. The House/Site Area Method

For the site area estimate, I make the following four assumptions. First, main-room area is taken as the most appropriate structural parameter for the projection of per capita space across Alutiiq, Koniag, and Kachemak time periods. This attempts to "correct" for changes in the use of domestic space between Kachemak and Koniag periods, even though it probably introduces a source of bias by excluding Koniag side rooms. As a corollary of this assumption, Kachemak people used their sleeping space for other domestic and social activities when not sleeping, while Koniag people did not. This may be only partially true, since main room areas may have been used for sleeping, especially by slaves and lower ranking kin.

Second, I assume the number of domestic structures at any site determines its area. This assumption discounts the possibility that sites of some, but not all, periods may incorporate non-domestic features (such as storage units, smoke houses, drying racks, ceremonial houses, or menstrual/birthing huts), or differing amounts of outdoor activity areas, that affect the total archaeologically-recorded site area. This assumption may exaggerate population estimates for later time periods, when large ceremonial structures came into use and food processing and storage were most intensively pursued. In contrast, Ocean Bay sites were more likely to expand horizontally (through time with shifting tent placement), as compared to later sites when semi-subterranean house-pits structured reoccupation. These two biases should tend to cancel each other out.

Third, for lack of further information, we assume that all domestic structures (used for estimating per capita main-room area and per capita site area) were occupied simultaneously. Because it disregards the likelihood that some houses in larger village sites were abandoned before others were occupied, this is a problematic assumption. As discussed in Chapter 4, some ethnohistoric accounts suggest that houses, or rooms, were sometimes abandoned when important residents died and were buried in them (Holmberg, 1985:53; Merck, 1980:108). Most ethnohistoric accounts, however, suggest that other locations, away from the house, were more often used to bury the dead (e.g., Merck, 1980:108; Gideon in Black, 1977:96). Also, availability of house-construction materials and labor probably minimized house abandonment, especially in the southern part of the archipelago where construction timbers were scarce. Additionally, archaeological data suggest that houses were often rebuilt in place during both the Late Kachemak and Koniag time periods (Jordan and Knecht, 1988; Knecht, 1995; Steffian, 1992b), indicating a degree of population continuity associated with house features through each phase. This assumption may foster over-representation in the later prehistoric periods, when villages were larger and the potential for social hierarchy was more pronounced (e.g., when house destruction may have been a form of costly signaling).

Finally, we assume that all Ocean Bay sites and all Kachemak, Koniag, and Alutiiq *villages* are correctly assigned to both temporal period and functional category; furthermore, the Kachemak and later *locations*, *camps*, *settlements*, and *refuges* represent seasonal or special activity sites for the use of *village* occupants. On this basis, *villages* then are the only categories included in population tallies for Kachemak and later periods. For Ocean Bay sites (where there are no *villages*), we assume that population duplication was counterbalanced by sample inadequacies due to erosional site loss and the difficulties in finding sites from this period.

While Appendix C presents the actual mathematical transformations, the procedure and results of the analysis are as follows. Lisianskii (1814; cited in Clark, 1987) noted that an average of 18 individuals lived in the Alutiiq multiroom houses he visited in AD 1803 and 1804. This finding concurs with Holmberg's (1985:43) estimate of 200–300 people in villages of 10–15 habitation structures. Lisianskii specifically made his estimate on the basis of visits to several villages on Southeast Kodiak and Sitkalidak Island; thus, it is reasonable to use these same villages (as well as villages he may not have visited) to derive an estimate of per capita house-floor area. Recall from Chapter 8 that the "Alutiiq (contact)" category of this study refers to sites inhabited during Lisianksii's tour of this part of Kodiak (he described and mapped the location of these villages; Clark, 1987), or sites believed to have been occupied at the same time, but missed in Lisianksii's inventory. The calculated average Alutiiq (contact) per-person main-room house area for each of six SAS Alutiiq (contact) villages is about 2 square meters per person (1.96 m^2). The total house area (including side-rooms) allotment is between 2.2 to 4.6 square meters per person (mean = 2.98 m^2).

Next, I multiplied the 2 m^2 per capita main-room estimate by the sum of main-rooms for each *village* site (*small* or *large*) with recorded house depressions. This generated population estimates for all sites with semi-subterranean features represented on the surface (i.e., Kachemak, Koniag, and Alutiiq), but it prevented any comparison with earlier phases. To include Ocean Bay samples in the analysis, I projected populations based on main-room area to total-site areas. Dividing the area of each Kodiak and Alutiiq (contact) *village* site with its population estimate based on main-room area totals, and determining the average of these outcomes, yielded an average per person site area of 29 m^2 and a median of 25 m^2. Using the average, population estimates for all sites were then calculated by dividing site areas by 29 m^2. Table 9.1 presents the period estimates.

Given the many assumptions involved, archaeological estimates of population are best viewed as relative measures of change through time, rather than as absolute estimates of population. Nevertheless, the estimated population numbers, in this

Table 9.1. Population estimates for Sitkalidak area cultural periods, based on house and site area estimates. See text for explanation.

Period	Sites	Population estimate	Average per site
Ocean Bay	14	396	28
Kachemak	9	819	91
Koniag	27	3611	134
Alutiiq (contact)	7	1502	214
Later Alutiiq	5	935	187

case, are in the appropriate range for their sites. Holmberg (1985:43) estimated 200–300 people per village in the Alutiiq Russian phase. The estimate of 214 for the Alutiiq (contact) phase is within that range. (Because the estimate started with a household population figure from Lisianskii that was consistent with Holmberg's village estimate, this confirmation is expected.) The best indication of a well-calibrated estimate relates to the site average for Ocean Bay *camps* and *settlements*, a calculation running very close to 25, the "magic number" anthropologists have long recognized as common for residentially mobile bands (Birdsell, 1953). This validation of the method suggests that measures for Kachemak and Koniag village sizes are reasonable approximations, at least for residential site populations. These measures offer a sense of the scale of change in village population from Kachemak to Koniag times.

Based on the Sitkalidak survey data, Figure 9.8 illustrates the trends in population calculated for period and phase. These estimates show overall population growth from the Ocean Bay period to the Koniag period as well as the dramatic drop following Russian conquest. The per-phase graph (Figure 9.8b), includes two methods for distributing phase-unspecific population. Components only attributed to period were distributed either evenly between phases (dark gray), or in proportion to components attributable to phase (light gray). The first method smoothes the differences between phases and may overestimate the relative population of the Early Kachemak, a phase with few confirmed components. The proportional bars better capture this anomaly. The dramatic increase in population of the Developed Koniag may also be exaggerated. While we have already reported the evidence against significant masking of older occupations by Developed Koniag sites, some overlay is nevertheless expected. Because of its relatively few distinct diagnostic features, the Early Koniag could hide more easily among Developed Koniag components, and may be particularly underrepresented. Nevertheless, the Sitkalidak region clearly reached a peak of population density during the few hundred years before Russian conquest. Previous analyses have already addressed the decline in the Alutiiq period population.

9.2.2. The Carbon-Date Frequency Method

Population estimates generated from spatial attributes of the Sitkalidak region can be compared to population estimates based on radiocarbon dates. This second method organizes regional radiocarbon dates into temporal intervals and plots them as proxy measures of population (e.g., Ames and Maschner, 1999; Erlandson et al., 1992); this technique is difficult to apply in localized contexts because it assumes (1) radiocarbon-dated sites represent the actual number of occupied sites in a time interval, and (2) each date represents a comparable unit of population. No account is taken of differences in site size or function. Nevertheless, biases generated by individual sites and unusual research strategies generally

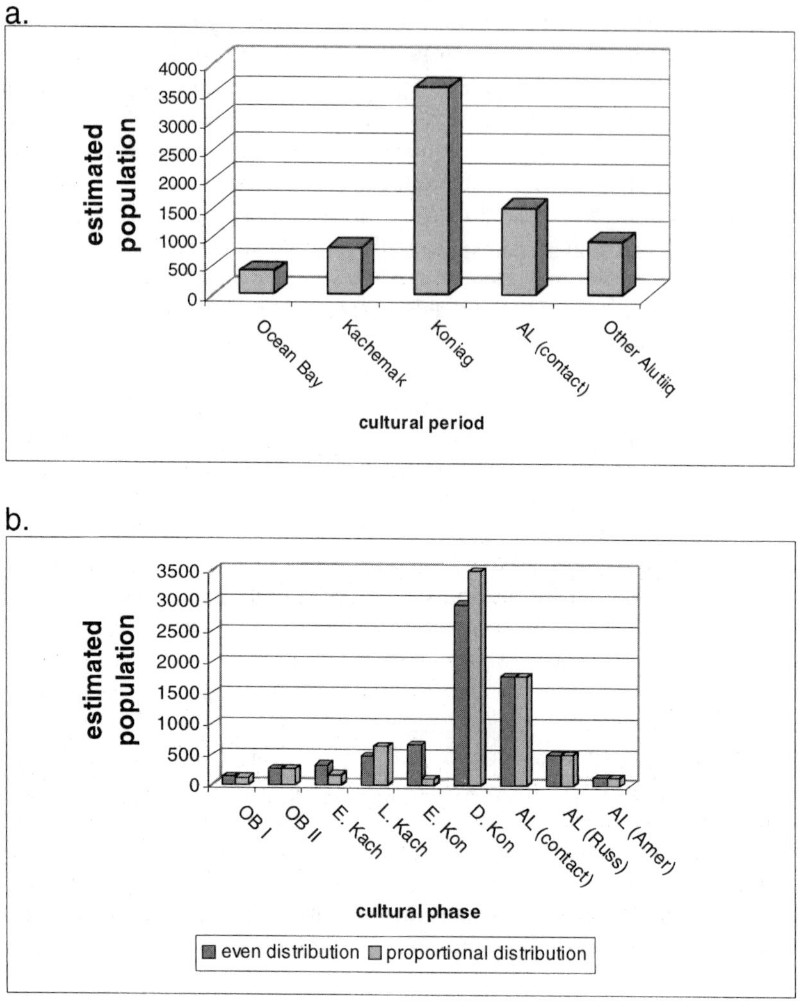

Figure 9.8. Population estimates for the Sitkalidak region based on house and site size data from the Sitkalidak Archaeological Survey: (a.) data plotted by period; (b.) data plotted by phase. In graph b, sites that were only dated to period but not phase were added to phase-dated components either evenly (dark gray bars) or in proportion to the phase-dated components (light gray bars). See text for explanation and assumptions.

cancel out each other when applied to large archaeological regions with numerous radiocarbon dates.

Using 169 dates, Erlandson and colleagues (1992) presented the first comprehensive paleodemographic analysis for the Gulf of Alaska and Prince William Sound region (GoA). To update the population estimates for this region, I have

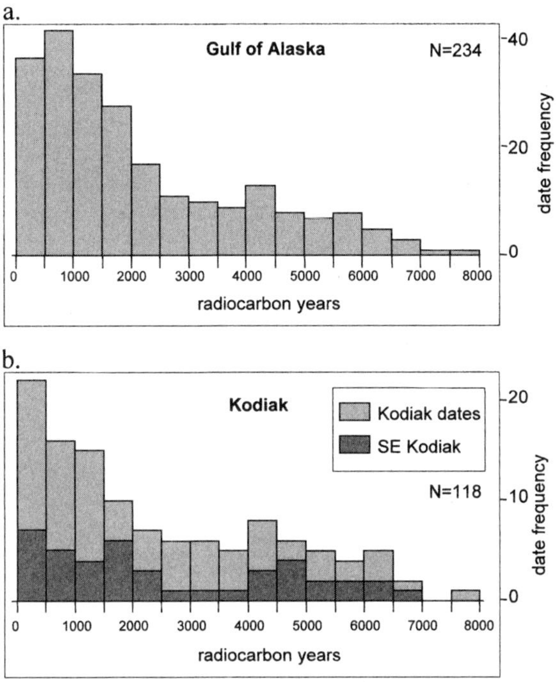

Figure 9.9. Chart of uncalibrated radiocarbon dates grouped per 500 year interval, used as proxy measures of population change through time (see text for explanation and assumptions): (a.) aggregated for the Gulf of Alaska, including the Kodiak Archipelago; (b.) Kodiak Archipelago dates only. Dates from Southeast Kodiak (the Sitkalidak vicinity) are shaded.

augmented their sample with more recently published and unpublished GoA dates (Appendix D). Dated sites are tallied per 500-year interval, counting each site only once per interval.

Figure 9.9 shows population patterns estimated by the radiocarbon method aggregated for the GoA region generally (Figure 9.9a) and for the Kodiak Archipelago specifically (Figure 9.9b). Considering the Gulf overall (Figure 9.9a), initial population growth commenses about 8000 bp and levels off soon after 6000 bp. A small surge appears in the 3500 to 4000 bp interval, soon returning to levels comparable to the previous 1500 years. With the exception of this spike, population grows slowly, but steadily, from 6000 bp to 2000 bp. After 2000 bp, population growth excels at an unprecedented rate, tripling its size by the time of Russian colonization; then it plummets, as known from other, more precise sources (Crowell and Lührmann, 2001).

Extracting the Kodiak dates from the GoA sample and plotting them independently (Figure 9.9b) shows the Kodiak-specific trends and unique aspects of the

Sitkalidak sample. The Kodiak pattern closely tracks the larger GoA trends, with gradual growth before, and accelerated growth after, 2000 bp. Sitkalidak dates (dark gray in Figure 9.9b), however, do not match the trends for the rest of the island and Gulf; removing the Sitkalidak dates shows the rest of Kodiak remaining stable from 6000 to 2000 bp. The Sitkalidak dates, taken alone, suggest that populations in this region fell dramatically between 4000 and 2500 bp, or, more likely perhaps, sites of this interval have been lost or obscured from detection. The significance of the apparent population decline in the Early Kachemak remains unknown. Though it could be real, recent research suggests one of several possible geological mechanisms may account for greater loss of sites in this interval (see Chapter 7). Even though additional dates from Sitkalidak might fill in the gap somewhat, the current sample should be statistically robust enough to hold its general shape.

Comparing the radiocarbon-date model (Figure 9.9) with the surface-area model for Sitkalidak (Figure 9.8) generates a synthetic overview of population change on Kodiak. This comparison is especially useful because each model depends on a different set of working assumptions. Assuming the radiocarbon-date model overestimates the degree of Ocean Bay population growth due to selective dating, the model's estimates are consistent with those of the site-area model: slow, but steady, population growth until the late Ocean Bay II phase, around 4000 bp. In the Sitkalidak portion of the radiocarbon-date plot and the site area data (both from the SAS project), population appears to have declined during the Early Kachemak phase, while elsewhere, Kachemak populations maintained themselves with little growth. Both models project an upswing from Late Kachemak through Koniag intervals. The site-area model does not capture growth in the Early Koniag (supporting the possibility of sampling error), while, in contrast, the radiocarbon-date model shows robust population growth from at least 2000 bp until Russian contact. The site-area model best represents the Alutiiq period, since the radiocarbon-date intervals are too coarse. These data confirm a dramatic decline in Sitkalidak-based population between the Developed Koniag phase and the Alutiiq-Russian phase and into the Alutiiq-American phase.

The Chapter 6 model predicted slow population growth through the Ocean Bay period. Factors hampering growth near the end of the period include: population packing, limits of subsistence potential prior to the development of technological intensification (mass harvesting), and labor intensification in processing. The drop in modeled population of the late Ocean Bay II and Early Kachemak phases may reflect this effect. Alternative explanations look more promising; these include either preferential site loss in the Sitkalidak region, or climate change and regional adjustments to Neoglacial winters. The model does *not* anticipate population decline at the Early Kachemak juncture, only deceleration or stasis.

Once mass-harvesting practices and processing intensification were established in the Early Kachemak phase, the population should have grown dramatically. In these data, growth starts again only in the Late Kachemak and continues to build to unprecedented levels through to the Developed Koniag phase. Consequently, we must reconsider the impact of the delayed-return economy on demography and social evolution on Kodiak. This issue will be discussed in the following chapter.

Radiocarbon date samples from Kodiak subregions and the surrounding regions remain too small to provide statistically representative estimates of regional variation in site preservation/population histories, and so they are not graphed. These data do, however, suggest that northern Kodiak (Chiniak Bay to Afognak Island) was a unique area with a higher frequency of sites recorded in the Early Kachemak phase than elsewhere. This pattern may derive from intense efforts in the past few years to investigate sites of this time period in this region (Clark, 1996, 1997, Patrick Saltonstall and Amy Steffian, personal communication, 2002). If true, additional dating should bring the northern distribution closer to the shape of the overall Kodiak and Gulf of Alaska patterns.

In a comparative study of relative sea level histories from the Outer Kenai and Katmai coasts, Crowell and Mann (1996) concluded that, unlike the Outer Kenai coast, the Katmai coast escaped serious submergence and site destruction after about 7000 radiocarbon years ago. Importantly, the Kodiak plot matches the shape of the plot for the Pacific coast of the Alaska Peninsula, including Katmai; this suggests that differential site preservation is not responsible for the overall trends observed in the aggregated data set. Nevertheless, local sea level histories vary and reconstructions are still needed to better understand regional variation in site preservation (Mann et al. 1998; Saltonstall and Carver, 2002).

9.3. SUMMARY

This chapter examined measures of social inequality and population change. The reported analyses suggest that social organization changed dramatically between Kachemak and Koniag times. Slight social differentiation is apparent in Kachemak house size variation, while marked variation is seen in Koniag houses. In addition, the rise of large non-residential structures in the Koniag period is linked to the practice of ethnohistorically-reported feasting. As in previous analyses, the Koniag period marks the most significant changes in social organization and in structures of social engagement. This is paralleled by the development of unprecedented population densities and mega-village aggregations. Surprisingly, the innovations of mass-harvesting and processing in the Early Kachemak do not appear to trigger significant population growth, perhaps because they may have been "counterbalanced" by increasing cold climates and more difficult winters.

Chapter 10

Reconciliation, Extension, and Implications

Anthropologists have concluded only recently that relatively complex hunting and gathering societies are relevant to questions about social evolution. To date, few archaeological studies have examined the actual evolutionary histories of markedly complex hunting and gathering societies. This book attempts to improve this record by exploring the development of complex hunter-gatherers on Kodiak.

Previous chapters reviewed social evolution on southern Alaska's Kodiak Archipelago and applied a sequential model (based on evolutionary ecological theory) to a case study of the Sitkalidak region. Several lines of evidence (based primarily on archaeological survey, mapping, and testing) were used to evaluate hypotheses drawn from the model. Core data included change through time in site/component frequencies, sizes, locations, and structural variation. Estimates of artifact variability, site function, and regional distributions presented a means to evaluate changes in land use and settlement pattern. House data from surface mapping offered examination of potential differences in social organization and rank. The measured house- and site-size data, as well as radiocarbon date frequencies, supported complementary estimates of demographic change at multiple spatial scales. Questions regarding the archaeological record's representativeness, especially for the early Holocene (for which we have no archaeological indications of occupation), and for the Early Kachemak phase (for which we have few), were discussed, but remain unanswered. In this chapter we evaluate the model, consider revisions in light of the case study findings, and situate the Kodiak case in a broader context.

10.1. INTERROGATING THE MODEL

Developed from an evolutionary/behavioral ecological point of view, the model presented in Chapter 6 combined a number of ideas about the dynamics of hunter-gatherer adaptation and social organization. At its simplest, the model considered the dynamic relationship between changing parameters and adaptive strategies as maritime hunter-gatherers colonized, expanded, and eventually packed ever more tightly into the finite Kodiak landscape. Patterns of residential and logistical mobility, subsistence pursuits, technology, and storage were predicted on the basis of seasonal resource distributions and population densities. Changes in social organization and political differentiation were predicted from expected changes in population packing, spatial variability in resource stability, and opportunities for exclusive control of particularly stable resources under circumstances of social circumscription. What distinguishes this model theoretically from previous "cultural ecological" arguments (and saves it from many of the functionalist/adaptationist criticisms that have challenged them) is the assumption that *individual* decision-making/behavior guides cultural process and motivates social change. Cultural process, then, emerges from thousands of decisions and ensuing actions by individuals seeking to adapt to local, and immediate, circumstances. While not always explicit, this focus underlies every behavioral ecological model used to guide the processual narrative.

How successfully did the model predict the archaeological sequence on Kodiak? Despite important unresolved issues, the model's predictions were supported. Table 10.1 summarizes the performance of each hypothesis, and Figure 10.1 illustrates the timing of key changes in the Kodiak/Sitkalidak sequence.

10.1.1. Ocean Bay I: The First Act—Aftermath of Colonization

From the model, I predicted hunter-gatherers on Kodiak (following colonization and immediate adjustments to a new environment) would gradually expand to fill the archipelago. During this expansion, these foragers would maintain small residential bands and relatively large foraging ranges. Occasional limitations of logistical foraging in winter and a primarily immediate-return economy would drive residential mobility.

Data from the Sitkalidak survey and excavations at the Tanginak Spring site support this understanding of early Kodiak prehistory. While people possibly occupied the archipelago for thousands of years before the currently known archaeological record begins, there are indications that early Ocean Bay I occupants at the Tanginak Spring site were adapting to an unfamiliar region. Tool and raw material categories suggest that early Ocean Bay residents initially depended on continued access to the mainland—a dependence that diminished with time. At this site (along with other Ocean Bay I sites in the survey area), camps were small, perhaps

Table 10.1. Comparison of model predictions from Chapter 6 against archaeological evidence presented in Chapters 7–9.

Act I predictions	Results	Ocean Bay I findings
1) . . . small campsites, in central places	Supported	Sitkalidak sites of Ocean Bay I and II are small. Average size appropriate for 25–30 individuals in no more than 3–5 structures (Figure 8.2; Table 9.1). Consistently in mid-bay locations (Table 8.5).
2) . . . few large aggregation sites	Supported tentatively	All but three Ocean Bay period sites are smaller than 1000 m^2. The others may have been aggregations or palimsests. None appears to represent permanent aggregation (village). Two larger sites are reported from northeast Kodiak (but are unpublished). These may have been aggregation sites or palimpsest deposits.
3) . . . low site densities and low variability in site functions.	Supported	Ocean Bay I sites have the lowest density reported (Figure 8.1). Only two functionally distinct site types were identified: *locations* and *camps/settlements* (where settlements were simply more frequently used camps; Table 8.2)
4) . . . portable residential structures.	Supported tentatively (for Sitkalidak data)	No semisubterranean structures were encountered. Excavations at the Tanginak Spring site reveal only flat occupation surfaces. One pit depression is noted at the Rice Ridge site (Don Clark, personal communication 1996).
5) . . . residential structures should exhibit low and normally distributed sizes variation.	Unevaluated	We need more excavation of Ocean Bay I structures
6) . . . portability in labor-intensive tools.	Supported	Labor intensive tools (especially lamps) are universally small. This has been demonstrated in excavation assemblages from the Afognak River sites and Sitkalidak Road Cut site (Clark 1979), Zaimka Mound (author's observation of collections in the Alutiiq Museum), and the Tanginak Spring site.
7) . . . relatively unspecialized hunting, fishing and gathering tools.	Unevaluated in Sitkalidak data	The tool categories known for Ocean Bay period are mostly generalized tools. These include projectile points, barbed harpoons, two-piece fishhooks, leister spears, and bone clam digging tools. Each could be used on a variety of resources.

(Continued)

Table 10.1. *(Continued)*

Act I predictions	Results	Ocean Bay I findings
8) . . . low proportions of processing tools.	Supported	Processing tools were at their lowest relative proportion of any phase/period in the archaeological samples evaluated (Figure 8.8).
9) . . . decrease through time in exotic raw materials, finished tools, and other off archipelago products.	Supported	Tanginak Spring site raw materials variability in blades from the 1994 excavations show decrease in basalt and increase in red chert through time (Figure 5.5).
10) . . . lithic technologies adapted to local and regional raw materials through time.	Supported	Ground slate technology in the later Ocean Bay I phase. Chipped stone technologies persist on Kodiak throughout prehistory, but ground slate assumes a central role in the production of hunting and processing implements (Clark 1982; Fitzhugh 2001).
11) . . . low species richness and low evenness in summer faunal assemblages; greater diversity and evenness in winter assemblages. Emphasis of larger bodied sea mammals and fish plus a wide variety of near camp resources. Bias against resources demanding high processing costs or specialized facilities.	Unevaluated	No faunal data available. Faunal preservation in the Sitkalidak region was limited to the Koniag period.

Act II predictions	Results	Ocean Bay II findings
1) . . . an increase in site density.	Supported	Site densities increase from Ocean Bay I to Ocean Bay II phases (Figure 8.1).
2) . . . no significant increase in site size.	Supported	Ocean Bay I and II sites could not be differentiated in terms of size.
3) . . . no increase in frequency of large aggregation sites.	Supported	No large Ocean Bay II phase sites were encountered in the Sitkalidak Survey. As discussed above, there may be candidates elsewhere on Kodiak.
4) . . . no significant change in site function variability or land use choices.	Supported	Ocean Bay II sites show no addition of site types from Ocean Bay I. Sites continue to be located in mid-bay regions. (Table 8.5)
5) . . . more durable constructions and thicker and denser site deposits.	Unevaluated	Proper evaluation requires more excavation and measures of artifact or feature density per unit time (with fine temporal resolution).

(Continued)

Table 10.1. *(Continued)*

Act II predictions	Results	Ocean Bay II findings
6) . . . increased use of non-portable technologies that could be left for reuse at camps integrated in a regular settlement cycle.	Contradicted	No change is apparent in artifact portability from Ocean Bay I. This prediction is fulfilled in the Kachemak period, suggesting that residential flexibility was still important in the Ocean Bay II phase.
7) . . . increased zooarchaeological taxonomic richness and evenness, with an increase in the relative dietary contribution of smaller prey of higher processing cost . . .	Unevaluated	No faunal data available.
8) . . . no significant evidence of technologies of mass harvesting, processing or storage.	Supported prior to model	Available evidence supports the prediction (mass harvesting and processing technologies appear in Kachemak times) but this evidence was used to calibrate the model.
9) . . . no change in mean or standard deviation of residential structure size.	Unevaluated	Need more excavation data
10) . . . no increase in frequency of military tools or defensive sites (minimal organized violence)	Supported weakly	Defensive sites are not apparent, although, reputedly, the Awa'uq site excavations revealed a component of this time period (with Arctic Small Tools tradition appearance; Rick Knecht, personal communication, 1994). Military tools are undifferentiated from hunting tools in the lithic assemblages of Kodiak making testing this prediction difficult.
11) . . . no increase in frequency of violent trauma in mortuary contexts or catastrophic site destruction.	Unevaluated	No Ocean Bay mortuary contexts have been investigated (to my knowledge) anywhere on the Kodiak Archipelago.
12) . . . no change in frequency of elaborate ornamentation, monuments, or labor-intensive crafts.	Supported	Absence of any evidence for monumental or prestige-based competition and low proportional representation of ornamental artifacts suggests that these activities were insignificant aspects of Ocean Bay period social interaction (Figure 8.8).

Act III predictions	Results	Kachemak findings
1) . . . increased efficiencies of harvest and processing technology.	Supported	The Sitkalidak artifact data (Figure 8.8) show the expected increase in processing tools corresponding to an increase in

(Continued)

Table 10.1. *(Continued)*

Act III predictions	Results	Kachemak findings
		fishing artifacts as expected. Toggling harpoons, net-hunting technology and more efficient processing tools (*ulus*) appear in the Kachemak period, but were used to calibrate the model.
2) . . . increases in the proportional representation of r-selected species and a reduction in the relative contribution of K-selected taxa.	Unevaluated	No faunal data available.
3) . . . increases in the proportional representation of processing tools in assemblages and processing facilities found in excavations.	Supported	Processing tools increase in importance in the Kachemak period (Figure 8.8). Processing facilities, per se, are suggested by recent excavations at the Outlet Site on the Buskin River (Kopperl personal communication 2002).
4) . . . larger and more structurally developed sites ("villages").	Supported	The first appearance of *villages* (Table 8.2), but fewer large villages than expected. Predominance of semi-subterranean house features is an important change.
5) . . . possible increase in mean house size but insignificant change in house variability.	Unevaluated	We need more Ocean Bay and Early Kachemak houses for comparison. Kachemak variability suggests minor inequity in residential space, population, and/or social status, probably a late Late Kachemak development (Figure 9.3).
6) . . . evidence of internalized storage (e.g., pits or rooms inside the dwellings) and greater site infrastructure (more feature types).	Supported elsewhere	Storage facilities are found in the form of clay lined pits in several excavated or tested Kachemak sites (e.g., Saltonstall 1995, 1996, 1997; Steffian 1992). Corner alcoves on Late Kachemak houses are described as storage facilities (Jordan and Knecht 1988).
7) . . . increase in functional site types, including seasonal occupation of mass-harvesting locations.	Supported	Kachemak sites include mid-bay *villages* and *settlements*, inner-bay/riverine *settlements* and *camps*, and *refuge* sites late in the period. (Table 8.5, Figure 8.7)
8) . . . increase in total site density (including sites of different functions).	Supported for Late Kachemak	Site densities increase moderately from the Ocean Bay to the Kachemak period, particularly in the Late Kachemak phase. Early Kachemak data if representative, contradict the prediction. (Figure 8.1)
9) . . . increase in prestige goods	Supported	Ornamental artifacts begin a trajectory of proportional increase in the Kachemak

(Continued)

Table 10.1. *(Continued)*

Act III predictions	Results	Kachemak findings
		period (Figure 8.8). Elsewhere, evidence includes Late Kachemak importation of coal for the production of labrets (Steffian and Saltonstall 2001) and the elaboration of labor-intensive decorations of stone lamps and other artifacts (Heizer 1956).
10) . . . increase in traumatic injury and death on skeletons.	Unevaluated	Late Kachemak mortuary research (e.g., Simon and Steffian 1994) suggests that treatment is highly variable, but violent death has not been monitored. An Ocean Bay mortuary sampled would be need to address this prediction.
11) . . . addition of defensive sites.	Supported	Late Kachemak *refuge* sites appear around 800 AD. They are small, located in protected waters and suggest local competition. More excavation and dating of *refuge* sites is needed (this finding is based on few sites and limited testing).

Act IV predictions	Results	Koniag findings
1) . . . greater development of infrastructure at defendable resource patches (patch defense through physical occupation).	Supported	The appearance of Koniag *settlements* at almost every stream outlet, no matter how small, suggests intensive exploitation and claims on patches. Many *small villages* were probably also seasonal resource sites by larger groups. (Table 8.2)
2) . . . evidence of economic and political integration into extended household or corporate groups.	Supported	Kachemak houses would not have accommodated as large gatherings as most Koniag structures (Figure 9.3). Some elites may not have lived in the most crowded houses (large main-rooms but few side-rooms;Figure 9.6), and factions must have been assembled from more than one household.
3) . . . increased variation between house size and/or house cluster size.	Supported	House size variability suggests differences in the ability/need to build and maintain large dwellings in the Koniag period. House clusters were not clearly observed in architectural layout, but the large number of small houses, suggests that many households were subordinate to a few larger ones. (Figures 9.3 and 9.4)

(Continued)

Table 10.1. (*Continued*)

Act IV predictions	Results	Koniag findings
4) . . . increased use of symbols of social identification.	Unevaluated or contradicted	The Sitkalidak artifact sample was too small and included too few artistic pieces appropriate to identifying regional affiliations. Steffian and Saltonstall (2001) identified regional clusters of labret styles linking people to adjacent sections of the mainland in the Late Kachemak. This and the discovery of similar "bird" designs on points from the Tanginak Bluff site and the Crag Point site (Figure 3.6) suggest this process was underway in the Late Kachemak phase.
5) . . . distinctive bodily ornaments and garments to demonstrate group affiliation and rank, found especially in larger houses.	Supported partially	Ornamental objects increase proportionally in the Koniag sample, which indicates greater circulation of prestige goods in general (Figure 8.8). Selective distribution could not be evaluated with survey data.
6) . . . increased frequencies of non-local materials and investment in labor intensive goods, especially in larger houses.	Unevaluated	Sitkalidak survey artifact samples were insufficient for this analysis. Systematic analysis of the Crag Point, Uyak, Karluk 31, and Karluk One assemblage could help to assess the relative importance of non-local and labor intensive prestige goods.
7) . . . increased construction of defensive facilities.	Supported	*Refuges* multiply in the Koniag period. (Table 8.2)
8) . . . enlargement and strategic reorientation of defensive sites.	Supported	We see the development of large, outer-coast *refuge* sites in the Koniag period by at least 500 years ago (Figure 8.11, Table 8.6). These site locations and sizes suggest coordinated defense from distant enemies.
9) . . . increased frequency of violent traumas.	Unevaluated	Mortuary data were neither sought nor encountered in the Sitkalidak survey.
10) . . . increased emphasis on military tools.	Supported	"Hunting" tools are highest on *refuges* and most refuges are Koniag (Figure 8.8).
11) . . . increases in genetic diversity (especially mtDNA compared to Y-chromosome)	Unevaluated	Biological evidence suggests some continuity and some genetic admixture (Scott 1992, 1994). Testing this prediction would require the interest and support of the Alutiiq community and the availability of human remains for testing.

(*Continued*)

Table 10.1. (*Continued*)

Act IV predictions	Results	Koniag findings
12) ... increase in adoption of non-local processing technologies but not hunting tools.	Unevaluated	Suggested by the appearance of pottery during the Koniag period.
13) ... greater diversity in mortuary treatment distinguishing slaves from non-slaves and high from low ranking members of society.	Unevaluated	Mortuary data were not sought nor encountered in the Sitkalidak survey. Other data is not specific, but Late Kachemak treatments are notably more elaborate (Simon and Steffian 1994; Urcid 1994; Workman 1992)

containing no more than 25–30 individuals. Probable tent use at the Tanginak Spring site and the small size of stone lamps (the most labor-intensive artifacts known for the period) suggest portable technologies and periodic residential mobility. Consistent placement of camps in intermediate locations between inner-bay/river locations and outer-coast zones, as well as the undifferentiated nature of Ocean Bay camps, further support the idea that logistical foraging and large foraging ranges predominated in daily subsistence activities. Low site densities suggest low population densities for this time period.

10.1.2. Ocean Bay II: The Second Act—Expansion to Density-Dependence

According to the model, gradual filling of the archipelago would foster compression of foraging ranges, limiting logistical, and eventually residential, mobility. I hypothesized that this would lead to resource depression and/or evasive behaviors in high-ranked prey species before population growth would be significantly hampered. With the increased variability of foraging success and greater exposure to the hazards of risk and unpredictability accompanying decreased access to prey, technological inventiveness would increase (Fitzhugh, 2001). This situation could persist indefinitely unless an innovation arose to improve foraging efficiencies and reduce risks. In cases of agricultural intensification, an effective innovation involved the development of strategies and technologies to increase yields in times of plenty and to extend seasonal surpluses into lean seasons. The ensuing shift to a delayed-return economy (focused on mass-harvesting and processing of r-selected resources such as herring and salmon) had broad implications for economic, residential, social, and political organization.

Data from the Ocean Bay II and Early Kachemak support many of these expectations. Judging from increased site densities, populations double from Ocean

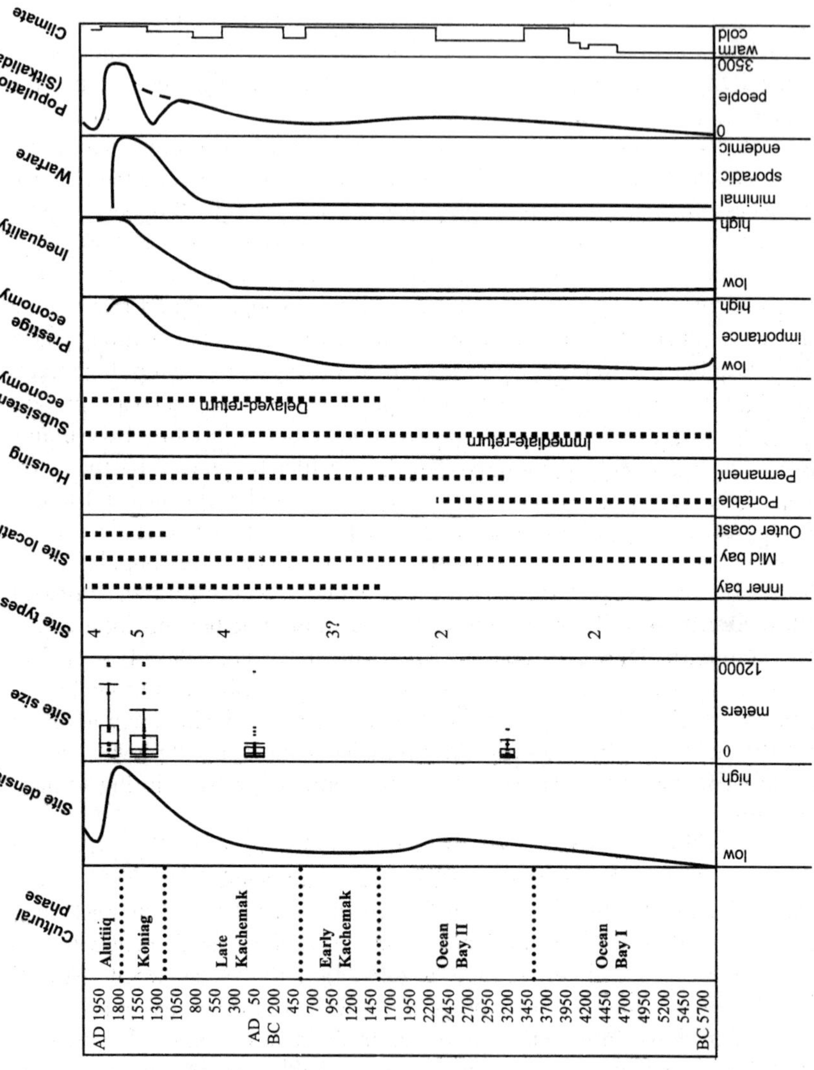

Figure 10.1. Comparative plot of 7500 years of economic, technological, social, and political changes in the Kodiak Archipelago as revealed through this study and other recent research.

Bay I to Ocean Bay II times. Ground-slate hunting technology may represent a technological intensification for improved marine mammal hunting (Hayden, 1981) as previously argued for the Scandinavian Younger Stone Age (Fitzhugh, 1974). Semi-subterranean housing developed in this phase, as evident in recent excavations at the Zaimka Mound site (Amy Steffian, personal communication, 2002) and elsewhere (although none were found around Sitkalidak). The abandonment of red ocher by the end of Ocean Bay II is interpreted as another indication of the shift to less portable housing.

While the resource-depression predictions have not been evaluated on the basis of Sitkalidak data, relevant information has been recently reported that can help address the model's subsistence predictions. For example, Kopperl's (2002) preliminary analyses from the Ocean Bay period Rice Ridge site (KOD 363) suggest a general increase in foraging pressure on marine mammals and cod fish between about 6000 and 3000 bp. In his analyses, Kopperl found that marine mammals dominated the faunal assemblage (relative to fish remains) through most of the Ocean Bay period occupation of the Rice Ridge site. In the top three levels (attributable to the later Ocean Bay II phase), remains of fish increase proportionately (to between 80% and 90% of identifiable bones [NISP]). This finding suggests that sea mammals may have become harder to capture by this time, perhaps due to overexploitation and/or the shift to colder climates and longer winters. Throughout the Rice Ridge sample, Kopperl found a gradual decrease in the average size of cod bones, suggesting depletion of near shore populations of codfish. I interpret this pattern as evidence for winter harvesting of cod, consistent with expectations of winter foraging close to camps. Whether this local depletion of cod stocks forced periodic residential mobility (as predicted) is unclear and will require careful analysis of associated artifact assemblages and site formation processes. Kopperl is expanding his faunal analysis to include samples from Kachemak and Koniag period sites. Once completed, his data will provide insight into subsistence patterns and the possibility of resource depression during Kodiak's prehistory.

The ultimate change to a mass-harvesting and storage economy is observed at the start of the Kachemak period. Nets, ulus, and toggling harpoons begin to appear in the initial stages of the Early Kachemak (Clark, 1997). Late Kachemak data (our only statistically meaningful sample for Kachemak settlement patterns) confirms an expansion of land use into inner-bay and river areas, where seasonal gluts of fish were captured and processed. Evidence from the Buskin River and shores of Chiniak Bay provide more information about Early Kachemak settlement patterns. These sites indicate intensive fishing and processing beginning during Ocean Bay II times and escalating in the Early Kachemak. While the number of sites remains small, the pattern appears to parallel the Late Kachemak settlement pattern from the Sitkalidak region (mid-bay occupations with river fishing and processing settlements and other resource extraction camps; Steffian et al., 1998;

Robert Kopperl, Patrick Saltonstall, and Amy Steffian, personal communications, 1999, 2000, 2001).

10.1.3. Early Kachemak: The Third Act—Density-Dependence or Near Extinction?

All else being equal, the development of a storage-based economy should have supported further population growth, increased population density, and aggregation into residentially-sedentary villages. In turn, these developments should have precipitated a series of social changes including the mobilization of larger domestic labor forces and increased political tension over the ownership and rights governing stored-food distribution. Tension should have developed surrounding the contradictory imperatives of storing for domestic use versus sharing resources with others.

Almost all of these predictions find support in archaeological records for the Late Kachemak and Koniag periods, but *not* the Early Kachemak. I conclude that all else was not equal in this time period. At the most general level, it appears that population growth continued slowly through the Early Kachemak and in some areas, such as around Sitkalidak Island, populations may even have declined during this period. Recent discoveries of Early Kachemak sites around Kodiak suggest that archaeologists may only now be learning how to recognize Early Kachemak processing sites (Amy Steffian, personal communication, 2002). That being the case, we can expect the predicted increase in population following technological intensification may yet be confirmed. Otherwise, we conclude that mass-harvesting and processing did not inspire the predicted demographic revolution, or, at least, we need to consider complicating factors as we seek to improve the model.

Two complicating factors are evident. First, already elaborated, an episode of selective site loss, perhaps due to sea-level fluctuations or tectonic processes, may have been especially devastating to the Early Kachemak record on the southeast side of the archipelago. Second, a reduction in site frequencies may indicate a near abandonment of the Sitkalidak area for nearly 2000 years. Possibly, the longer and less productive winters of neoglaciation facilitated the change to an intensive, storage-based economy while preventing the economic reorganization from translating into increased rates of population growth. With a new economy based on fishing, the Sitkalidak region (with its preponderance of small streams) may have been less attractive than areas with larger, and more predictable, anadromous fish streams. It is intriguing that most of the documented Early Kachemak sites are located on substantial salmon rivers or near their outlets (Afognak River, Buskin River, Karluk Lagoon, Kiavak). Interestingly, there is a significant movement of Kachemak people up the Kenai River in approximately the same interval (Reger, 1998; Reger and Boraas, 1996). If this pattern holds, it becomes reasonable to characterize the Late Kachemak as a phase of expansion, backing out of the river zones to reoccupy the more open coasts. Climate change may underlie both developments.

Given these observations, it is necessary to modify the evolutionary model to better account for the Early and Late Kachemak developments. It is simply impossible to claim that intensive storage strategies (delayed-return economy) automatically precipitated population growth and social complexity, although it may have prevented local extinction. For now, we can conclude that low population density and minimal socio-political change characterized the majority of the Early Kachemak phase. Future research should be devoted to explaining this anomalous interval.

Population growth began in earnest about 2000 years ago, near the beginning of the Late Kachemak period. The reasons for this growth remain uncertain. An ameliorating climate may have provided opportunities for the previously developed delayed-return economy to fund a genuine population explosion. I find no evidence to suggest an influx of outside populations to Kodiak beginning 2000 years ago, or that immigration significantly influenced the steady growth in population through the transition to the Koniag period (but see Dumond, 1987, 1988a).

As predicted, once significant population growth was underway, it stimulated increased political competition, as evident in the incipient house-size asymmetries and local defensive constructions. This population growth did not, however, precipitate large-scale village aggregation in the Late Kachemak; possibly families retained a strong commitment to self-sufficiency and resisted the pressures to become enmeshed in larger and more complex social configurations. In an improved climate, greater resource availability would support continued equality, as each family retained the ability to be self-sufficient. Such changes would, however, have entailed greater *social* risk and uncertainty (see below).

10.1.4. Late Kachemak and Koniag: The Fourth Act—Competition and Cooperation in a New World Order

Increased population density, a storage economy, and the emergence of a more patchy resource environment (in which dominant individuals maintained exclusive control of resource areas—a despotic environment) should have encouraged spatial reorganization around corporate kin-groups. In turn, these circumstances should have fostered asymmetrical control over the most reliable/consistently productive patches, rendering a shift from scramble to contest competition and political factionalization. From this the following circumstances were predicted: amplified economic and political competition, patron–client relationships between more and less successful families, and prestige economies driven by the need of newly powerful families to advertise political and economic strength (to potential subordinates, allies, and enemies). Endemic warfare should have become an important component of social and political checks and balances. Initially driven by challenges to control resources, warfare (raiding) could eventually have become a means for establishing reputations, building prestige wealth, and increasing productive labor through war-captive slavery.

Because changes from Late Kachemak to Koniag happened so rapidly (in the scale of resolution available in this study), teasing out micro-developmental sequences and determining what initiated the cascade of transformations remains difficult. The Sitkalidak data contributes the following information. First, population growth was continuous from early in the Late Kachemak period until Russian contact; this implies that the causes of population growth cannot be attributed to changes in the organization of production at the inception of the Koniag phase, though these developments probably contributed momentum to population growth. Instead, population growth likely influenced social changes through its effect on resource access and structure, crowding, and altered options for resolving subsistence and social crises. Second, social stress developed in the Late Kachemak, when social identification and ritual behavior emerged as accentuated parts of social life. Incipient social differentiation (indicated by a small degree of house variation) probably dated to the terminal Late Kachemak, when small, locally-oriented defensive facilities were first constructed. These facilities suggest inter-family feuds or localized warfare at the time. This implies that violence and changes in social organization followed (rather than preceded) population growth and increased crowding.

In the Koniag period, houses became larger, perhaps as clusters of nuclear-family dwellings joined to an enclosed common patio turned into sleeping chambers joining a common living room (Saltonstall, 1995). Multi-roomed houses appeared around the archipelago as early as 1300 A.D. (Partlow, 2000). Around Sitkalidak, house data suggest a definitive development of differentiation in dwelling size and the number of side rooms, indicative of the ranking noted in ethnohistoric documents. Partlow (2000) finds evidence for increasingly intensive salmon harvesting following 1300 AD in the quantities of processed fish remains stored in winter houses. Around Sitkalidak, the increased frequency of settlements at stream mouths reinforces this notion of intensified salmon production.

Large villages appeared in the Sitkalidak region during the Koniag period, although at least two reasonably large Kachemak villages were identified (others might have been lost to erosion). Mega-village formation in outer-coast regions suggests a shift from local to regional social interaction (potlaches and warfare); this shift may have followed the development of an intensive whale-hunting strategy, allowing competitive hunters to contribute large amounts of meat and blubber to village feasts and stores and to fund long-distance trade in whale oil and other products (Figure 10.2). Massive defensive sites developed at the same time, indicating an expansion in the scale of aggressive competition and inter-village alliances.

These data are insufficient to demonstrate that a despotic environment with limited numbers of stable and defendable resource patches directly caused social

Figure 10.2. A Fox Aleut whale hunt by Henry Elliot. Courtesy of the National Anthropological Archives, Smithsonian Institution (SI-NAA 7119–9).

inequality. For that we need fine-grained evidence relating population density to the heterogeneity of resource patches themselves. However, with growing population densities through the Late Kachemak and Koniag periods (as well as the onset of the Little Ice Age around 1350 AD), the number of stable patches should have declined and most foragers would have experienced greater risk (variability) in their production of staple resources. This, in turn, would have created a despotic environment encouraging factionalization and defense of better patches.

The Koniag data support expectations for corporate control of resources in a despotic environment. For example, infrastructural investments at fixed resource locations (like salmon streams) facilitated claims of territorial ownership and defense. While streamside settlements became important in Kachemak period, settlements were located at virtually every stream only in the Koniag period. Partlow's (2000) data indicate an increase in the amount of salmon stored inside Koniag, as compared to Kachemak, dwellings. This internalization of storage facilities suggests increased intention to control the use, and distribution, of stores for potlatch feasts and other political purposes.

A majority of small houses in Koniag times (compared to a minority of large ones) suggests that relatively few households held a position of dominance. Elite control would have been supported, in part, by the more productive labor-potential of larger households (more spouses, slaves, children, and followers) and disproportionate extension of kinship ties to other households and communities. These ties would expand the allegiance base available for aggressive and defensive warfare as well as increase the demand for shared defensive facilities. Large central rooms (some in residential structures and some standing alone), suggest the importance of social aggregations for feasts and ceremonies. This is consistent with ethnohistoric accounts of Koniag ceremonialism (summarized in Chapter 4), and supported by data from the New Karluk site and elsewhere (Donta, 1993).

10.2. OVERTURES TO EMERGENT PROPERTIES

Despite some important unresolved issues, the evolutionary model, fundamentally, performed well. The model provides a reasonable account of the mechanisms and processes underlying the Kodiak sequence and the emergence of a limited degree of inequality and social complexity. But, readers should not finish this book with the impression that biological and economic principles alone sufficiently explain social evolution, or that I advocate such a position. Though grounded in a materialist and ecological frame of reference, I find it necessary to consider other dimensions of social life to develop a more comprehensive understanding of the past. In so doing I depart from (but do not violate) the ecological materialism of the evolutionary model at the core of this book, as I consider the more proximate (conscious, immediate, emergent) features of Kodiak's later prehistory—the period of time witnessing significant social upheaval.

Synthesized with other evidence from Kodiak archaeology, the Sitkalidak findings document how the Late Kachemak phase represents a critical transition in the development of Kodiak's non-egalitarian society. Ocean Bay and Early Kachemak people had vibrant social customs, decorated their artifacts, performed ceremonies, and engaged in social rituals for a wide variety of reasons (as most people do). But, the material culture of the Late Kachemak and Koniag suggests a qualitative change in ritual activity; examples include increased specialization of bodily decoration (Steffian and Saltonstall, 2001) and labor-intensive treatment of some stone lamps (Heizer, 1956). Mortuary evidence offers, perhaps, the most striking change (Simon and Steffian, 1994; Urcid, 1994; Workman, 1992). As previously discussed, Late Kachemak sites often contained perforated human bones; some had drill-holes to preserve bodily articulation (presumably part of a mummification procedure), and other bones were drilled for independent suspension. Several burials have been found lacking head, hands, and feet.

These unique Late Kachemak developments strongly imply concerns regarding spiritual forces. Why would these practices develop prior to, and in tandem with, the emergence of institutional social competition, inequality, and warfare? I propose this Late Kachemak complex arises from the social anxiety brought about by the processes of culture change engendered by the shift to a delayed-return economy (with its requisite dependence on hoarding) that eventually led to unprecedented crowding and a transition from egalitarian to non-egalitarian systems of interaction. If equal access to resources and equal exposure to economic risk maintain egalitarian relations, and if these balances become jeopardized, people likely will resist overt competitiveness or selfishness. Demand-sharing and other forms of status-leveling might retard initial moves towards inequality (however ineffectual they may ultimately prove to be).

In the warming environment of the early Late Kachemak, when it became possible for some households to grow relatively independent of others (despite close proximity), occurrences of unprecedented jealousy and small-scale skirmishes are expected. These tensions would elevate anxieties as age-old egalitarian traditions started to unravel. Powerful magic was called for.

> "Shamans are healers (chasing away evil spirits that cause sickness), priests (honoring ancestors and conducting ceremonies), and ambassadors (negotiating with the Masters of the Animals and other spirits). They travel to the land above and the land below with the help of their drum, belts, rattles, helping spirits, and occasionally hallucinogenics. Shamans divine the future, identify spirits, and insure that the community follows rules and taboos." (Chaussonnet, 1995:73)

A common aspect of hunter-gatherer cultures, shamanism encompasses a system of ritual and belief. Shamans are individuals experienced in the art of metaphysical transcendence, able to send themselves through dreams and trances into the realm of spirits. Through their work, shamans divine spiritual causes for earthly afflictions, anxieties, and deviations from the expected. Through the process they offer advice and remedies, pacify angry spirits, and establish harmony between the spiritual and earthly worlds. These individuals gain deep respect, and great fear, for their abilities to help and harm. I suspect the Late Kachemak experienced increased shamanistic activity, inspired by a palpable sense of anxiety surrounding instability of social relationships and uncertainty of the future in a time of remarkable, if yet subtle, change; a time when some families found it advantageous to hoard (rather than share) and to compete for exclusive rights to productive patches.

Practicing their art, shamans of the North American and Eurasian arctic and subartic regions often wore garments embellished with varied bangles and charms (e.g., Pavlinskaya, 1994). Human bones hanging from shamans' clothing or house beams, flickering oil lamps with stone figures "floating" in glassy pools of oil (Figure 10.3), and mummified ancestors watching over the proceedings of divination rituals, all conveyed tremendous power to the shaman's performance. The magic of the ceremony, while far from alleviating the causes of cultural anxiety (associated with the change from egalitarian to non-egalitarian society), helped address its symptoms (by establishing an ideology sanctifying hierarchy); in the process, the shaman's role gradually evolved with the social ritual of non-egalitarian

Figure 10.3. The imagined scene of a Late Kachemak shamanic ritual. Artistic rendition by Laada Bilaniuk.

competition. (See Donta, 1993 for a discussion of how status was embedded in Koniag ceremonialism.) By the contact period, shamanism was institutionalized in the competitive performance of feasting events throughout the Northwest Coast (de Laguna, 1988), and the shamanistic activities on Kodiak (Crowell and Leer, 2001) suggest similar roles. Shamans' supernatural powers endowed competitive feasts with spiritual legitimacy. As mediators of the supernatural, shamans helped reinforce existing structures of power. I propose that the artistic elaboration and treatment of the dead in Late Kachemak times relate, in part, to the emergence of this kind of shamanism. In the Late Kachemak, shamans helped hold together a world in spiritual turmoil. By Koniag times, shamans became part of the ritual of social competition itself (Desson, 1995).

The practice of whaling was imbued with shamanism and spirituality; whaling cults are extensively described in ethnohistorical accounts (see Chapter 4; Desson, 1995, Donta, 1993; Lantis, 1984). Whalers from the Gulf of Alaska and the Aleutians stole dead bodies from graves and preserved them as mummies in caves. They often dissected these bodies for human fat to rub on whaling lance blades; sometimes they ate or wore pieces of flesh from dead whalers to give them spiritual power to kill whales. Some of the mortuary patterns from the Late Kachemak suggest the beginnings of such practices.

Data from the Sitkalidak survey indicate that whale hunting became more important in the very Late Kachemak (noted by movement of settlements and villages to the outer-coast zone at this time). The earliest radiocarbon date from a village site in the outer-coast zone (KOD-106) dates to this interval (930 ± 70 bp), and whale bones cover the site (although most probably date to Koniag occupations). The diagnostic Three-Saints point (commonly marked with engraved designs that researchers sometimes interpret as "ownership marks") also indicates possible whaling intensification in the Late Kachemak phase. Some decorated lamps from the Late Kachemak feature whaling imagery, suggesting the animals' growing symbolic importance.

Whale hunting possibly expanded from a low-intensity activity to a high-intensity pursuit late in the Late Kachemak for two reasons: (1) increased economic stress (higher population densities and renewed resource depression), and (2) increased economic and political competitiveness (an avenue for sanctioned competition). Hunters were encouraged by political rewards of prestige and power derived from whaling success. Significant dangers involved in the hunt fostered the use of magical interventions. Shamanistic rituals served as appropriate sources for that magic, and successful hunters eventually acquired shaman-like mystique for their involvement with, and power over, the life and death of whales and people (Crowell, 1994).

Near the beginning of Koniag times, the decline in durable, labor-intensive art forms (e.g., decorated stone lamps and labrets of exotic raw materials) may reflect a new status quo in social competition and hierarchy. Dominant families now held relatively stable positions of economic and political control in the community. Turmoil was replaced by a new social structure with rules of rank and status that everyone understood. Competition became expected, sanctioned, channeled into predictable avenues, and spiritually reinforced. People no longer saw social asymmetry as a source of ideological confusion, but as a predictable dimension of social interaction. Perishable artistic elaboration (especially in wood) continued through the Koniag period, used for shamanistic practices (such as dance masks, dolls, puffin beak rattles, and drums) as well as for broader contexts of ceremony and gambling. Unfortunately, comparative samples of perishable artifacts are lacking from earlier times, making it impossible to establish the relative novelty of the Koniag assemblages.

Previously discussed Late Kachemak developments indicate a shift in the underlying character of social interaction. People jockeyed for position, recognition, and affiliation in an uncertain world. Less interdependent than they had been for millennia, they were unsure of how to deal with the realities of the new social environment. Within this context, early signs of defensive constructions appear, suggesting localized feuding between families and/or households. As competition intensified, some families found it necessary to build durable structures for refuge in defendable locations. These refuges only work when people use them in anticipation of attack, not in response to attacks. (Warriors tried to launch surprise attacks on unsuspecting villages.) Accordingly, certain groups expected their enemies to attack, perhaps in retribution for prior aggression or insult. From the beginning, the refuge sites may have been designed as shelters for mothers, children, and elderly family members during warriors' and hunters' prolonged absences. Possibly, these Late Kachemak refuge sites were established by the first families of the region to openly claim exclusive ownership of production locales, defending against the challenge of rivals jealous of their audacious claims. Also, it is possible that aggression proved an effective mechanism for leveling social competitiveness in this phase. Defensive sites of this time may have ineffectively protected upstarts from angry mobs of people unhappy with the inequality imposed by the aggrandizing behaviors of a few families.

Physical aggression and defense would have been insufficient to stem the tide of jealous competitors. Successful elites needed to be effective politicians, using generosity to draw their neighbors into interdependent relationships. A key feature of "transegalitarian" societies (those moving from egalitarian to non-egalitarian social organization) is a shift in the nature of sharing (rather than wholesale abandonment of sharing). Whereas egalitarian hunter-gatherers share for insurance against future failures or their inability to utilize the entire carcass before it spoils, elites share to ensure good will and engender obligations (such as labor, political support, and tribute). Often elites' sharing behavior is competitive. As de Laguna (1988:277) notes, Kwakiutl (Kwakwaka'wakw) chiefs were said to "fight with property." For the most audacious and sufficiently-endowed chiefs, extreme generosity in gifting and potlatch ceremony could ruin a rival (unable to reciprocate). Egalitarian people generally downplay their generosity while non-egalitarian elites often exaggerate theirs. This shift occurred in Koniag times, paralleling the development of: dedicated ceremonial houses and/or super-sized living rooms; the expansion of military alliances and endemic warfare; enlarged co-residential labor pools; expanded trade in subsistence and prestige goods; and, more definitive signs of inequality between households. Kachemak structures were still too small to host large crowds of supporters and guests.

10.3. SUMMARY AND CONCLUSION

The Sitkalidak data support the socio-ecologically based model in most dimensions. With a better understanding of the timing of critical changes afforded by the data, we can further envision some of the environmental, social, and ideological processes that facilitated change during the Late Kachemak and Koniag periods, from egalitarian to non-egalitarian or hierarchical societies with slavery, endemic warfare and large scale prestige trade. Enhanced understanding of the Early and Late Kachemak phases will foster appreciation of the critical transition from immediate to delayed-return foraging and the relationship between this change and the initial development of semi-sedentary villages and socio-political reorganization.

The evolutionary history of hunting and gathering societies on Kodiak makes important contributions to the understanding of Kodiak, Alaskan, and North Pacific cultural history and heritage; it sheds light on the emergent complexity of hunter-gatherer societies as well as the more general puzzle of human social evolution. To the people of Kodiak, this book relates a history of social development, a history of ancestral traditions; it highlights some of the definitive challenges and successes of people living for millennia on the Kodiak Archipelago. While some healthy scientific uncertainty remains, the evidence points strongly towards cultural continuity on the islands from the beginning of the archaeological sequence. This has implications for the cultural origins of the Alutiiq people, and possibly for the historical relationships between EskAleut biology, language, and culture (Clark, 1992a, 1996, 1998; Dumond, 1988a, 2000). Even if evidence someday turned against cultural continuity, Kodiak's history of cultural change would be no less informative, interesting, or valuable to the current inhabitants.

Today, Alutiiq people (like other Native Alaskan and Native American groups) remain immersed in a struggle to define themselves in the modern, post-colonial, multi-national, corporate, capitalist world. While they conduct board meetings, run businesses, watch CNN, order supplies on the internet, and travel widely, they also look to the past to claim their place in the modern world, demonstrating their cultural dignity and endurance. These efforts respond to a colonial experience in which outsiders sought to deprive Alutiiq people of their identity and legitimacy. Archaeology helps to reveal a dimension of that cultural heritage (obscured by centuries and millennia of sedimentation and erosion, conquest and oppression, abuse and neglect). Negotiating the past two centuries of subjugation, disease, and assimilation, the Alutiiq find renewed unity and power in their common history and heritage. Aside from the technical aspects of this book, I hope my Alutiiq friends find insights into forgotten dimensions of the past, helping them bring more perspective to their history.

In the anthropological and scientific tradition, this book goes beyond history by seeking to understand cultural development as something more than a unique series of sequential events and to improve the basis for comparative anthropology of hunter-gatherer societies. In this book, we see how Kodiak society persisted, with little apparent change, for thousands of years. Occasional developments in technology and settlement mark time in an otherwise conservative social trajectory. Some technological developments rendered longer-term consequences than others. Possibly, the delayed-return "revolution" saved Kodiak populations from extinction during neoglaciation, and only later facilitated population growth and social change. Climatic fluctuations probably had significant impacts and also provided unforeseen opportunities. But, major and rapid social changes in the direction of social and political complexity did not occur until a series of conditions aligned themselves together. Population growth was one of the major catalysts, possibly spurred on by improved climate and new possibilities afforded by the delayed-return economy. Social changes and dynamic tensions are observed during this Late Kachemak interval, but only incipient inequality can be discerned. (These observations are reasonable as long as conditions are good and everyone maintains unrestricted access to resource locations.) When the Little Ice Age began, after about AD 1300, population densities and social changes made it impossible to revert to simpler ways of life. Some families found themselves differentially dependent on others. More fortunate families became patrons of the less fortunate, providing an unprecedented form of social welfare, while at the same time benefiting from the labor and political support they received. Social inequalities emerged around the concept of wealth (ownership of resources and prestige goods as well as ability to control the labor of others). Elite status-holders established, and justified, their roles by means of competition in whale hunting, gambling, feasting, and warfare. Elaborate ceremonials, perhaps inspired by shamanism, developed to support the new world-view. With the increased intensification of whale hunting and fishing, populations continued to grow through this transition to socio-political inequality, possibly encouraged by pulses of climatic deterioration and improvement.

This new system of social competition and political ferment was only a few hundred years old when the Russians arrived with guns, ships, diseases, missionaries, and an unquenchable passion for sea otter furs. The people of Kodiak (and other inhabitants of southern Alaska) found themselves pawns in a capitalist world system, orders of magnitude larger than the regional exchange systems in which they were previously engaged (Crowell, 1997). Though the natives' resistance was initially successful, the Russian's relentless push to expand the fur territories eventually overwhelmed the capacity of indigenous political organizations to repel them. The previously discussed post-contact data provide a glimpse of the scale of reorganization and demographic collapse (also documented in historic accounts and by oral traditions). Archaeological data, however, also document the persistence of many dimensions of technology and culture despite colonialism. In spite of it all, Alutiiq culture has adapted and survived.

With respect to popular accounts of the emergence of complex hunter-gatherers, I offer two conclusions. First, the evidence reported here does not validate the idea that the North Pacific Rim was an inherently resource-rich environment that could support unlimited population growth. The pattern of demographic change through time, in fact, suggests periods when resource availability, seasonality, and harvesting technologies severely limited population growth. Claims to the contrary merely project modern (or proto-historic) social and technological capacities into the past, a procedure of dubious value, as archaeology has demonstrated time and time again. Not only did the environment fluctuate substantially, but economic productivity is always critically dependent on evolving cultural parameters.

The second conclusion purports that population density (not necessarily population pressure, though possibly) plays an important role in increased complexity, at least in this case. In the Sitkalidak area and across Kodiak (even at the broader Gulf of Alaska scale), population growth accelerates only a few centuries before we find evidence of social change, increased political competition, and incipient inequalities. Legros' (1985) account of social hierarchy among low-density Tchutchone populations demonstrates that inequality is not necessarily an issue of absolute population densities, but rather, of relative ones; this important study shows how small populations of Tchutchone Indians (of the post-contact period in interior Yukon Territories) took control of salmon fishing hot-spots and gained exclusive rights to trade relations with the Northwest Coast. Again, the relative exclusivity of access to resources conditioned the emergence of hierarchy. On Kodiak (and the Pacific Rim more generally), relatively more productive environments meant asymmetrical control would not emerge until population densities were considerably higher than in the Tchutchone case.

I draw these conclusions to argue strongly against the claim (made by Hayden, 1994; Clark and Blake, 1994; Ames and Maschner, 1999, and many others) that complex hunter-gatherers emerged on the Northwest Coast *because of* the environment's productivity; I wager rather that complex hunter-gatherers emerged *in spite of* this productivity. Open access to unlimited critical resources would not give anyone leverage to claim positions of power over others. I argue, instead, that what prevents hierarchy among low-density hunter-gatherers is not their low density, but circumstances associated with the following factors: flexibility of movement and alliance; economic interdependence or shared vulnerability to risk and uncertainty; difficulties controlling resource patches; and, the generally low utility of surplus resources (Spielmann, 1986). Circumstances reverse with development of a heterogeneous environment where some people are excluded from equal access to critical resources; this can happen in impoverished, or productive, environments. Asymmetrical dependencies seldom develop in productive ecosystems at low population levels, which is why high population levels generally correlate with institutional inequality (cf. Keeley, 1988). Economic inequality in impoverished environments, while it can occur, is less likely to be sustainable, simply because resources in this context tend towards unpredictability and instability.

A brief comparison to other complex hunter-gatherer cases affords additional consideration of these circumstances. Horizontal social complexity increases with population density as a result of increased interactions and tensions arising from crowding; but social hierarchy (vertical complexity) sometimes emerges with increased horizontal complexity, and sometimes not. For example, groups from Kodiak, the Northwest Coast (Ames and Maschner, 1999), and the Channel Islands (Arnold, 1992, 2000) developed definitive social inequalities, based on differences in the ability to control productive fishing streams, sea mammal rookeries, shellfish beds, trade routes, and productive technologies. From these foundations, elites expanded their political influence and power by controlling the labor of non-elites. Perhaps the Florida Calusa illustrate the most extreme case of social asymmetry documented for a hunter-gatherer group. Calusa paramount chiefs commanded vast armies, controlled large networks of villages, and acquired massive tribute in goods and labor from subsidiary chiefs and commoners (Marquardt, 1986, 1988; Widmer, 1988). Massive construction projects (such as canals and shell mounds in mangrove swamps and islands) improved tributary commerce, military deployment, and visibility, as well as elevated the social prominence of chiefs and subchiefs (Luer, 1989; Marquardt, 1992). Warfare and slavery were common to all of these cases.

By contrast, other groups appear to have maintained egalitarian relations while engaging in complex economies and massive constructions projects. For example, Gibson (2001) finds the composition of the 3700 year old Poverty Point Culture included an egalitarian core population; these people voluntarily moved almost one million cubic yards of sediment, in less than 300 years, to erect a massive network of earthworks at the Poverty Point site. This feat, in and of itself, required unusually complex coordination and planning. These Late Archaic hunter-fisher-gatherers sustained themselves on fish, deer, nuts, and other quarry while engaged in a complex trade system for raw materials, many located far from their homes. Gibson sees the need for exotic raw materials and the subsequent ability to deploy effective technologies for on-demand fish harvesting from the productive Mississippi bayous as critical elements of the complexity; if his thoughts are correct, the lack of significant social asymmetry may be attributed to people's unrestricted opportunities of access to productive resources, especially fish (assuming they could acquire stone materials for fashioning net weights).

Similarly, Habu (n.d.) argues that the organizationally complex Middle Jomon population (occupying the Sannai Maruyama site in northern Honshu, Japan) lacked significant social inequalities or hierarchies. Nevertheless, its occupants constructed a massive town with over 700 small dwellings (not all contemporaneous), several non-domestic long-houses, and monumental structures utilizing immense chestnut posts. Despite the possibility of cultivated chestnuts and barnyard millet, products of hunting, fishing, and gathering dominate subsistence. Habu believes residents only used this site seasonally, and she suggests the site may have functioned as an important seasonal trade center. Exotic materials such as jade, amber, asphalt, and obsidian were found at the site, supporting a role for long-distance exchange. Social inequality, identified in mortuary treatments (Nakamura, 1999), only appeared in Late and Final Jomon contexts, after the Sannai Maruyama site fell into disuse.

A central difference between the complex hunter-gatherer cases with and without evidence of inequality involves the presence of social and physical circumscription, limiting the opportunities for potential subordinates to escape the influence of dominating factions. From the Gulf of Alaska to the Northwest Coast (and in the Aleutians), interlocking social developments occurring throughout the region reinforced patterns of rank and inequality. Because everyone was caught in the same process, movement from one coastal location to another could not provide relief from political and economic competition. The trajectories of Kodiak and the Channel Islands were also affected by relative insularity (which might have initially protected islanders from mainland networks but later amplified dependencies on defendable productive hotspots and mainland trade in times of increased circumscription). By contrast, Poverty Point and Middle Jomon hunter-gatherers lacked significant circumscription; mobility remained possible and aggrandizing strategies would have been less sustainable. It is notable that these cases lack evidence of warfare and slavery. Importantly, as the Tchutchone case illustrates, circumscription does not result from population density or environmental productivity, but from a relationship between population density, environmental productivity, and ecological structure (patchiness and stability).

Fundamental to understanding social evolution, ecological perspectives need not be considered contradictory to more proximate perspectives focused on social and ideological dimensions of organization and change. To the extent that humans approach common problems in predictable ways (e.g., through social reorganization and expressions of ideological anxiety), we can make inferences about prehistoric processes of change that include gender and kinship relations, shamanism, and other aspects of cultural interaction (sometimes considered antagonistic to evolutionary and ecological perspectives). These extensions of ecological theory benefit from a holistic view; a view that understands human behavior and social organization to be derived from individual fitness-seeking propensities, set within an environment that is partially constructed by emergent and ever-changing symbolic systems, themselves driven by the uniquely human capacity for linguistic communication and abstract thought.

Near the conclusion of a major study of the Calusa, Widmer (1988:277) notes "the fact that the particular adaptation in question relies on nonagricultural resources is not important." He goes on to argue that the specific mode of economic production is not as important to a society's complexity as the energetic basis upon which it is built. Impoverished environments usually do not support high populations that build mounds, dig canals, and wage wars; naturally-productive aquatic ecosystems and artificially-enriched agricultural ones often do. I have argued that environmental *structure* is more important than environmental productivity in many of the dimensions of complexity we have considered (such as rank and hierarchy). Examining emergent complexity among hunter-gatherers, we have an opportunity to challenge ingrained assumptions about the importance of agriculture in social evolution. We reduce the complexity of our analytical task and consider a greater range of variables in an evolutionary context. Perhaps, most importantly, through these kinds of studies we better appreciate the diversity of paths possible in hunter-gatherer social evolution; we free ourselves from simplistic notions of progress and unilineal development as well as enrich our understanding of social change.

As one of the first attempts to develop a comprehensive view of social evolution on the Kodiak Archipelago, this research joins a small number of similar studies about other areas of the North Pacific Rim. As many issues remain unresolved, theory development and additional research will, no doubt, improve our understanding of the processes and mechanisms of Kodiak cultural development. Though we may later question some of the assumptions and conclusions drawn here, these analyses and interpretations provide a solid foundation for future research. As we continue to seek more empirically-grounded understandings of human social development, the archaeological history and heritage of the North Pacific holds great promise.

Appendix A

Site Data from the Sitkalidak Archaeological Survey

Appendix A

AHRS # (KOD)	SAS #	Locus/ Comp	Period	Location	Cultural components	Distinct cultural strata	Midden preserved	Site length	Site width	Site area (sq. m)	houses obsvd	TotHSarea	avgHSarea	stdev HS	TotMRarea	AvMRar	stdev MR	Site function
472	1		KO	outer	1	1	calc	20	12	240	3				40			settlement
473	2		AL (contact)	outer	1	1	y	15	10	150	0							location
474	3	L.1	HI	outer	2	1	?	10	10	100	1							settlement
474	3	L.2	KO	outer	2	1	?	45	20	900	0							settlement
475	4		?	outer	1		y	8	6	48	0							camp
381	5		?	outer	1		n	4	1	4	0							camp
476	6	C.1	?	outer	2		y	30	8	240	0							camp
476	6	C.2	HI	outer	2		y	30	8	240	0							camp
477	7	C.1	?	outer	2		y	100	15	1500	0							camp
477	7	C.2	HI	outer	2		y	100	15	1500	0							camp
442	8		DKO	outer	1	1	y	110	30	3300	26				306	18	12	refuge
479	9	L.1,2,4	DKO	outer	1	2	y	78	25	1950	5	227			138			small village
479	9	L.3	KA?	outer	1	1	?	4	3	12	1	4	4	0	4	4	0	settlement
480	10		AL (contact)	outer	1	1	y	60	30	1800	6	240			187			small village
116	11	C.1	AL (contact)	outer	2		y	310	30	9300	23	1074			737			large village
116	11	C.2	DKO	outer	2	10	y	310	30	9300	23	1074			737			large village
481	12		OB1	mid	1	7	n	30	30	900	0							settlement

114	13	C.1	AL (contact)	mid	1	1	y	200	20	4000	9	541	45		431			small village
114	13	C.2	DKO?	mid	1	1	y	200	20	4000	9	541	45		431			small village
482	14		?	mid	1	1	n	5	5	25	0							location
483	15		OB	mid	1	1	n	1	1	1	0							location
484	16		?	mid	1	1	?	5	5	25	0							camp
485	17		DKO	mid	1	1	y	6	5	30	0							settlement
486	18	L.1	EKO?	mid	1	2	calc	25	10	250	1							settlement
486	18	L.2	KA	mid	1	2	y	15	15	225	1							settlement
486	18	L.3	?	mid	1	1	n	5	5	25	0							location
487	19		?	mid	1	1	n	5	5	25	1							camp
488	20		?	mid	1	1	n	5	3	15	0							location
489	21	D	AL?	mid	1	2	n	25	10	250	2	43	21		28	14		settlement
489	21	D	KO?	mid	1	2	n	25	10	250	2	43	21		28	14		settlement
490	22		DKO	mid	1	3	y	10	10	100	1	12	12	0	12	12	0	settlement
491	23		DKO	mid	1	6	y	25	15	375	1	65	65	0	42	42	0	settlement
491	24		DKO	mid	1		y	20	15	300	2	66	33		37			settlement
110	25	C.1	AL (contact)	mid	2		y	290	40	11600	34	1270	47		841			large village
110	25	C.2	DKO	mid	2	4	y	290	40	11600	34	1270	47		841			large village
492	26		?	mid	1	3	y	28	5	280	2							settlement
493	27		AL (Amer)?	mid	1	2	n	20	15	300	2				67			settlement
494	28		HI(EU)	mid	1	1	n	12	8	96	1				52	52	0	settlement

(*Continued*)

Appendix A (*Continued*)

AHRS # (KOD)	SAS #	Locus/ Comp	Period	Location	Cultural components	Distinct cultural strata	Midden preserved	Site length	Site width	Site area (sq_ m)	houses obsvd	TotHSarea	avgHSarea	stdev HS	TotMRarea	AvMRar	stdev MR	Site function
111	29	L.1	AL (contact)	mid	2		y	70	50	3500	11				194			small village
111	29	L.1	DKO	mid	2	5	y	70	50	3500	11				194			small village
111	29	L.2	KO	mid	1	4	?	36	20	720	6							settlement
392	30	D	EKO?	outer	1	4	y	50	20	1000	2				24	12	0	settlement
392	30	D	KA?	outer	1	4	y	50	20	1000	2				24	12	0	settlement
112	31	C.1	KO?	inner	2	3	n				0							settlement
112	31	C.2	KA	inner	2		n				0							settlement
112	31	C.3	OB?	inner	2		n				0							unknown
495	32		?	mid	1		n			49	0							location
496	33		OB?	mid	1		n	8	5	40	0							location
379	34	C.1	AL (Russ)?	mid	2		y	80	40	3200	5							small village
379	34	C.2	KO	mid	2		y	80	40	3200	5							small village
379	34	C.3	KA	mid	2		y	80	40	3200								settlement
565	35	L.1	LKA	mid	1		n	40	20	800	4	143			129			settlement
384	35	L.1	OB?	mid	2		y	20	10	200								camp
384	35	L.2, C.2	LKA	mid	3		y	90	20	1800	8	91			91	11		small village
384	35	L.2–2	EKO	mid	3	3	y	40	25	1000	3							unknown

384	35	L.2, C.1	HI	mid	3		y											unknown
384	35	L.3	LKA	mid	1	2	y	60	20	1200	10	68			67			small village
384	35	L.4	LKA	mid	1		?	15	10	150		12	12	0	12	12	0	settlement
497	36		OB2	mid	1	3	n	15	8	120	0							settlement
115	37		EKA?	mid	1		y	80	20	1600	10	99			99			small village
498	38		?	inner	1		y				0							unknown
499	39		EKO	mid	1		y	30	12	360	2				110			settlement
101	40	C.1	AL (contact)	outer	2		y	230		11900								small village
101	40	C.2	DKO	outer	2		y	230		11900								small village
101	41		DKO?	outer	1		n	40	20	800	2	140	70		85			settlement
500	42		KA	mid	1		?	50	10	500	6				45	9	3	small village
501	43		?	mid	1		?	10	5	50								settlement
502	44		?	mid	1		n	17	15	255	4	12	6	0	12	6	0	settlement
107	45		KA?	inner	1		y	30	15	450								settlement
503	46		?	inner	1		calc	10	10	100	0							camp
108	47		?	inner	1		n				0							unknown
504	48		DKO	inner	1		n	6	6	36	1							camp
564	49	C.1	EKA	mid	3		calc											camp
564	49	C.2	LKA	mid	3		calc	60	8	480	5	21			21			settlement
564	49	C.3	OB1	mid	3		calc				0							camp
564	49	C.4	OB2	mid	3		calc	60	8	480	0							camp

(*Continued*)

Appendix A (*Continued*)

AHRS # (KOD)	SAS #	Locus/ Comp	Period	Location	Cultural components	Distinct cultural strata	Midden preserved	Site length	Site width	Site area (sq. m)	houses obsvd	TotHSarea	avgHSarea	stdev HS	TotMRarea	AvMRar	stdev MR	Site function
505	50	L.1, D	EKO?	mid	2		n	25	10	250	2	37			37			settlement
505	50	L.1, D	LKA?	mid	2		n	25	10	250	2	37			37			settlement
505	50	L.2,3,4	?	mid	1		n				0							camp
506	51		KA	mid	1		?	35	10	350								settlement
507	52	L.1, D	AL?	mid	2		calc	25	11	225	2							settlement
507	52	L.1, D	KO	mid	2		calc	25	11	225	2							settlement
507	52	L.2	OB?	mid	1		n	20	10	200	0							camp
507	52	L.3	?	mid	2		?	10	10	100	0							settlement
508	53	L.1	KA?	mid	1		calc	20	20	400								settlement
508	53	L.2	KO	mid	1		?	40	40	1600	10	219			205			small village
509	54		EKO	inner	1		y	60	15	900	8	78			73			small village
510	55		DKO (, KA?)	inner	2		y	100	25	2500	10							small village
511	56		?	inner	1		n	50	15	750								camp
512	57		?	inner	1		n	10	10	100	1							settlement
513	58		HI?	inner	1		n				2							settlement
514	59		?	mid	1		n	50	15	750								settlement
515	60		HI(EU)	inner	1		n	10	10	100	1							ranch
102	61		EKO?	inner	1		n	40	10	400	4							settlement

516	62		LKA	?	1		n	15	10	150								camp
517	63		?	?	1		n	10	5	50								camp
518	64		?	?	1		n	10	5	50	1							camp
519	65		?	?	1		n	5	5	25	1							camp
520	66		DKO	inner	1		n	15	15	225	1	37	37		35	35	0	settlement
521	67		?	?	1		calc	10	10	100	1							camp
522	68		OB2	?	1		calc	5	5	25	0							camp
523	69		?	mid	2		calc											settlement
118	70		KO	outer	1		calc	50	30	1500	3							settlement
122	71	C.1	AL (Russ)	outer	1		y	125	65	8125	9							large village
122	71	Poss C.2	KO?	outer	1		y	125	65	8125	9							large village
121	72		KA?	outer	1		y	75	50	3750								small village
385	73		LKA	outer	1		y				0							camp
120	74		?	outer	1		y				0							camp
119	75	C.1	OB1	?	2		n	30	30	900	0							camp
119	75	C.2	OB2	?	2		n	30	30	900	0							camp
524	76		KO	?	1		n	120	20	2400	6	144	18		118			small village
525	77		DKO	?	1		?	130	20	2600	7	239			173			small village
526	78		?	?	?		?											settlement
527	79		?	?	1		?											settlement
528	80		?	?	1		?	20	20	400								camp

(Continued)

Appendix A (*Continued*)

AHRS # (KOD)	SAS #	Locus/ Comp	Period	Location	Cultural components	Distinct cultural strata	Midden preserved	Site length	Site width	Site area (sq. m)	houses obsvd	TotHSarea	avgHSarea	stdev HS	TotMRarea	AvMRar	stdev MR	Site function
529	81		KO	?	1		?	90	20	1800	6	150			119			small village
530	82		OB	?	1		n	20	10	200	0							camp
531	83		?	?	1		n	5	5	25	0							camp
532	84		KO?	mid	1		y	40	10	400	1							settlement
450	85	C.1	AL (contact)	outer	2		?											refuge
450	85	C.2	DKO	outer	2		?											refuge
566	86		LKA	mid	1		calc	70	15	1050	3	73			70			settlement
382	87	C.1	KO	mid	2		y	140	20	2800	9	219			142			small village
382	87	C.2	KA?	mid	2		y	140	20	2800	1							settlement
533	88		?	mid	1		n	10	5	50	0							camp
534	89		KO	mid	1		n	40	10	400	3	88			51			settlement
535	90		HI(EU)	mid	1		nr	10	10	100								settlement
117	91		DKO	outer	1		y	125	30	3750	8				283			small village
536	92		EU (contact)	inner	1		n				0							location
537	93		?	inner	1		n	4	4	16	0							camp
538	94		?	inner	1		n	10	5	50	0							camp

113	95		KO?	outer	1	y	80	50	4000	13			507	39		small village
391	96		KO?	outer	1	y	68	20	1360	6						small village
539	97	L.1	LKA?	outer	2	?	50	25	1250	10						small village
539	97	L.2	DKO	outer	2	?	78	25	1950	8	286		213			small village
383	98		AL (Russ)	outer	2	y	10	30	300	2	78	39	50	25		settlement
378	99	L.1	LKA?	mid	2	n	33	20	660	5	68	17	53			settlement
378	99	L.2	KA?	mid	2	n	10	7	70	1	15	15	15	15	0	settlement
378	99	L.3	KO?	mid	2	y	44	35	1540	5						settlement
540	100		LKA	?	1	n	7	5	35	0						camp
541	101		LKA	?	1	calc	10	10	100	0						settlement
542	102		LKA?	?	1	n	22	15	330	4						settlement
543	103		?	mid	1	n	3	3	9	0						settlement
478	104	C.1	DKO	mid	2	y	40	20	800	2						settlement
478	104	C.2	KA?	mid	2	y	10	5	50	2						settlement
544	105		KO	inner	1	n	70	12	840	8	239		222		11	small village
88	106		?	inner	1	n										camp
545	107		?	mid	1	n	10	10	100	1						settlement
546	108		?	mid	1	n	10	5	50	0						settlement
547	109		HI	mid	1	n				0						camp
548	110	C.1	AL?	mid	2	n	50	17	850	5	71		60	30	0	settlement
548	110	C.2	OB2?	mid	2	n				0						unknown

(Continued)

Appendix A (*Continued*)

AHRS # (KOD)	SAS #	Locus/ Comp	Period	Location	Cultural components	Distinct cultural strata	Midden preserved	Site length	Site width	Site area (sq. m)	houses obsvd	TotHSarea	avgHSarea	stdev HS	TotMRarea	AvMRar	stdev MR	Site function
549	111		KA?	inner	1		calc	25	15	375	4							settlement
92	112		KO	inner	1		y	94	25	2350	11	348			233			small village
550	113	L.1	?	inner	2		n	20	15	300	2							settlement
550	113	L.2	?	inner	2		n	15	15	225	1							settlement
550	113	L.3	LKA	inner	2		y	60	30	1800	4	44			44			settlement
550	113	L.4	?	inner	2		n	17	10	170	2							settlement
93	114	L.1, C.2	OB?	mid	1		n				0							camp
93	114	L.1,2, C.1	AL (Amer)?	mid	1		y	80	50	4000	8	200	50		178			small village
551	115		?	mid	2		n	80	15	1200	5	77	11	3	66	11		settlement
86	116	C.1	KO	outer	2		y	35	25	875	0							refuge
86	116	C.2	LKA	outer	2		y	35	25	875	0							refuge
552	117		LKA	mid	1	2	n	7	5	16	1	9	9	0	9	9	0	refuge
94	118	C.1	KO	mid	1		y	130	30	3700	4	104	26		78		11	small village
94	118	C.2	AL?	mid	1		y	130	30	3700	4							settlement
553	119		KA	mid	2		y	180	60	10800	19	226			226			large village
554	120		OB2	mid	1		calc	150	15	2250	0							settlement
555	121		OB1	mid	1		n	132	12	1637	0							settlement
556	122		HI	inner	1		n	15	10	150	2							location

557	123	L.1	KO	inner	2	calc	25	20	500	5							settlement
557	123	L.2	KA?	inner	2	calc	30	10	300	3							settlement
558	124		EKO	mid	1	n	10	5	50	1	36	36	0	36	36	0	refuge
559	125		KO?	mid	1	y	45	25	1125	6				133			small village
106	126	C.1,L.1,2,4,5	AL (contact)	outer	2	y	40	25	1000	8				227			small village
106	126	C.2,L.1,2,4,5	DKO	outer	2	y	40	25	1000	8				227			small village
106	126	L.3	EKO	outer	1	y	25	10	250	4				78			settlement
89	127	C.1	AL (Russ)	mid	2	y	100	35	3500								small village
89	127	C.2	OB2	mid	2	y	100	35	3500	0							settlement
90	128		?	mid	?	?											unknown
91	129		?	mid	?	?											unknown
194	130		?	mid	?	?											unknown
	131	unused	•	•		•											
567	132		KA?	inner	1	n	40	30	1200	6	179			179			small village
568	133	D	EKO?	inner	1	n	70	20	1400	6	181	37		181	37		small village
568	133	D	LKA?	inner	1	n	70	20	1400	6	181	37		181	37		small village
569	134		?	mid	?	?											unknown
570	135		?	inner	?	?											settlement
571	136		?	inner	1	n				0							settlement

(Continued)

Appendix A (*Continued*)

AHRS # (KOD)	SAS #	Locus/ Comp	Period	Location	Cultural components	Distinct cultural strata	Midden preserved	Site length	Site width	Site area (sq_ m)	houses obsvd	TotHSarea	avgHSarea	stdev HS	TotMRarea	AvMRar	stdev MR	Site function
572	137	C.1	AL(Russ)	mid	1		y	200	40	8000	7							large village
572	137	Poss C.2	DKO?	mid	1		y	200	40	8000	7							large village
573	138		?	mid	1		?	15	7	105	2							settlement
574	139		?	mid	1		?	10	7	70	1							settlement
575	140		DKO	outer	1		y	100	60	6000	14	529			399	43		large village
576	141		DKO	outer	1		y	24	12	288	0							refuge
577	142		?	?	1		n	5	5	25	1							settlement
578	143		EKA?	?	1		n	50	15	750	2							settlement
579	144		?	?	1		n	5	5	25	0							camp
580	145		?	mid	1		n	5	5	25	0							unknown
581	146		?	mid	1		n				0							unknown
582	147		?	mid	2		n	100	10	1000								unknown
583	148		?	mid	1		n	30	15	450	1							settlement
584	149		?	mid	1		n	4	4	16	0							camp
585	150		?	mid	1		n	100	15	1500	0							settlement
586	151	L.1	?	mid	1		n	60	10	600	5							settlement
587	151	L.2	?	mid	?		n	30	15	450	3							settlement
588	152		?	mid	1		n	10	5	50	1							settlement

Appendix B

House/Structure Data from the Sitkalidak Archaeological Survey

Appendix B

SAS Site #	Structure	Period/ Phase	Main room area (m^2)	Total structure area (m^2)	Side rooms	Site area (m^2)
1	H1	KO	12	12	0	240
1	H2	KO	15.9	15.9	0	240
1	H3	KO	17.5	17.5	0	240
3	H1	KO	14	14	0	900
8	H1	DKO	16	16	0	3300
8	H2	DKO	26.25	38.25	2	3300
8	H3	DKO	14	14	0	3300
8	H4	DKO	16	43	4	3300
8	H5	DKO	56	74	1	3300
8	H6	DKO	13.5	17	1	3300
8	H7	DKO	27.5	31.5	1	3300
8	H8	DKO	13.5	13.5	0	3300
8	H9	DKO	9	18	1	3300
8	H10	DKO	16.5	17.5	1	3300
8	H11	DKO	18	18	0	3300
8	H12	DKO	24	29	1	3300
8	H13	DKO	5	7	1	3300
8	H14	DKO	6	9	1	3300
8	H15	DKO	16	19	1	3300
8	H16	DKO	24	38	3	3300
8	H17	DKO	7.5	9.5	1	3300
8	H18	DKO	7.5	10	1	3300
8	H19	DKO	8	10	1	3300
8	H20	DKO	9	12	1	3300
8	H21	DKO	12	17	2	3300
8	H22	DKO	6	6	0	3300
8	H23	DKO	10	10	0	3300
8	H24	DKO	7	7	0	3300
8	H25	DKO	6	12	1	3300
8	H26	DKO	16	16	0	3300
8	H27	DKO	9	9	0	3300
8	H28	DKO	9	9	0	3300
8	H29	DKO	16	16	0	3300
8	H30	DKO	15	19	1	3300
8	H31	DKO	16	16	0	3300

Appendix B (*Continued*)

SAS Site #	Structure	Period/ Phase	Main room area (m^2)	Total structure area (m^2)	Side rooms	Site area (m^2)
8	H32	DKO	8	8	0	3300
8	H33	DKO	8	8	0	3300
8	H34	DKO	6	6	0	3300
8	H35	DKO	8	8	0	3300
8	H36	DKO	10	15	1	3300
8	H37	DKO	6	6	0	3300
9 L.1	H1	DKO	29	47	3	1950
9 L.1	H2	DKO	22.5	22.5	0	1950
9 L.2	H3	DKO	40	74.75	4	1950
9 L.3	H6	KA?	4	4	0	12
9 L.4	H4	DKO	33	53	3	1950
9 L.4	H5	DKO	17	30	2	1950
10	H1	AL	13.5	17.5	1	1800
10	H2	AL	88	146	5	1800
10	H3	AL	36	36	0	1800
10	H4	AL	26	34	3	1800
10	H5	AL	15	15	0	1800
10	H6	AL	10	10	0	1800
10	H7	AL	10	10	0	1800
11	H1	DKO	30	67	5	9300
11	H2	DKO	54	81	5	9300
11	H3	DKO	15	15	0	9300
11	H4	DKO	54	54	0	9300
11	H5	DKO	24	40	3	9300
11	H6	DKO	70	107	5	9300
11	H7	DKO	49	108	4	9300
11	H8	DKO	44	81	4	9300
11	H9	DKO	35	69	1	9300
11	H10	DKO	54	153	7	9300
11	H11	DKO	72	114	3	9300
11	H12	DKO	22	26	1	9300
11	H13	DKO	27	41	1	9300
11	H14	DKO	54	104	5	9300
11	H15	DKO	45	53	1	9300

(*Continued*)

Appendix B (*Continued*)

SAS Site #	Structure	Period/ Phase	Main room area (m^2)	Total structure area (m^2)	Side rooms	Site area (m^2)
11	H16	DKO	80	95	1	9300
21	H1	KO?	24	30	1	250
21	H2	KO?	4	4	0	250
22	H1	KO	12	12	0	250
23	H1	DKO	42	65	5	375
24	H2	DKO	30	48	3	300
24	H3	DKO	9	15	1	300
25	H1	DKO	40	44	1	11600
25	H2	DKO	72	72	0	11600
25	H4	DKO	24	24	0	11600
25	H5	DKO	56	56	0	11600
25	H7	DKO	32	32	0	11600
25	H8	DKO	30	33	1	11600
25	H9	DKO	27	31	1	11600
25	H10	DKO	25	28	1	11600
25	H12	DKO	9	9	0	11600
25	H15	DKO	24	24	0	11600
25	H16	DKO	40	40	0	11600
25	H19	DKO	48	48	0	11600
25	H21	DKO	35	35	0	11600
25	H22	DKO	35	50	1	11600
25	H23	DKO	17	17	0	11600
25	H24	DKO	63	78	2	11600
25	H25	DKO	72	86	1	11600
25	H26	DKO	40	41	0	11600
25	H28	DKO	66	82	3	11600
25	H29	DKO	54	54	0	11600
25	H30	DKO	49	49	0	11600
25	H31	DKO	48	48	0	11600
25	H32	DKO	30	38	1	11600
27	H1	HI	42	42	0	300
27	H2	HI	11	11	0	300
28	H1	EA	50	50	0	96
30	H1	KA?	12	12	0	1000
30	H2	KA?	12	12	0	1000

Appendix B (*Continued*)

SAS Site #	Structure	Period/ Phase	Main room area (m^2)	Total structure area (m^2)	Side rooms	Site area (m^2)
34	H3	KA	48	48	0	3200
34	H4	DKO	36	45	2	3200
34	H5	DKO	36	49	3	3200
35 L.1	H20	KA	12	12	0	800
35 L.1	H21	KA	25	25	0	800
35 L.1	H22	KA	64	64	0	800
35 L.1	H23	KA	14	14	0	800
35 L.1	H24	KA	5	5	0	800
35 L.2-1	H14	KA	9	9	0	1800
35 L.2-1	H15	KA	12	12	0	1800
35 L.2-1	H17	KA	9	9	0	1800
35 L.3	H2	KA	6	6	0	1200
35 L.3	H3	KA	9	9	0	1200
35 L.3	H4	KA	9	9	0	1200
35 L.3	H5	KA	9	9	0	1200
35 L.3	H6	KA	9	9	0	1200
35 L.3	H6	KA	16	16	0	1200
35 L.3	H7	KA	9	9	0	1200
35 L.4	H1	KA?	12	12	0	150
37	H1	EKA?	36	36	0	1600
37	H2	EKA?	18	18	0	1600
37	H3	EKA?	30	30	0	1600
37	H4	EKA?	9	9	0	1600
39	H1	EKO	50	50	0	360
41	H1	KO?	55	92	2	800
41	H2	KO?	30	48	2	800
42	H1	KA	11	19	1	500
42	H2	KA	6	6	0	500
42	H3	KA	12	12	1	500
42	H4	KA	9	9	0	500
44	H1	?	6	6	0	255
44	H2	?	6	6	0	255
49	H1	KA	12	12	0	480
49	H2	KA	6	6	0	480

(*Continued*)

Appendix B (*Continued*)

SAS Site #	Structure	Period/ Phase	Main room area (m^2)	Total structure area (m^2)	Side rooms	Site area (m^2)
50 L.1	H1	KA?	25	27	1	250
50 L.1	H2	KA?	14	14	0	250
53 L.2	H1	EKO?	19	19	0	1600
53 L.2	H2	EKO?	25	29	1	1600
53 L.2	H3	EKO?	12	16	0	1600
53 L.2	H4	EKO?	24	28	2	1600
53 L.2	H5	EKO?	20	20	0	1600
53 L.2	H6	EKO?	25	25	0	1600
53 L.2	H7	EKO?	30	30	0	1600
53 L.2	H8	EKO?	20	20	0	1600
53 L.2	H9	EKO?	16	20	1	1600
54	H1	EKO	16	16	0	900
54	H2	EKO	18	18	0	900
54	H3	EKO	19	19	0	900
54	H4	EKO	15	19	2	900
54	H5	EKO	9	9	0	900
66	H1	DKO	35	37	1	225
76	H1	DKO	18	40	4	2400
76	H2	DKO	22	22	0	2400
76	H3	DKO	14	14	0	2400
77	H1	DKO	25	29	1	2600
77	H2	DKO	25	25	0	2600
77	H3	DKO	29	33	1	2600
77	H4	DKO	28	50	3	2600
77	H5	DKO	21	21	0	2600
77	H6	DKO	30	41	2	2600
81	H1	KO	21	21	0	1800
81	H2	KO	17	22	1	1800
81	H3	KO	25	31	1	1800
81	H4	KO	21	32	1	1800
81	H5	KO	88	88	0	1800
81	H6	KO	36	36	0	1800
86	H1	LKA	35	35	0	1050
86	H2	LKA	11	11	0	1050
86	H3	LKA	4	4	0	1050

Appendix B (*Continued*)

SAS Site #	Structure	Period/ Phase	Main room area (m^2)	Total structure area (m^2)	Side rooms	Site area (m^2)
86	H4	LKA	20	20	0	1050
87	H1	DKO	35	51	3	2800
87	H2	DKO	12	26	4	2800
87	H3	DKO	9	14	2	2800
87	H4	DKO	35	56	3	2800
87	H5	DKO	16	16	0	2800
87	H6	DKO	16	21	1	2800
87	H7	DKO	36	51	2	2800
89	H1	KO	35	46	1	400
89	H2	KO	16	16	0	400
91	H2	DKO	54	n.r.	3	3750
91	H3	DKO	42	42	0	3750
91	H4	DKO	42	n.r.	4	3750
91	H5	DKO	25	n.r.	3	3750
91	H6	DKO	30	30	0	3750
91	H7	DKO	54	n.r.	4	3750
91	H8	DKO	24	n.r.	1	3750
95	H1	KO?	70	n.r.	1?	4000
95	H2	KO?	64	n.r.	0–1?	4000
95	H3	KO?	32	n.r.	0–2?	4000
95	H4	KO?	42	n.r.	1?	4000
95	H5	KO?	35	35	0	4000
95	H6	KO?	40	n.r.	1–3?	4000
95	H7	KO?	35	35	0	4000
95	H8	KO?	35	n.r.	0–2?	4000
95	H9	KO?	56	n.r.	?	4000
95	H10	KO?	44	44	0	4000
95	H12	KO?	25	n.r.	0–4?	4000
95	H13	KO?	12	12	0	4000
97 L.2	H1	DKO	32	61	6	1950
97 L.2	H2	DKO	34	41	3	1950
97 L.2	H3	DKO	28	30	1	1950
97 L.2	H4	DKO	15	25	2	1950
97 L.2	H6	DKO	14	23	1	1950

(*Continued*)

Appendix B (*Continued*)

SAS Site #	Structure	Period/ Phase	Main room area (m^2)	Total structure area (m^2)	Side rooms	Site area (m^2)
97 L.2	H7	DKO	14	23	1	1950
97 L.2	H8	DKO	14	18	1	1950
97 L.2	H9	DKO	10	10	0	1950
98	H1	AL	30	56	4	300
98	H2	AL	20	20	0	300
99 L.1	H1	KA?	25	30	1	660
99 L.1	H2	KA?	12	16	1	660
99 L.1	H3	KA?	4	4	0	660
99 L.1	H4	KA?	16	16	0	660
99 L.2	H7	KA?	15	15	0	70
99 L.3	H5	EKO?	31.5	31.5	0	70
99 L.3	H6	EKO?	18	18	0	70
105	H1	EKO?	15	15	0	840
105	H2	EKO?	18	27	2	840
105	H3	EKO?	18	18	0	840
105	H4	EKO?	35	35	0	840
105	H5	EKO?	35	35	0	840
105	H6	EKO?	49	49	0	840
105	H7	EKO?	25	25	0	840
105	H8	EKO?	30	30	0	840
110	H1	AL?	30	35	1	850
110	H2	AL?	25	30	0	850
112	H2	DKO	20	23	2	2350
112	H3	DKO	20	29	2	2350
112	H4	DKO	16	18	3	2350
112	H5	DKO	25	41	4	2350
112	H6	DKO	30	45	3	2350
112	H7	DKO	20	27	1	2350
112	H8	DKO	30	42	2	2350
112	H9	DKO	32	50	2	2350
112	H10	DKO	35	64	5	2350
113 L.3	H1	LKA	9	6	0	1800
113 L.3	H2	LKA	6	9	0	1800
114	H1	AL	48	72	4	4000
114	H2	AL	37	45	2	4000

Appendix B (*Continued*)

SAS Site #	Structure	Period/ Phase	Main room area (m^2)	Total structure area (m^2)	Side rooms	Site area (m^2)
114	H3	AL	32	49	2	4000
114	H4	AL	22	31	2	4000
114	H5	AL	7	10	1	4000
115	H1	KA?	12	12	0	1200
115	H2	KA?	11	1	0	1200
115	H3	KA?	12	12	0	1200
115	H4	KA?	6	6	0	1200
115	H5	KA?	13	13	0	1200
115	H6	KA?	8	8	0	1200
117	H1	LKA	9	9	0	16
118	H1	KO	35	59	4	3700
118	H2	KO	18	18	0	3700
118	H3	KO	16	16	0	3700
118	H4	KO	9	9	0	3700
119	H1	KA	8	8	0	10800
119	H2	KA	9	9	0	10800
119	H3	KA	9	9	0	10800
119	H4	KA	9	9	0	10800
119	H5	KA	9	9	0	10800
119	H6	KA	9	9	0	10800
119	H7	KA	10	10	0	10800
119	H8	KA	12	12	0	10800
119	H9	KA	12	12	0	10800
119	H10	KA	12	12	0	10800
119	H11	KA	12	12	0	10800
119	H12	KA	6	6	0	10800
119	H13	KA	14	14	0	10800
119	H14	KA	14	14	0	10800
119	H15	KA	15	15	0	10800
119	H16	KA	15	15	0	10800
119	H17	KA	16	16	0	10800
119	H18	KA	20	20	0	10800
119	H19	KA	25	25	0	10800
124	H1	EKO	36	36	0	50

(*Continued*)

Appendix B (*Continued*)

SAS Site #	Structure	Period/ Phase	Main room area (m^2)	Total structure area (m^2)	Side rooms	Site area (m^2)
126 L.1	H1	DKO/AL	36	n.r.	5	1000
126 L.1	H2	DKO/AL	56	56	0	1000
126 L.1	H3	DKO/AL	56	56	0	1000
126 L.1	H13	DKO/AL	6	6	0	1000
126 L.1	H14	DKO/AL	6	6	0	1000
126 L.3	H4	EKO	16	16	0	250
126 L.3	H5	EKO	28	28	0	250
126 L.3	H7	EKO	12	12	0	250
126 L.3	H8	EKO	9	9	0	250
126 L.3	H9	EKO	13	13	0	250
126 L.4	H10	DKO/AL	39	n.r.	6	1000
126 L.5	H11	DKO/AL	16	16	0	1000
126 L.6	H12	DKO/AL	12	12	0	1000
132	H1	KA?	15	15	0	1200
132	H2	KA?	22	22	0	1200
132	H3	KA?	25	25	0	1200
132	H4	KA?	16	16	0	1200
132	H5	KA?	48	48	0	1200
132	H6	KA?	49	49	0	1200
133	H1	EKO?	32	32	0	1400
133	H2	EKO?	44	44	0	1400
133	H3	EKO?	28	28	0	1400
133	H4	EKO?	34	34	0	1400
133	H5	EKO?	35	35	0	1400
140	H1	DKO	48	n.r.	1	6000
140	H2	DKO	56	84	3	6000
140	H3	DKO	24	35	2	6000
140	H4	DKO	50	65	1	6000
140	H5	DKO	44	66	2	6000
140	H6	DKO	43	57	3	6000
140	H7	DKO	30	45	1	6000
140	H8	DKO	35	56	0	6000
140	H10	DKO	27	36	1	6000
140	H11	DKO	33	42	1	6000
140	H12	DKO	30	47	2	6000
140	H13	DKO	30	56	3	6000

Appendix C

Population Estimates Based on House and Site Area Data from the Sitkalidak Archaeological Survey

Appendix C

SAS	Locus	Period	Site area (m2)	Houses obsvd	TotalHSarea	MeanHSarea	TotalMRarea	MeanMRarea	Site function	M^2/ person by mean MR area	M^2/ person by total HS area	Pop_ estimate (MR)	M^2 Site Area/ person	Pop_ estimate (site area)
9	L.1,2,4	DKO	1950	5	227	34.8	138	27.6	small village	—	—	69.00	28.26	68
10		AL (contact)	1800	6	240	39.9	187	31.2	small village	1.73	2.22	93.50	19.25	63
11	C.1	AL (contact)	9300	23	1074	82.6	737	43.4	large village	2.41	4.59	368.50	25.24	324
11	C.2	DKO	9300	23	1074	82.6	737	43.4	large village	—	—	368.50	25.24	324
12		OB1	900	0	—	—	—	—	settlement	—	—	—	—	31
13	C.1	AL (contact)	4000	9	541	45	431	35.9	small village	1.99	2.50	215.50	18.56	139
13	C.2	DKO?	4000	9	541	45	431	35.9	small village	—	—	215.50	18.56	139
15		OB	1	0	—	—	—	—	location	—	—	—	—	0
25	C.1	AL (contact)	11600	34	1270	47	841	33.6	large village	1.87	2.61	420.50	27.59	404
25	C.2	DKO	11600	34	1270	47	841	33.6	large village	—	—	420.50	27.59	404
29	L.1	AL (contact)	3500	11	—	—	194	38.8	small village	2.16	—	97.00	36.08	122
29	L.1	DKO	3500	11	—	—	194	38.8	small village	—	—	97.00	36.08	122
33		OB?	40	0	—	—	—	—	location	—	—	—	—	1
34	C.1	AL(Russ)?	3200	5	—	—	—	—	small village	—	—	—	—	111
34	C.2	KO	3200	5	—	—	—	—	small village	—	—	—	—	111
35	L.1	OB?	200	—	—	—	—	—	camp	—	—	—	—	7
35	L.2, C.2	LKA	1800	8	91	10.7	91	11	small village	—	—	45.50	39.56	63
35	L.3	LKA	1200	10	68	9.71	67	9.57	small village	—	—	33.50	35.82	42
36		OB2	120	0	—	—	—	—	settlement	—	—	—	—	4
37		EKA?	1600	10	99	24.8	99	24.8	small village	—	—	49.50	32.32	56

40	C.1	AL (contact)	11900	—	—	—	—	—	small village	—	—	—	—	415
40	C.2	DKO	11900	—	—	—	—	—	small village	—	—	—	—	415
42		KA	500	6	46.8	11.7	45	9	small village	—	—	22.50	22.22	17
49	C.4	OB2	480	0	—	—	—	—	camp	—	—	—	—	17
52	L.2	OB?	200	0	—	—	—	—	camp	—	—	—	—	7
53	L.2	KO	1600	10	219	21.9	205	20.1	small village	—	—	102.50	15.61	56
54		EKO	900	8	78	16.6	73	14.6	small village	—	—	36.50	24.66	31
55		DKO(, KA?)	2500	10	—	—	—	—	small village	—	—	—	—	87
68		OB2	25	0	—	—	—	—	camp	—	—	—	—	1
71	C.1	AL(Russ)	8125	9	—	—	—	—	large village	—	—	—	—	283
71	Poss C.2	KO?	8125	9	—	—	—	—	large village	—	—	—	—	283
72		KA?	3750	—	—	—	—	—	small village	—	—	—	—	131
75	C.1	OB1	900	0	—	—	—	—	camp	—	—	—	—	31
75	C.2	OB2	900	0	—	—	—	—	camp	—	—	—	—	31
76		KO	2400	6	144	18	118	14.8	small village	—	—	59.00	40.68	84
77		DKO	2600	7	239	34.1	173	29.3	small village	—	—	86.50	30.06	91
81		KO	1800	6	150	37.5	119	28.7	small village	—	—	59.50	30.25	63
82		OB	200	0	—	—	—	—	camp	—	—	—	—	7
87	C.1	KO	2800	9	219	36.5	142	23.7	small village	—	—	71.00	39.44	98
91		DKO	3750	8	—	—	283	35.4	small village	—	—	141.50	26.50	131
95		KO?	4000	13	—	—	507	39	small village	—	—	253.50	15.78	139
96		KO?	1360	6	—	—	—	—	small village	—	—	—	—	47
97	L.1	LKA?	1250	10	—	—	—	—	small village	—	—	—	—	44
97	L.2	DKO	1950	8	286	23.8	213	17.8	small village	—	—	106.50	18.31	68
105		KO	840	8	239	29.9	222	27.8	small village	—	—	111.00	7.57	29

(Continued)

Appendix C *(Continued)*

SAS	Locus	Period	Site area (m2)	Houses obsvd	TotalHSarea	MeanHSarea	TotalMRarea	MeanMRarea	Site function	M^2/ person by mean MR area	M^2/ person by total HS area	Pop-estimate (MR)	M^2 Site Area/ person	Pop-estimate (site area)
112		KO	2350	11	348	43.5	233	25.9	small village	—	—	116.50	20.17	82
114	L.1,2, C.1	AL (Amer)?	4000	8	200	50	178	29.7	small village	—	—	89.00	44.94	139
118	C.1	KO	3700	4	104	26	78	19.5	small village	—	—	39.00	94.87	129
119		KA	10800	19	226	10.6	226	10.6	large village	—	—	113.00	95.58	376
120		OB2	2250	0	—	—	—	—	settlement	—	—	—	—	78
121		OB1	1637	0	—	—	—	—	settlement	—	—	—	—	57
125		KO?	1125	6	—	—	133	66.5	small village	—	—	66.50	16.92	39
126	C.1,L.1,2,4,5	AL (contact)	1000	8	—	—	227	28.4	small village	1.58	—	113.50	8.81	35
126	C.2,L.1,2,4,5	DKO	1000	8	—	—	227	28.4	small village	—	—	113.50	8.81	35
127	C.1	AL(Russ)	3500	—	—	—	—	—	small village	—	—	—	—	122
127	C.2	OB2	3500	0	—	—	—	—	settlement	—	—	—	—	122
132		KA?	1200	6	179	22.5	179	22.5	small village	—	—	89.50	13.41	42
133	D	EKO?	1400	6	181	37	181	37	small village	—	—	90.50	15.47	49
133	D	LKA?	1400	6	181	37	181	37	small village	—	—	90.50	15.47	49
137	C.1	AL(Russ)	8000	7	—	—	—	—	large village	—	—	—	—	279
137	Poss C.2	DKO?	8000	7	—	—	—	—	large village	—	—	—	—	279
140		DKO	6000	14	529	52.9	399	43	large village	—	—	199.50	30.08	209

Appendix D

Radiocarbon Dates from the Central Gulf of Alaska

Appendix D

Site name	AHRS no.	Lab no.	^{14}C date	Material	Culture	Reference	Reject
			KODIAK ARCHIPELAGO				
Afognak Region							
Tsunami	AFG 215	Beta 165141	880 ± 40	CH	LKA	D. Clark,p.c. 2002	
Tsunami	AFG 215	Beta 165140	1750 ± 60	CH	LKA	D. Clark,p.c. 2002	
Tsunami	AFG 215	Beta 165139	1320 ± 80	CH	LKA	D. Clark,p.c. 2002	
Salmon Bend	AFG-010	Beta 170060	1400 ± 80	CM	LKA	D. Clark,p.c. 2002	
Salmon Bend	AFG-010	Beta 170061	1330 ± 60	CM	LKA	D. Clark,p.c. 2002	
Aleut Town	AFG 004	Beta 150810	920 ± 50	CM	LKA	D. Clark,p.c. 2002	
Aleut Town	AFG 004	Beta 150811	1090 ± 80	CM	LKA	D. Clark,p.c. 2002	
Malina Creek	AFG 005	Beta 42073	620 ± 70	?	KO	Haggarty et al. 1991; Mills 1994	
Malina Creek	AFG 005	?	500 ± 50	?	KO	Knecht 1995:39	
	AFG 008	GaK 3801	5750 ± 240	CH lens	OB I	Clark 1979; Mills 1994	
	AFG 008	GaK 3802	4150 ± 200	CH	OB II	Clark 1979; Mills 1994	
	AFG 011	S 1418	4480 ± 160	?	OB II	Clark 1969; Mills 1994	
	AFG 011	S 1419	4475 ± 125	?	OB II	Clark 1969; Mills 1994	
	AFG 011	GaK 3804	4200 ± 140	?	OB II	Clark 1979; Mills 1994	
	AFG 011	GaK 3803	3890 ± 110	?	OB II	Clark 1979; Mills 1994	
	AFG 012	Beta 101916	450 ± 60	?	KO	Partlow 2000	
	AFG 012	Beta 101915	420 ± 60	?	KO	Partlow 2000	
	AFG 012	Beta 101914	310 ± 40	?	KO	Partlow 2000	
	AFG 012	Beta 101917	280 ± 60	?	KO	Partlow 2000	
Settlement Point	AFG 015	Beta 101551	620 ± 50	?	KO	P. Saltonstall, p.c. 2002	

Settlement Point	AFG 015	Beta 118300	570 ± 60	?	KO	P. Saltonstall, p.c. 2002
Settlement Point	AFG 015	Beta 114204	450 ± 50	?	KO	P. Saltonstall, p.c. 2002
Settlement Point	AFG 015	Beta 101912	440 ± 50	?	KO	P. Saltonstall, p.c. 2002
Settlement Point	AFG 015	Beta 114202	440 ± 60	?	KO	P. Saltonstall, p.c. 2002
Settlement Point	AFG 015	Beta 114205	440 ± 60	?	KO	P. Saltonstall, p.c. 2002
Settlement Point	AFG 015	Beta 101913	390 ± 50	?	KO	P. Saltonstall, p.c. 2002
Settlement Point	AFG 015	Beta 114096	370 ± 80	?	KO	P. Saltonstall, p.c. 2002
Settlement Point	AFG 015	Beta 114097	350 ± 70	?	KO	P. Saltonstall, p.c. 2002
Settlement Point	AFG 015	Beta 114098	340 ± 60	?	KO	P. Saltonstall, p.c. 2002
Settlement Point	AFG 015	Beta 114203	330 ± 60	?	KO	P. Saltonstall, p.c. 2002
Settlement Point	AFG 015	Beta 101552	300 ± 50	?	KO	P. Saltonstall, p.c. 2002
Shuyak	AFG 098	GX 17327	950 ± 65	?	KO	Reger et al. 1992
Shuyak	AFG 098	GX 17328	625 ± 60	?	KO	Reger et al. 1992
Shuyak	AFG 098	GX 17331	570 ± 60	?	KO	Reger et al. 1992
Shuyak	AFG 098	GX 17325	500 ± 105	?	KO	Reger et al. 1992
Shuyak	AFG 098	GX 17326	500 ± 100	?	KO	Reger et al. 1992
Shuyak	AFG 098	GX 17332	360 ± 125	?	KO	Reger et al. 1992
Shuyak	AFG 098	GX 17330	1175 ± 110	?	KO	Reger et al. 1992
Shuyak	AFG 098	GX 17324	1055 ± 105	?	KO	Reger et al. 1992
Shuyak	AFG 098	GX 17329	1040 ± 105	?	KO	Reger et al. 1992
	AFG 119	Beta 42074	1000 ± 80	?	KA	Haggarty et al. 1991; Mills 1994
Chiniak Bay Region						
Blisky	KOD 210	Beta 77806	340 ± 70	WD/CH	KO	Clark, p.c. 2002
Blisky	KOD 210	Beta 77805	410 ± 80	GR	KO	Clark, p.c. 2002

(Continued)

Appendix D (*Continued*)

Site name	AHRS no.	Lab no.	^{14}C date	Material	Culture	Reference	Reject
Blisky	KOD 210	Beta 77804	2010 ± 80		KA	Clark, p.c. 2002	
Blisky	KOD 210	Beta 113164	2880 ± 120	CH	KA	Steffian p.c. 2002	
Blisky	KOD 210	Beta 113163	3050 ± 60	CH	KA	Steffian p.c. 2002	
Outlet	KOD 562	Beta 160042	240 ± 50	?	KO	P. Saltonstall, p.c. 2002	
Outlet	KOD 562	Beta 160047	310 ± 60	?	KO	P. Saltonstall, p.c. 2002	
Outlet	KOD 562	Beta 160043	320 ± 60	?	KO	P. Saltonstall, p.c. 2002	
Outlet	KOD 562	Beta 132444	1140 ± 60	WD	KA	P. Saltonstall, p.c. 2002	
Outlet	KOD 562	Beta 160045	1230 ± 60	?	KA	P. Saltonstall, p.c. 2002	
Outlet	KOD 562	Beta 145863	1360 ± 60	?	KA	P. Saltonstall, p.c. 2002	
Outlet	KOD 562	Beta 132443	1880 ± 60	?	KA	P. Saltonstall, p.c. 2002	
Outlet	KOD 562	Beta 160046	2650 ± 50	?	KA	P. Saltonstall, p.c. 2002	
Outlet	KOD 562	Beta 145865	3070 ± 70	?	KA	P. Saltonstall, p.c. 2002	
Outlet	KOD 562	Beta 145864	3140 ± 70	?	KA	P. Saltonstall, p.c. 2002	
Outlet	KOD 562	Beta 160044	3350 ± 150	?	OB II	P. Saltonstall, p.c. 2002	
Array	KOD 561	Beta 145862	4480 ± 80	?	OB II	P. Saltonstall, p.c. 2002	
Chiniak River Village	KOD 449	?	4300 ±	?	OB II	Haggarty et al. 1991; Mills 1994	
Rice Ridge	KOD 363	Beta 43135	3850 ± 80	?	OB II	Knecht 1995:33	
Rice Ridge	KOD 363	Beta 43134	3860 ± 90	?	OB II	Haggarty et al. 1991; Mills 1994	
Rice Ridge	KOD 363	Beta 26230	4310 ± 60	?	OB I	Haggarty et al. 1991; Mills 1994	
Rice Ridge	KOD 363	GX 14674	5030 ± 250	CH	OB I	Haggarty et al. 1991; Mills 1994	
Rice Ridge	KOD 363	GX 14673	6080 ± 90	?	OB I	Knecht 1995:33	
Zaimka	KOD 013	Beta 130190	3890 ± 70	CH	EKA	A. Steffian, p.c. 2002	R

Zaimka	KOD 013	Beta 130191	5530 ± 130	CH	OB I	A. Steffian, p.c. 2002
Zaimka	KOD 013	Beta 130188	6250 ± 70	CH	OB I	A. Steffian, p.c. 2002
Zaimka	KOD 013	Beta 130189	6390 ± 70	CH	OB I	A. Steffian, p.c. 2002
Marmot Bay Region						
Crag Pt.	KOD 044	Beta 20122	910 ± 60	CH	LKA	Haggarty et al. 1991; Mills 1994
Crag Pt.	KOD 044	Beta 45944	910 ± 70	CH	LKA	Mills 1994
Crag Pt.	KOD 044	B 835	1110 ± 100	COR2	KO	Mills 1994; Oeschger et al. 1970
Crag Pt.	KOD 044	Beta 20533	1890 ± 90	CH	KA	Haggarty et al. 1991; Mills 1994
Crag Pt.	KOD 044	Beta 92094	1940 ± 60	WD		Clark, p.c. 2002
Crag Pt.	KOD 044	Beta 48044	2000 ± 70	CH	KA	Mills 1994
Crag Pt.	KOD 044	P 1057	2033 ± 52	CH	LKA	Clark p.c. 2002
Crag Pt.	KOD 044	Beta 48043	2190 ± 90	?	KA	Mills 1994
Crag Pt.	KOD 044	Beta 45943	2380 ± 70	CH	KA	Mills 1994
Crag Pt.	KOD 044	Beta 94894	3150 ± 80	BO	EKA	Clark, p.c. 2002
Crag Pt.	KOD 044	Beta 45942	3160 ± 70	CH	KA	Mills 1994
Crag Pt.	KOD 044	Beta 66656	3190 ± 50	CH	?	Mills 1994
Crag Pt.	KOD 044	Beta 66655	3290 ± 50	CH	?	Mills 1994
Crag Pt.	KOD 044	Beta 45945	3340 ± 60	CH	KA	Mills 1994
Crag Pt.	KOD 044	Beta 20123	7790 ± 620	CH	PA?/OB I	Jordan 1992; Mills 1994
Kizhuyak	KOD 043	B 836	600 ± 100	CH	KO	Mills 1994; Oeschger et al. 1970
Kizhuyak Bay	KOD 324	Beta 14497	2700 ± 90	CH	EKA	Crozier 1986, 1987; Mills 1994
Kizhuyak Bay	KOD 324	Beta 8186	3520 ± 60	CH	OB II	Crozier 1986, 1987; Mills 1994
Kizhuyak Bay	KOD 324	Beta 14500	3630 ± 80	CH	OB II	Crozier 1986, 1987; Mills 1994
Kizhuyak Bay	KOD 324	Beta 14499	3850 ± 270	CH	OB II	Crozier 1986, 1987; Mills 1994

(Continued)

Appendix D (*Continued*)

Site name	AHRS no.	Lab no.	^{14}C date	Material	Culture	Reference	Reject
Kizhuyak Bay	KOD 324	Beta 14498	3920 ± 150	CH	OB II	Crozier 1986, 1987; Mills 1994	
Kizhuyak Bay	KOD 324	Beta 14501	4620 ± 110	CH	OB II	Crozier 1986, 1987; Mills 1994	
Kizhuyak Bay	KOD 324	Beta 8185	6620 ± 60	CH	OB I	Crozier 1986, 1987; Mills 1994	
Monashka Bay	KOD 026	P 1049	298 ± 44	CH		Clark 1966b; Mills 1994; Stuckenrath et al. 1966	
Monashka Bay	KOD 026	Beta 33545	1570 ± 60	CH lens	LKA	C. Donta p.c. 1992 to Mills 1994	
Monashka Bay	KOD 026	Beta 34832	1680 ± 50	CH lens	LKA	C. Donta p.c. 1992 to Mills 1994	
Sitkinak Region							
Sitkinak	XTI 052	Beta 7325	200 ± 50	?	KO	Haggarty et al. 1991; Mills 1994	
Sitkinak	XTI 052	Beta 7326	750 ± 80	?	KO	Haggarty et al. 1991; Mills 1994	
Southeast (Sitkalidak) Region							
Kiliuda Bay 1	KOD 077	Beta 42079	280 ± 80	CH lens	KO	Haggarty et al. 1991; Mills 1994	
Kiavak	KOD 099	P 1044	280 ± 44	CH	KO	Clark 1966b; Mills 1994; Stuckenrath et al. 1966	
Kiavak	KOD 099	P 1045	391 ± 48	CH	KO	Clark 1966b; Mills 1994; Stuckenrath et al. 1966	
Kiavak	KOD 100	P 1041	937 ± 49	COR2	KO	Clark 1966b; Mills 1994; Stuckenrath et al. 1966	R
Kiavak	KOD 100	S 2996	1960 ± 75	CH	EKA	D. Clark p.c. 1992 to Mills 1994	
Kiavak	KOD 100	S 2998	2400 ± 235	CW	EKA	D. Clark p.c. 1992 to Mills 1994	
Kiavak	KOD 100	S 3488	2750 ± 130	CH	EKA	D. Clark p.c. 1993 to Mills 1994	
Kiavak	KOD 100	P 1039	3263 ± 71	CH	EKA	Clark 1974a; Stuckenrath 1966	

Kiavak	KOD 100	S 2997	3365 ± 225	CH	EKA	D. Clark p.c. 1992 to Mills 1994
Kiavak	KOD 100	P 1038	4698 ± 71	CH lens	OB I	Clark 1966b; Mills 1994; Stuckenrath et al. 1966
Roadcut	KOD 119	P 1036	3929 ± 65	CH lens	OBII	Clark 1979; Mills 1994; Stuckenrath 1966
Roadcut	KOD 119	P 1034	5503 ± 78	CH2	OB I	Clark 1979; Mills 1994; Stuckenrath 1966
Rolling Bay	KOD 101	P 1048	353 ± 44	CH	KO	Clark 1966b; Mills 1994; Stuckenrath et al. 1966
Rolling Bay	KOD 101	P 1047	393 ± 40	CH	KO	Clark 1966b; Mills 1994; Stuckenrath et al. 1966
SAS 02	KOD 473	Beta 78508	120 ± 50	CH	ALc	Fitzhugh 1996
SAS 08	KOD 442	Beta 71089	610 ± 60	CH	DKO	Fitzhugh 1996
SAS 10	KOD 480	Beta 78501	90 ± 70	CH	ALc	Fitzhugh 1996
SAS 25	KOD 110	Beta 78502	480 ± 60	CH	ALc; DKO	Fitzhugh 1996
SAS 35	KOD 384	Beta 78520	1490 ± 70	MS	LKA	Fitzhugh 1996
SAS 36	KOD 497	Beta 71091	4390 ± 70	CH	OB II	Fitzhugh 1996
SAS 48	KOD 504	Beta 78503	360 ± 60	CH	DKO	Fitzhugh 1996
SAS 49	KOD 564	Beta 71092	1750 ± 60	CH	LKA	Fitzhugh 1996
SAS 49	KOD 564	?	1890 ± 70	CH	LKA	Fitzhugh 1996
SAS 49	KOD 564	Beta 78509	5300 ± 110	CH	OB I	Fitzhugh 1996
SAS 54	KOD 509	Beta 78505	820 ± 90	CH	EKO	Fitzhugh 1996
SAS 55	KOD 510	Beta 78521	400 ± 50	CH	DKO	Fitzhugh 1996
SAS 62	KOD 516	Beta 78506	1720 ± 60	CH	LKA	Fitzhugh 1996
SAS 68	KOD 522	Beta 78507	4850 ± 120	CH	OB II	Fitzhugh 1996
SAS 82	KOD 530	?	4560 ± 50	CH	OB	Fitzhugh 1996

(*Continued*)

Appendix D (*Continued*)

Site name	AHRS no.	Lab no.	^{14}C date	Material	Culture	Reference	Reject
SAS 86	KOD 566	Beta 78510	1610 ± 60	CH	LKA	Fitzhugh 1996	
SAS 100	KOD 540	Beta 78511	1500 ± 60	CH	LKA	Fitzhugh 1996	
SAS 101	KOD 541	Beta 78512	1530 ± 60	CH	LKA	Fitzhugh 1996	
SAS 116	KOD 086	Beta 78513	1240 ± 60	BO	LKA	Fitzhugh 1996	
SAS 117	KOD 552	Beta 78514	1090 ± 60	CH	LKA	Fitzhugh 1996	
SAS 120	KOD 554	Beta 78516	4330 ± 90	CH	OB II	Fitzhugh 1996	
SAS 121	KOD 555	Beta 78519	6340 ± 100	CH	OBI	Fitzhugh 1996	
SAS 124	KOD 558	Beta 78517	960 ± 90	CH	EKO	Fitzhugh 1996	
SAS 126	KOD 106	Beta 78518	930 ± 70	CH	EKO	Fitzhugh 1996	
SAS 127	KOD 089	Beta 83510	4450 ± 100	CH	OB II	Fitzhugh 1996	
SAS 127	KOD 089	Beta 83511	4610 ± 100	CH	OB II	Fitzhugh 1996	
SAS 143	KOD 578	?	2110 ± 50	CH	EKA	Fitzhugh 1996	
Tanginak Spring	KOD 481	Beta 134790	5370 ± 60	CH	OB I	This report	
Tanginak Spring	KOD 481	Beta 161617	5630 ± 70	CH	OB I	This report	
Tanginak Spring	KOD 481	Beta 134789	5710 ± 70	CH	OB I	This report	
Tanginak Spring	KOD 481	Beta 134791	5880 ± 80	CH	OB I	This report	
Tanginak Spring	KOD 481	Beta 161616	6070 ± 60	CH	OB I	This report	
Tanginak Spring	KOD 481	Beta 161618	6250 ± 60	CH	OB I	This report	
Tanginak Spring	KOD 481	Beta 161613	6280 ± 60	CH	OB I	This report	
Tanginak Spring	KOD 481	Beta 71714	6380 ± 50	CH	OB I	This report	
Tanginak Spring	KOD 481	Beta 161619	6430 ± 60	CH	OB I	This report	
Tanginak Spring	KOD 481	Beta 161615	6480 ± 70	CH	OB I	This report	
Tanginak Spring	KOD 481	Beta 161614	6490 ± 60	CH	OB I	This report	

Tanginak Spring	KOD 481	Beta 161612	6579 ± 60	CH	OB I	This report	
Tanginak Spring	KOD 481	Beta 76738	6600 ± 230	BCH	OB I	Fitzhugh 1996	
Three Saints	KOD 083	P 1043	1119 ± 49	CH	LKA	Clark 1966b; Mills 1994; Stuckenrath et al. 1966	
Three Saints	KOD 083	P 1042	2028 ± 55	CH	LKA	Clark 1966b; Mills 1994; Stuckenrath et al. 1966	
Shelikof Strait Region							
Karluk Lake Outlet	KAR 023	?	215 ± 140	?	?	source misplaced	
New Karluk	KAR 001	Beta 15014	290 ± 60	WD	KO	Jordan and Knecht 1988; Mills 1994	
New Karluk	KAR 001	Beta 8942	370 ± 50	CH	KO	Mills 1994	
New Karluk	KAR 001	Beta 15013	410 ± 70	WD	KO	Mills 1994	R
New Karluk	KAR 001	Beta 15015	480 ± 80	WD	KO	Jordan and Knecht 1988; Mills 1994	
New Karluk	KAR 001	GX 14670	545 ± 70	BB	KO	Mills 1994	R
New Karluk	KAR 001	Beta 25599	630 ± 50	WD	KO	Mills 1994	
New Karluk	KAR 001	Beta 15016	740 ± 80	WD	EKO	Jordan and Knecht 1988; Mills 1994	
New Karluk	KAR 001	Beta 25600	780 ± 60	WD	EKO	Mills 1994	
Nunakakhnak	KAR 037	GX 14672	5115 ± 250	CH	OB I	Mills 1994	
Old Karluk	KAR 031	Beta 15017	320 ± 60	CH	KO	Mills 1994	
Old Karluk	KAR 031	Beta 15690	430 ± 60	CH	KO?	Mills 1994	
Old Karluk	KAR 031	Beta 15691	980 ± 60	WD	LKA	Jordan 1992; Mills 1994	
Old Karluk	KAR 031	Beta 8946	2010 ± 70	WD	LKA	Jordan and Knecht 1988; Mills 1994	

(Continued)

Appendix D (*Continued*)

Site name	AHRS no.	Lab no.	^{14}C date	Material	Culture	Reference	Reject
Old Karluk	KAR 031	Beta 8945	2540 ± 60	CH	?	Jordan and Knecht 1988; Mills 1994	
Old Karluk	KAR 031	Beta 11245	4900 ± 100	CH	OB I	Jordan and Knecht 1988; Mills 1994	
	KAR 039	Beta 8943	2650 ± 60	CH	KA?	Haggarty et al. 1991; Mills 1994	
	KAR 048	Beta 8944	3050 ± 70	CH	KA?	Mills 1994	
Uganik I.	KOD 224	UGa 2823	1080 ± 90	?	KA	Haggarty et al. 1991; Mills 1994	
Uganik I.	KOD 224	UGa 2820	3130 ± 85	?	Takli-like	Clark 1984a; Mills 1994	
Uganik I.	KOD 224	UGa 2822	3365 ± 70	?	Takli-like	Clark 1984a; Mills 1994	
Uganik I.	KOD 224	UGa 1931	5065 ± 135	CH lens	OB I	Clark 1984a; Mills 1994; Nowak 1978	
Uganik I.	KOD 224	DIC 1236	6220 ± 70	CH	OB I	Clark 1984a; Mills 1994	
Chief Cove Island	KOD 172	UGa-?	2075 ±	?	KA?	Nowak 1978; Mills 1994	
Chief Cove Island	KOD 172	UGa-?	2180 ±	?	KA?	Nowak 1978; Mills 1994	
Larsen Bay	KAR 029	Beta 23767	450 ± 70	CS	KO	Crozier 1989; Mills 1994	
Larsen Bay	KAR 029	Beta 23769	620 ± 50	CS	KO	Crozier 1989; Mills 1994	
Larsen Bay	KAR 029	Beta 23768	870 ± 70	CH	EKO	Crozier 1989; Mills 1994	
Larsen Bay	KAR 029	Beta 23765	990 ± 60	CH	LKA	Crozier 1989; Mills 1994	
Larsen Bay	KAR 029	Beta 23766	1000 ± 110	CH	LKA	Crozier 1989; Mills 1994	
Larsen Bay	KAR 029	Beta 23771	1290 ± 80	CH	LKA	Crozier 1989; Mills 1994	
Larsen Bay	KAR 029	Beta 23770	1310 ± 70	CS	LKA	Crozier 1989; Mills 1994	
Slab Grave Site	KOD 157	Beta 20128	100 ± 90	CH	AL	Haggarty et al. 1991; Mills 1994	
Uyak	KOD 145	Beta 25601	460 ± 50	WD	LKA	Steffian 1992b; Mills 1994	R
Uyak	KOD 145	Beta 34281	1130 ± 70	CH	LKA	Steffian 1992b; Mills 1994	
Uyak	KOD 145	Beta 25603	1140 ± 90	WD	LKA	Steffian 1992b; Mills 1994	

Uyak	KOD 145	Beta 34283	1270 ± 100	CH	LKA	Steffian 1992b; Mills 1994
Uyak	KOD 145	Beta 25602	1310 ± 70	CH	LKA	Steffian 1992b; Mills 1994
Uyak	KOD 145	Beta 34282	1320 ± 70	WD	LKA	Steffian 1992b; Mills 1994
	KOD 336	Beta 20127	100 ± 3	CH	AL	Haggarty et al. 1991; Mills 1994
	KOD 235	UGa 1935	1170 ± 60	CH	KA?	Nowak 1978; Mills 1994
		CENTRAL ALASKA PENINSULA–PACIFIC COAST				
Aniakchak Region						
Aniakchak Bay House	SUT 016	?	375 ± 40	CH	?	R. Vanderhoek, p.c. 2002
Aniakchak Bay House	SUT 016	Beta 109924	870 ± 90	CH/WD	?	R. Vanderhoek, p.c. 2002
Aniakchak Bay House	SUT 016	Beta 108731	1600 ± 40	CH	?	R. Vanderhoek, p.c. 2002
Aniakchak Bay House	SUT 016	Beta 143672	1600 ± 40	CH	?	R. Vanderhoek, p.c. 2002
	SUT 052	Beta 143671	260 ± 40	CH	?	R. Vanderhoek, p.c. 2002
	SUT 033	Beta 143678	270 ± 30	CH	?	R. Vanderhoek, p.c. 2002
	SUT 039	Beta 149048	330 ± 60	CH/WD	?	R. Vanderhoek, p.c. 2002
	SUT 033	Beta 138995	380 ± 40	CH	?	R. Vanderhoek, p.c. 2002
	SUT 051	Beta 143670	380 ± 40	CH	?	R. Vanderhoek, p.c. 2002
	SUT 039	Beta 149051	390 ± 50	CH	?	R. Vanderhoek, p.c. 2002
	SUT 037	Beta 128613	490 ± 40	CH	?	R. Vanderhoek, p.c. 2002
	SUT 036	Beta 128612	510 ± 40	CH/WD	?	R. Vanderhoek, p.c. 2002
	SUT 043	WSU 5049	510 ± 60	CH	?	R. Vanderhoek, p.c. 2002
	SUT 036	Beta 149056	520 ± 60	CH	?	R. Vanderhoek, p.c. 2002
	SUT 011	Beta 109899	620 ± 60	CH/WD	?	R. Vanderhoek, p.c. 2002
	SUT 039	Beta 149049	690 ± 40	CH	?	R. Vanderhoek, p.c. 2002

(Continued)

Appendix D (*Continued*)

Site name	AHRS no.	Lab no.	^{14}C date	Material	Culture	Reference	Reject
	SUT 050	Beta 149055	880 ± 50	CH	?	R. Vanderhoek, p.c. 2002	
	SUT 041	Beta 143669	930 ± 80	CH	?	R. Vanderhoek, p.c. 2002	
	SUT 050	Beta 143677	940 ± 40	CH	?	R. Vanderhoek, p.c. 2002	
	SUT 040	Beta 128616	970 ± 30	CH	?	R. Vanderhoek, p.c. 2002	
	SUT 013	Beta 157012	1090 ± 50	CH	?	R. Vanderhoek, p.c. 2002	
	SUT 030	Beta 138998	1160 ± 80	CH	?	R. Vanderhoek, p.c. 2002	
	CHK 059	Beta 149044	1190 ± 70	CH	?	R. Vanderhoek, p.c. 2002	
	SUT 025	Beta 110266	1190 ± 40	CH	?	R. Vanderhoek, p.c. 2002	
	SUT 039	Beta 149052	1230 ± 40	CH	?	R. Vanderhoek, p.c. 2002	
	SUT 013	Beta 109896	1250 ± 50	CH	?	R. Vanderhoek, p.c. 2002	
	SUT 038	Beta 149046	1250 ± 60	CH	?	R. Vanderhoek, p.c. 2002	
	SUT 039	WSU 5048	1250 ± 60	CH	?	R. Vanderhoek, p.c. 2002	
	SUT 022	Beta 109895	1260 ± 50	CH	?	R. Vanderhoek, p.c. 2002	
	SUT 014	Beta 157013	1270 ± 40	CH (uncl.)	?	R. Vanderhoek, p.c. 2002	
	SUT 022	WSU 5047	1275 ± 60	CH	?	R. Vanderhoek, p.c. 2002	
	SUT 024	Beta 108733	1290 ± 40	CH	?	R. Vanderhoek, p.c. 2002	
	SUT 022	Beta 108729	1300 ± 40	CH	?	R. Vanderhoek, p.c. 2002	
	SUT 027	Beta 128609	1310 ± 50	CH	?	R. Vanderhoek, p.c. 2002	
	SUT 022	Beta 157014	1320 ± 40	CH	?	R. Vanderhoek, p.c. 2002	
	SUT 013	Beta 143675	1350 ± 40	CH	?	R. Vanderhoek, p.c. 2002	
	SUT 039	Beta 149050	1360 ± 40	CH	?	R. Vanderhoek, p.c. 2002	
	SUT 024	Beta 157015	1380 ± 40	CH	?	R. Vanderhoek, p.c. 2002	

	SUT 038	Beta 128614	1380 ± 40	CH	?	R. Vanderhoek, p.c. 2002
	SUT 039	Beta 149047	1380 ± 60	CH	?	R. Vanderhoek, p.c. 2002
	SUT 013	Beta 143674	1390 ± 30	CH	?	R. Vanderhoek, p.c. 2002
	SUT 027	Beta 108734	1450 ± 40	CH	?	R. Vanderhoek, p.c. 2002
	SUT 024	WSU 5045	1460 ± 75	CH	?	R. Vanderhoek, p.c. 2002
	SUT 024	WSU 5046	1470 ± 80	CH	?	R. Vanderhoek, p.c. 2002
	SUT 013	Beta 157011	1490 ± 40	CH	?	R. Vanderhoek, p.c. 2002
	SUT 024	Beta 128608	1490 ± 50	CH	?	R. Vanderhoek, p.c. 2002
	SUT 027	Beta 109898	1490 ± 50	CH	?	R. Vanderhoek, p.c. 2002
	SUT 027	Beta 128611	1530 ± 50	WD	?	R. Vanderhoek, p.c. 2002
	SUT 013	Beta 143673	1550 ± 70	CH	?	R. Vanderhoek, p.c. 2002
	SUT 013	Beta 157010	1550 ± 40	CH (uncl.)	?	R. Vanderhoek, p.c. 2002
	SUT 039	Beta 128615	1590 ± 40	CH	?	R. Vanderhoek, p.c. 2002
	SUT 024	Beta 128985	1600 ± 40	CH	?	R. Vanderhoek, p.c. 2002
	SUT 039	Beta 149054	1600 ± 40	CH	?	R. Vanderhoek, p.c. 2002
	SUT 024	Beta 109897	1620 ± 50	CH	?	R. Vanderhoek, p.c. 2002
	SUT 039	Beta 149053	1630 ± 50	CH	?	R. Vanderhoek, p.c. 2002
	SUT 027	Beta 128610	1660 ± 40	CH	?	R. Vanderhoek, p.c. 2002
	SUT 024	Beta 138996	1740 ± 60	CH	?	R. Vanderhoek, p.c. 2002
	SUT 048	Beta 143676	2060 ± 40	CH	?	R. Vanderhoek, p.c. 2002
	SUT 014	Beta 108730	2110 ± 40	CH	?	R. Vanderhoek, p.c. 2002
Katmai Region						
Aguchik Island Cove	XMK 115	Beta 74664	3560 ± 80	CH	?	Crowell and Mann 1996
Aguchik Island Tombolo	XMK 116	Beta 74673	2970 ± 60	CH	?	Crowell and Mann 1996

(Continued)

Appendix D *(Continued)*

Site name	AHRS no.	Lab no.	^{14}C date	Material	Culture	Reference	Reject
Cape Gull	XMK 058	GX 17008	510 ± 105	CH	?	Haggarty et al. 1991; Crowell and Mann 1996	
Cape Gull	XMK 058	GX 17006	525 ± 60	CH	?	Haggarty et al. 1991; Crowell and Mann 1996	
Cape Gull	XMK 058	GX 17009	550 ± 85	CH	?	Haggarty et al. 1991; Crowell and Mann 1996	
Cape Gull	XMK 058	GX 17007	590 ± 105	CH	?	Haggarty et al. 1991; Crowell and Mann 1996	
Cape Gull	XMK 058	GX 17005	730 ± 120	CH	?	Haggarty et al. 1991; Crowell and Mann 1996	
Cape Gull	XMK 058	GX 17004	750 ± 110	CH	?	Haggarty et al. 1991; Crowell and Mann 1996	
Hook Point	XMK 020	I-1942	1680 ± 100	CH	?	Clark 1977; Mills 1994	
Hook Point	XMK 020	I-1943	3470 ± 110	CH	?	Clark 1977; Mills 1994	
Island E. of Takli	XMK 027	GX 17218	3810 ± 160	CH	?	Dekin et al. 1993; Crowell and Mann 1996	
Island E. of Takli	XMK 027	GX 17219	4645 ± 220	CH	?	Dekin et al. 1993; Crowell and Mann 1996	
Island E. of Takli	XMK 030	GX 17216	5815 ± 85	CH	?	Dekin et al. 1993; Crowell and Mann 1996	
Island E. of Takli	XMK 030	GX 17217	5990 ± 85	CH	?	Dekin et al. 1993; Crowell and Mann 1996	
Kaflia River Mouth	XMK 120	Beta 75321	460 ± 70	CH	?	Crowell and Mann 1996	
Kaflia River Mouth	XMK 119	Beta 75320	3350 ± 90	CH	?	Crowell and Mann 1996	
Kinak Bay Islet	XMK 112	Beta 75316	230 ± 80	CH	?	Crowell and Mann 1996	

Kinak River Wet	XMK 113	Beta 74851	210 ± 60	CH	?	Crowell and Mann 1996
Kukak	XMK 006	Beta 97002	720 ± 70	CH	MD	D. Dumond, p.c. 2002
Kukak	XMK 006	I-1636	775 ± 110	CH	?	Clark 1977; Mills 1994
Kukak	XMK 006	I-505	775 ± 95	CH	?	Clark 1977; Mills 1994
Kukak	XMK 006	I-1638	1075 ± 100	CH	?	Clark 1977; Mills 1994
Kukak	XMK 006	I-1637	1450 ± 130	CH	?	Clark 1977; Mills 1994
Kukak	XMK 006	I-1944	1460 ± 95	CH	?	Clark 1977; Mills 1994
Kukak Bay Refuge	XMK 059	Beta 74856	360 ± 60	CH	?	Crowell and Mann 1996
Kukak isolated HP	XMK 006a	I-1945	5830 ± 120	CH	?	Clark 1977; Mills 1994
Kukak Point Village	XMK 118	Beta 75319	900 ± 60	CH	?	Crowell and Mann 1996
Kukak River Wet	XMK 113	Beta 74852	960 ± 60	CH	?	Crowell and Mann 1996
Russian Anchorage	XMK 056	Beta 74853	690 ± 60	CH	?	Crowell and Mann 1996
Russian Anchorage	XMK 056	Beta 75318	1890 ± 70	CH	?	Crowell and Mann 1996
Russian Anchorage	XMK 056	Beta 74855	4940 ± 60	CH	?	Crowell and Mann 1996
Russian Anchorage	XMK 056	Beta 74854	5060 ± 90	CH	?	Crowell and Mann 1996
Sukoi Bay Terrace	AFG 207	Beta 74849	2020 ± 80	CH	?	Crowell and Mann 1996
Sukoi Bay Terrace	AFG 207	Beta 74850	3570 ± 60	CH	?	Crowell and Mann 1996
Takli	XMK 018	I-3733	2810 ± 100	CH	?	Clark 1977; Mills 1994
Takli	XMK 018	I-1941	2910 ± 105	CH	?	Clark 1977; Mills 1994
Takli	XMK 018	I-1639	4110 ± 160	CH	?	Clark 1977; Mills 1994
Takli	XMK 018	I-1940	5650 ± 115	CH	?	Clark 1977; Mills 1994
Takli Island	XMK 075	GX 17213	2020 ± 180	CH	?	Dekin et al. 1993; Crowell and Mann 1996
Takli Island	XMK 075	GX 17212	2175 ± 205	CH	?	Dekin et al. 1993; Crowell and Mann 1996

(Continued)

Appendix D (*Continued*)

Site name	AHRS no.	Lab no.	^{14}C date	Material	Culture	Reference	Reject
Takli Island	XMK 022	GX 17515	4245 ± 165	CH	?	Dekin et al. 1993; Crowell and Mann 1996	
Takli Island	XMK 022	GX 17520	4300 ± 165	CH	?	Dekin et al. 1993; Crowell and Mann 1996	
Takli Island	XMK 022	GX 18195	4410 ± 135	CH	?	Dekin et al. 1993; Crowell and Mann 1996	
Takli Island	XMK 022	GX 17514	5205 ± 165	CH	?	Dekin et al. 1993; Crowell and Mann 1996	
Takli Islet	XMK 072	GX 17214	3605 ± 150	CH	?	Dekin et al. 1993; Crowell and Mann 1996	
Takli Islet	XMK 072	GX 17215	3875 ± 175	CH	?	Dekin et al. 1993; Crowell and Mann 1996	
Tiny Island Passage	XMK 111	Beta 75315	3270 ± 70	CH	?	Crowell and Mann 1996	
Tiny Island Village	XMK 106	Beta 74857	1530 ± 80	CH	?	Crowell and Mann 1996	
Tiny Island Village	XMK 106	Beta 74859	5630 ± 60	CH	?	Crowell and Mann 1996	
	XMK 047	Beta 75314	640 ± 90	CH	?	Crowell and Mann 1996	
KATM-99 (NPS #)	XMK-089	Beta 130082	3840 ± 40	CH	?	Lake Clark Katmai Cultural Resources Program, (M. Hilton, p.c. 2002)	
KATM-99 (NPS #)	XMK-089	Beta 130083	5500 ± 40	CH	?	Lake Clark Katmai Cultural Resources Program, (M. Hilton, p.c. 2002)	
			CHIRIK OF ISLAND				
Chirikof 9	XTI 032	P 1050	4029 ± 63	CH(2)	Takli-like	Mills 1994; Workman 1966	

LOWER KENAI PENINSULA

Kenai Region							
Merrill	KEN 029	Gx 16523	2135 ± 130	?	RKA	W. Workman, p.c. 2002	
Merrill	KEN 029	S 1040	2560 ± 300	?	RKA	W. Workman, p.c. 2002	
Merrill	KEN 029	S 1041	2245 ± 115	?	RKA	W. Workman, p.c. 2002	
Moose River	KEN 043	12.171	1960 ± 120	?	RKA	W. Workman, p.c. 2002	
Moose River	KEN 043	Beta 6680	2050 ± 70	?	RKA	W. Workman, p.c. 2002	
Moose River	KEN 043	Beta 6682	1650 ± 60	?	RKA	W. Workman, p.c. 2002	
Moose River	KEN 043	Beta 6683	2010 ± 60	?	RKA	W. Workman, p.c. 2002	
Moose River	KEN 043	Beta 6684	1910 ± 60	?	RKA	W. Workman, p.c. 2002	
Moose River	KEN 043	GX 5039	1515 ± 125	?	RKA	W. Workman, p.c. 2002	
Moose River	KEN 043	I 12	500 ± 80	?	DN	Reger and Boraas 1996	
Moose River	KEN 043	WSU 1888	1495 ± 70	?	RKA	W. Workman, p.c. 2002	
Moose River	KEN 043	WSU 2226	2180 ± 70	?	RKA	W. Workman, p.c. 2002	
Moose River	KEN 043	WSU 2227	2330 ± 70	?	RKA	W. Workman, p.c. 2002	
Moose River	KEN 043	WSU 2228	2180 ± 90	?	RKA	W. Workman, p.c. 2002	
Clam Gulch	KEN 045	Beta 6685	1210 ± 50	?	DN	Reger 1987; Reger and Boraas 1996	R
Tust Cmp.	KEN 065	Beta 23385	1986 ± 130	?	RKA	W. Workman, p.c. 2002	
Nilnunqa	KEN 066	Beta 6691	1690 ± 90	?	RKA	W. Workman, p.c. 2002	
Nilnunqa	KEN 066	GX 14166	1020 ± 21	?	RKA	W. Workman, p.c. 2002	
Nilnunqa	KEN 066	WSU 2109	660 ± 65	?	DN	Reger and Boraas 1996	
Nilnunqa	KEN 066	WSU 2947	1150 ± 130	?	RKA	W. Workman, p.c. 2002	
Nilnunqa	KEN 066	WSU 2949	1675 ± 75	?	RKA	W. Workman, p.c. 2002	
Nilnunqa	KEN 066	WSU 2950	2140 ± 80	?	RKA	W. Workman, p.c. 2002	

(Continued)

Appendix D (*Continued*)

Site name	AHRS no.	Lab no.	^{14}C date	Material	Culture	Reference	Reject
Nilnunqa	KEN 066	WSU 2951	1620 ± 60	?	RKA	W. Workman, p.c. 2002	
Nilnunqa	KEN 066	WSU 2952	585 ± 50	?	DN	Reger and Boraas 1996	
Nilnunqa	KEN 066	WSU 3105	870 ± 70	?	DN	Reger and Boraas 1996	
Nilnunqa	KEN 066	WSU 3107	895 ± 85	?	DN	Reger and Boraas 1996	
Nilnunqa	KEN 066	WSU 3108	2755 ± 160	?	RKA	W. Workman, p.c. 2002	
	KEN 147	WSU 2944	1480 ± 50	?	RKA	W. Workman, p.c. 2002	
	KEN 214	WSU 3898	1940 ± 160	?	RKA	W. Workman, p.c. 2002	
	KEN 214	WSU 3899	740 ± 50	?	DN	Reger and Boraas 1996	
Pelch	KEN 233	WSU 4147	540 ± 90	?	DN	Reger and Boraas 1996	
Pelch	KEN 233	WSU 4149	645 ± 60	?	DN	Reger and Boraas 1996	
Round Mtn	SEW 214	WSU 3089	565 ± 65	?	DN	Reger and Boraas 1996	
Bearbones Rockshelter	SEW 517	Beta 42081	3035 ± 55	?	?	source misplaced	
?	SEW 756	?	1000 ± 60	?	DN	Corbett 1998	
?	SEW 756	?	1720 ± 60	?	RKA	W. Workman, p.c. 2002	
?	SEW 756	?	1760 ± 70	?	RKA	W. Workman, p.c. 2002	
?	SEW 756	?	1820 ± 70	?	RKA	W. Workman, p.c. 2002	
?	SEW 756	?	2110 ± 70	?	RKA	W. Workman, p.c. 2002	
Kachemak Bay Region							
Chugachik Island	SEL 033	UGa 2344	1475 ± 70	CH	KA II/subIII	Mills 1994; Workman and Workman 1988	
Chugachik Island	SEL 033	S 1063	1705 ± 65	CH	KA II/subIII	Mills 1994; Workman 1977	
Chugachik Island	SEL 033	UGa 2342	1940 ± 90	CH	KA II/subIII	Mills 1994; Workman and Workman 1988	

Chugachik Island	SEL 033	S 1062	2310 ± 65	BB	KA II/subIII	Mills 1994; Workman 1977	
Chugachik Island	SEL 033	UGa 2343	2740 ± 75	WD	KA II/subIII	Mills 1994; Workman and Workman 1988	
Chugachik Island	SEL 033	WSU 4303	4005 ± 100	CH	ASTt-like	Workman p.c. to Mills 1994	
Chugachik Island	SEL 033	WSU 4302	18910 ± 250	CH	?	Workman p.c. to Mills 1994	R
Cottonwood Creek	SEL 030	S 1054	1555 ± 75	CH	KA subIII/III	Mills 1994; Workman 1977	
Cottonwood Creek	SEL 030	S 1055	1630 ± 65	CH	KA subIII/III	Mills 1994; Workman 1977	
Cottonwood Creek	SEL 030	S 1042	1745 ± 65	CW	KA subIII/III	Mills 1994; Workman 1977	
Cottonwood Creek	SEL 030	S 1043	1750 ± 125	CW	KA subIII/III	Mills 1994; Workman 1977	
Faulkner	SEL 009	Beta 158402	6670 ± 90	CM	?	Zollars and Klein 2002	
Faulkner	SEL 009	Beta 152924	6790 ± 70	CM	?	Zollars and Klein 2002	
Faulkner	SEL 009	Beta 152923	6830 ± 90	CM	?	Zollars and Klein 2002	
Faulkner	SEL 009	Beta 158403	7070 ± 90	CM	?	Zollars and Klein 2002	
Faulkner	SEL 009	Beta 152922	7150 ± 100	CM	?	Zollars and Klein 2002	
Halibut Cove 1 (Pt. West)	SEL 010	WSU 3812	510 ± 60	CH	DN	Mills 1994	
Halibut Cove 1 (Pt. West)	SEL 010	WSU 3810	775 ± 60	CH	DN	Mills 1994	
Halibut Cove 1 (Pt. West)	SEL 010	WSU 3859	1100 ± 60	CW	KA III	Mills 1994	
Halibut Cove 1 (Pt. West)	SEL 010	WSU 3811	1940 ± 70	CH	KA III	Mills 1994	
Port Graham	SEL 027	Beta 099312	570 ± 80	?	post-KA?	Workman and Workman 1997	
Port Graham	SEL 027	Beta 099310	610 ± 60	?	post-KA?	Workman and Workman 1997	
Seal Beach (Indian I.)	SEL 079	UGa 3638	410 ± 55	CH	DN	Mills 1994; Workman and Workman 1988	
Seal Beach (Indian I.)	SEL 079	UGa 3637	680 ± 55	CH	?	Mills 1994; Workman and Workman 1988	R
Seal Beach (Indian I.)	SEL 079	UGa 3634	1030 ± 205	CH	?	Mills 1994; Workman and Workman 1988	R

(Continued)

Appendix D (*Continued*)

Site name	AHRS no.	Lab no.	^{14}C date	Material	Culture	Reference	Reject
Seal Beach (Indian I.)	SEL 079	UGa 3635	1685 ± 100	CH	KA III	Mills 1994; Workman and Workman 1988	
Seal Beach (Indian I.)	SEL 079	UGa 3636	2050 ± 60	CH	KA III	Mills 1994; Workman and Workman 1988	
Sylva	SEL 245	Beta 58166	4440 ± 90	CH	OB-II	Mills 1994; Reger and Boraas 1996	
Fox Farm and Bluff	SEL 041	UGa 2339	1090 ± 195	CH	NOR	Mills 1994; Workman et al 1980	
Fox Farm and Bluff	SEL 041	UGa 2340	1130 ± 120	CH	NOR	Mills 1994; Workman et al 1980	
Fox Farm and Bluff	SEL 041	UGa 2341	1315 ± 250	CH	NOR	Mills 1994; Workman et al 1980	
Great Midden	SEL 001	P 138	1369 ± 102	ANT	KA III	Mills 1994, Reger and Boraas 1996	R
Yukon I. Great Midden	SEL 001	P 139	2706 ± 118	ANT	KA I/ YI 1?	Mills 1994, Reger and Boraas 1996	R
Kenai Fjiords Region							
	?	Beta 85205	80 ± 70	?	?	M. Yarborough	
MacArthur Pass	SEL 188	Beta 39475	620 ± 50	CH	?	Erlandson et al. 1992; Schaaf and Johnson 1990	
MacArthur Pass	SEL 188	Beta 39476	560 ± 50	CH	?	Erlandson et al. 1992; Schaaf and Johnson 1990	
MacArthur Pass	SEL 188	Beta 39477	710 ± 50	CH	?	Erlandson et al. 1992; Schaaf and Johnson 1990	
MacArthur Pass	SEL 188	Beta 39478	700 ± 90	CH	?	Erlandson et al. 1992; Schaaf and Johnson 1990	
MacArthur Pass	SEL 188	Beta 39479	1350 ± 70	CH	?	Erlandson et al. 1992; Schaaf and Johnson 1990	
MacArthur Pass	SEL 188	GX 17226	825 ± 65	CH	?	Dekin et al. 1993; Crowell and Mann 1996	

MacArthur Pass	SEL 188	GX 17227	660 ± 60	CH	?	Dekin et al. 1993; Crowell and Mann 1996
MacArthur Pass	SEL 188	GX 17228	1690 ± 140	CH	?	Dekin et al. 1993; Crowell and Mann 1996
MacArthur Pass	SEL 188	GX 17229	1710 ± 120	CH	?	Dekin et al. 1993; Crowell and Mann 1996
MacArthur Pass	SEL 188	GX 17230	1555 ± 180	CH	?	Dekin et al. 1993; Crowell and Mann 1996
MacArthur Pass	SEL 188	GX 17232	855 ± 115	CH	?	Dekin et al. 1993; Crowell and Mann 1996
MacArthur Pass	SEL 188	GX 17234	1005 ± 65	WD	?	Dekin et al. 1993; Crowell and Mann 1996
MacArthur Pass	SEL 188	GX 17235	1210 ± 65	WD	?	Dekin et al. 1993; Crowell and Mann 1996
MacArthur Pass	SEL 188	GX 17236	585 ± 105	CH	?	Dekin et al. 1993; Crowell and Mann 1996
MacArthur Pass	SEL 188	GX 17237	670 ± 105	CH	?	Dekin et al. 1993; Crowell and Mann 1996
MacArthur Pass	SEL 188	GX 17238	770 ± 65	CH	?	Dekin et al. 1993; Crowell and Mann 1996
MacArthur Pass	SEL 188	GX 17239	925 ± 105	CH	?	Dekin et al. 1993; Crowell and Mann 1996
MacArthur Pass	SEL 188	GX 17240	1105 ± 120	CH	?	Dekin et al. 1993; Crowell and Mann 1996
MacArthur Pass	SEL 188	GX 17241	800 ± 140	CH	?	Dekin et al. 1993; Crowell and Mann 1996
Northwest Lagoon	XBS 020	Beta 23382	320 ± 50	CH	?	Dotter 1988; Mills 1994
Northwest Lagoon	XBS 020	Beta 23383	140 ± 60	CH	?	Dotter 1988; Mills 1994

(Continued)

Appendix D (*Continued*)

Site name	AHRS no.	Lab no.	^{14}C date	Material	Culture	Reference	Reject
Northwest Lagoon	XBS 020	Beta 67267	580 ± 80	CH	?	Crowell and Mann 1996	
Northwest Lagoon	XBS 020	Beta 67268	830 ± 70	CH	?	Crowell and Mann 1996	
Northwest Lagoon	XBS 020	Beta 67269	660 ± 90	CH	?	Crowell and Mann 1996	
Northwest Lagoon	XBS 020	Beta 67270	690 ± 90	CH	?	Crowell and Mann 1996	
Northwest Lagoon	XBS 020	Beta 67271	610 ± 90	CH	?	Crowell and Mann 1996	
Northwest Lagoon	XBS 020	Beta 67272	240 ± 70	CH	?	Crowell and Mann 1996	
Verdant Cove Pond	XBS 028	Beta 67278	180 ± 60	CH	?	Crowell and Mann 1996	
Verdant Cove Village	XBS 029	Beta 74860	180 ± 60	CH	?	Crowell and Mann 1996	
Bear Cove Village	XBS 030	Beta 67273	590 ± 50	CH	?	Crowell and Mann 1996	
Bear Cove Village	XBS 030	Beta 67274	640 ± 110	CH	?	Crowell and Mann 1996	
Verdant Cove South Midden	XBS 031	Beta 67277	570 ± 50	CH	?	Crowell and Mann 1996	

Material: ANT = Antler; BB = Birch bark; BO = Bulk organic; CH = Charcoal (wood?); CM = Charred material; CW = Charred wood; GR = Grass; MS = Marine shell; WD = Wood.

Culture Codes: OB = Ocean Bay (I/II); KA = Kachemak (R = Riverine/E = Early/L = Late or I/II/III); KO = Koniag (E = Early/D = Developed); AL = Alutiiq (c = contact); DN = Denaina; MD = Mound Phase; NOR = Norton; PA = Paleoarctic.

Endnotes

Notes for Chapter 1

1. The term *Alutiiq* refers to the language and people of the "Pacific Eskimo" region of the north-central Gulf of Alaska. The term is relatively recent and is preferred by the Native people of this region over either "Pacific Eskimo" or "Koniag" (both used in varying contexts in the past). The term "Suqpiaq" is synonymous with Alutiiq and is preferred by some communities. The term "Aleut" is also a common synonym but should not be confused with Aleutian Islanders. Kodiak Alutiiq are a subset of the more widespread population inhabiting the southeastern Alaska Peninsula, Kodiak, sections of Kenai Peninsula, and Prince William Sound. In this book, the term "Alutiiq" refers primarily to the Kodiak Alutiiq.

Notes for Chapter 2

1. The description of Kodiak ecology is based on a combination of personal experience, discussions with local inhabitants, and the published literature. While citations are included where appropriate, in many cases observations are not directly attributable to a single source and are the result of several years of adapting "book" knowledge and personal experience of Kodiak conditions.
2. A more detailed description of the Kodiak terrestrial ecosystem is found in Russell (1991:5–6). She breaks the archipelago into five ecozones based on vegetation: the coastal spruce-hemlock forest; the shrub-thicket zone; alpine tundra; moist tundra; and wet tundra.
3. In addition to shorter frequency and lower amplitude variation, longer interval cycle (approximately 15 years) is apparent in the freshwater discharge data (Wilson and Overland, 1986:figure 2–22).
4. In this book, I adopt the convention of noting calibrated radiocarbon dates with capital 'BP' (years before present) or 'BC' and 'AD.' Uncalibrated radiocarbon dates are indicated by uncapitalized 'bp.' Period date ranges are derived from Mills' (1994) calibration of dates associated with independently attributed sites.
5. These estimates are extrapolated from Gilpin's stratigraphy diagrams (1995:216). Approximate ages are based on calibrated radiocarbon dates indicated for peat sections underlying or capping tephra deposits.

Notes for Chapter 3

1. Throughout this book the term "period" refers to a major culture-historical unit (e.g., Ocean Bay, Kachemak, Koniag and Alutiiq) and "phase" refers to subunits of periods (e.g., Ocean Bay I and Ocean Bay II). The term "tradition" is sometimes used in the literature as a synonym for period, as it is occasionally used here.
2. As of September 1998, 17 confirmed and 5 tentative Ocean Bay sites were listed on the Kodiak AHRS site register. The SAS project added 10–15 additional Ocean Bay period sites.
3. Also see Aigner (1977) for a suggestion of a pit-house in an excavation profile at the Anangula site in the Eastern Aleutians dating to between 8000 and 9000 BP.
4. Don Dummond (personal communication 1998) sees greater similarity between the Tanginak Spring site artifacts dating to greater than 5500 bp and northern ASTt characteristics (of ca 4000 bp) than he does between ASTt and the contemporaneous Rice Ridge material he has observed. This suggests that so-called ASTt assemblages on Kodiak may actually be poorly dated, or out-of-context early Ocean Bay I materials.
5. A problem exists in Knecht's (1995) presentation of data from the New Karluk site that makes it difficult to draw quantitative conclusions about changes in assemblage composition through time. Knecht reports several classes of artifacts by numbers of specimen, grouped for each floor layer and two excavated midden layers. No attempt is made to correct for possible differences in rates of deposition, occupation duration, excavation area, and volume. Based on the schematic profile and floor plans reproduced in Knecht (1995:138–151), it appears that a major change in the extent of excavation and volume of deposit occurs exactly where he identifies major shifts in Koniag organization (AD 1400—between House floor 8 and 9). Correction for these biases (e.g., for volume of excavation) may even out the relative frequency of many of the items Knecht sees surging after this juncture. Nevertheless, Knecht's conclusion of continuity with gradual evolution in artifact classes is supported, even if the quantitative aspects of this change need revision.
6. Born of a noble family on Kodiak, Qaspeq (meaning 'nobleman') was captured and enslaved by Fox Aleuts in a raid (Holmberg, 1985:59).
7. In an effort to dramatize the danger to his group and to glorify his victory, Shelikhov's number of 4000 is probably an overestimate. Other sources (cited in Black 1992; Britiukov, 1788; Holmberg, 1985; and Izmailov, 1790), suggest that the number may have been as few as 1000. A refuge population of 1500 to 2500 is probably a reasonable estimate based on a note made by one of Shelikhov's crewmen (Anonymous, n.d., cited in Black, 1992:171).
8. Reference to the continued practice of shamanism (and its repression by missionaries and other colonists) suggests that the conversion to Christianity did not immediately signal a rejection of traditional systems of belief. See Brody (2001) for a compelling perspective on this process and on the relationship to colonialization, the social imperatives supporting missionization, and reasons for success in missionizing hunter-gatherers.
9. Brody's (2001) book "The Other Side of Eden" makes a passionate and informed argument for the rationale behind colonial native education policies in the United States, Canada, Australia, and elsewhere. Policies often advocated the systematic destruction of native languages, and their associated culture, knowledge, and claim to the land.

Notes for Chapter 4

1. Davydov visited Kodiak in AD 1802, 40 years after first contact between Russians and Kodiak Alutiiq and 18 years after Shelikhov brutally conquered the archipelago. Father Gideon, whose testimony contradicts several of Davydov's points, visited Kodiak in approximately the same period. Holmberg's account, made almost 50 years later in AD 1851, includes both original observation and testimony from survivors of the conquest. It also contains descriptions and interpretations drawn directly from Davydov's account.
2. Interestingly, the sharing of property is not uncommon among middle-range societies with territoriality and wealth accumulation. This practice may appear to contradict the dramatic distinction between egalitarian sharing and the property concept as I outline here. In fact, "sharing property" is decidedly non-egalitarian and potentially competitive, as Marcell Mauss illustrates in*The Gift: The Form and Reason for Exchange in Archaic Societies* (1969). The emphasis on generosity (as a medium for social competition in complex hunting and gathering societies and between small-scale food producers), however, may emerge directly from antecedents in egalitarian-sharing systems.
3. Chumash elites from the California Channel Islands apparently also capitalized on trade with the mainland in a way similar to that envisioned for the contact era Alutiiq. In the Chumash case, the power was even more concentrated because of the chiefs' exclusive ability to leverage labor for boat construction and control the flow of goods on and off the islands (Arnold, 1993, 1995).
4. Inter-tribal warfare has been noted throughout coastal Alaska all the way to the Arctic Ocean (Burch, 1988a). However, my focus in this discussion is on a more endemic warfare, organized by the elites for the purpose of accumulating prestige, wealth, and power. It is what Mitchell (1984) calls "predatory warfare," engaged as one of many "business strategies" with the primary goal of capturing wealth objects, food stores, and slaves. Some Bering Strait groups might have practiced a similar form of warfare in the past, but the evidence for this is unclear (see Burch, 1988a; Mason, 1998).

Notes for Chapter 5

1. The date itself is not problematic because the analysis compensated for a small sample size by extending the normal count time four times to measure the rate of radiocarbon decay. This method results in a wide margin of error for the actual target date, and it is this range (close to 2000 years at 95% probability limit) that renders the date nearly useless as a measure of early occupation.
2. Fine-grained volcanics are sometimes found in glacial deposits around Kodiak (Donald Clark, personal communication, 1996) making it possible that tools made from these materials were derived from Kodiak quarries, not ones on the Alaska Peninsula. We would then expect more continuous use of fine-grained igneous rocks throughout the Tanginak sequence than we see, as well as the presence of macrocores. None were found.
3. This pattern needs to be confirmed with a larger sample of artifacts. A preliminary analysis of debitage raw materials provides less support for the colonization hypothesis, though taphonomic issues remain to be resolved (Widner, 2002).

Notes for Chapter 6

1. In this book I do not derive or test the formal models. Rather, I use them as a theoretical backdrop for generating predictions of linked changes in variables related to subsistence and symbolic economies, social organization, residential and logistical mobility, technological change, demographic variability, and political differentiation.

Notes for Chapter 7

1. As of 1998, only 18 definitive Ocean Bay sites were recorded on the AHRS register for the Kodiak Archipelago, not including the Sitkalidak sites. Five other sites were listed as probable, or possible, Ocean Bay sites.

Notes for Chapter 8

1. Minor differences in these plots from similar analyses presented in Fitzhugh (2002a), are based on reclassifying some borderline site types to be consistent with the procedures of site classification noted earlier in this chapter.
2. Composite, socketted, bone fish harpoons are well represented in the New Karluk (Koniag period) assemblage, and may have been used earlier; but bone preservation is poor for the older sites. Interestingly, at the New Karluk site and in the Uyak Bay region, netsinkers are found at the mouths of salmon streams (author's personal observation, 1987; Crowell, 1986; Knecht, 1995). Also, bone fish harpoons are not identified in excavated Kachemak sites where they should have been preserved, and alternative methods of capture are implied (Amy Steffian, personal communication, 1996).
3. A large refuge site situated at the southern opening of the Sitkalidak Strait is not included in Table 8.5. Wind and the islet's exposure to sea swells prevented close investigation. Observing from the air, Rick Knecht (personal communication, 1993) reports seeing numerous house-pits on this island.

Notes for Chapter 9

1. Total house size includes the area of all rooms, measured from the top of each sod wall, but not entrance and room passageways.

References

Ackerman, Robert E., 1968, The Archaeology of the Glacier Bay Region, Southeastern Alaska. *Laboratory of Anthropology Report of Investigations* 44. Washington State University, Pullman.

1988, Early Subsistence Patterns in Southeast Alaska. In *Diet and Subsistence: Current Archaeological Perspectives*, edited by B. V. Kennedy and G. M. LeMoine, pp. 175–189. University of Calgary, Calgary.

Ackerman, Robert E., K. C. Reid, J. D. Gallison, and M. E. Roe, 1985, *Archaeology of Haceta Island: A Survey of Sixteen Timber Harvest Units in the Tongass National Forest, Southeast Alaska*. Center for Northwest Anthropology, Washington State University, Pullman.

Aigner, Jean S., 1970, A Unifacial Core and Blade Site on Anangula Islands, Aleutians. *Arctic Anthropology* 7(2):59–88.

1976a, Asiatic New World Continuum and the Origin and Development of Bering Sea Mongoloids, with Special Emphasis on the Aleuts. *Anthropologie* 14(1–2):101–103.

1976b, Dating the Early Holocene Maritime Village of Anangula. *Anthropological Papers of the University of Alaska* 18(1):51–62.

1977, Anangula: An 8,500 B. P. Coastal Occupation in the Aleutian Islands. *Sonderdruck aus "Quartar"*, Band 27/28. pp. 65–104.

Aigner, Jean S. and Terry Del Bene, 1982, Early Holocene Maritime Adaptation in the Aleutian Islands. In *Peopling of the New World*, edited by J. E. Ericson, R. E. Taylor, and R. Berger. Ballena Press, Los Altos, Ca.

Alaska Department of Fish and Game, 1985, *Alaska Habitat Management Guide, Southcentral Region Vol. 1: Life Histories and Habitat Requirements of Fish and Wildlife*. State of Alaska Department of Fish and Game Division of Habitat, Juneau.

Alcock, John and P. Sherman, 1994, The Utility of the Proximate-Ultimate Dichotomy in Ethology. *Ethology* 96(1):58–62.

Aldenderfer, Mark, 2002, Explaining Changes in Settlement Dynamics across Transformations of Modes of Production: From Hunting to Herding in the South-Central Andes. In *Beyond Foraging and Collecting: Evolutionary Change in Hunter-Gatherer Settlement Systems*, edited By B. Fitzhugh and J. Habu, pp. 387–412. Kluwer Academic/Plenum, New York.

Ames, Kenneth M., 1981, The Evolution of Social Ranking on the Northwest Coast of North America. *American Antiquity* 46:789–805.

1985, Hierarchies, Stress and Logistical Strategies among Hunter-Gatherers in Northwestern North America. In *Prehistoric Hunter-Gatherers: the Emergence of Cultural Complexity*, edited by T. D. Price and J. Brown, pp. 155–180. Academic Press, New York.

1991, The Archaeology of the *Longue Durée*: Temporal and Spatial Scale in the Evolution of Social Complexity on the Southern Northwest Coast. *Antiquity* 65:935–945.

1994, The Northwest Coast: Complex Hunter-Gatherers, Ecology, and Social Evolution. *Annual Review of Anthropology* 23:209–229.

1995, Chiefly Power and Household Production on the Northwest Coast. In *Foundations of Social Inequality*, edited by T. D. Price and G. M. Feinman, pp. 155–187. Plenum Press, New York.

1998, Economic prehistory of the northern British Columbia coast. *Arctic Anthropology* 35(1): 68–87.

2001, Slaves, Chiefs and Labour on the Northern Northwest Coast. *World Archaeology* 33(1):1–17.

2002, Going by Boat: The Forager-Collector Continuum at Sea. In *Beyond Foraging and Collecting: Evolutionary Change in Hunter-Gatherer Settlement Systems*, edited By B. Fitzhugh and J. Habu, pp. 19–52. Kluwer Academic/Plenum, New York.

Ames, Kenneth M. and Herbert D. G. Maschner, 1999, *Peoples of the Northwest Coast: Their Archaeology and Prehistory*. Thames & Hudson, London.

Amorosi, Thomas, 1986, First Preliminary Report of an Archaeofauna from Uyak Bay and Karluk, Kodiak Island, Alaska. Manuscript on file at the Hunter Bioarchaeology Facility, Department of Anthropology, Hunter College, CUNY.

1987, The Karluk and Uyak Archaeofaunas: An Approach towards the Paleoeconomy of Kodiak Island, Alaska. Paper presented at the 52nd Annual Meeting of the Society for American Archaeology, Toronto.

1988, The Use of Faunal Remains in the Interpretation of Kodiak Island's Past. Paper presented at the First Kodiak Island Cultural Heritage Conference, Kodiak, Alaska.

Anonymous, n.d., A Ledger for 1784–1788 [containing individual accounts for the *promyshlenniki* from the vessel *Three Saints*]. Manuscript from the Yudin Collection, Box 1, Folder 3. Manuscript Division, Library of Congress, Washington, D.C. (Cited in Black 1992)

Arnold, Jeanne E., 1992, Complex Hunter-Gatherer-Fishers of Prehistoric California: Chiefs, Specialists, and Maritime Adaptations of the Channel Islands. *American Antiquity* 57:60–84.

1993, Labor and the Rise of Complex Hunter-Gatherers. *Journal of Anthropological Archaeology* 12: 75–119.

1995, Transportation Innovation and Social Complexity among Maritime Hunter-Gatherer Societies. *American Anthropologist* 97:733–747.

1996a, Understanding the Evolution of Intermediate Societies. In *Emergent Complexity: The Evolution of Intermediate Societies*, edited by J. E. Arnold, pp. 1–12. International Monographs in Prehistory, Archaeological Series 9, Ann Arbor.

1996b, Organizational Transformations: Power and Labor among Complex Hunter-Gatherers and Other Intermediate Societies. In *Emergent Complexity: The Evolution of Intermediate Societies*, edited by J. E. Arnold, pp. 59–73. International Monographs in Prehistory, Archaeological Series 9, Ann Arbor.

1996c, The Archaeology of Complex Hunter-Gatherers. *Journal of Archaeological Method and Theory* 77–126.

1997, Bigger Boats, Crowded Creekbanks: Environmental Stresses in Perspective. *American Antiquity* 62(2):337–339.

2000, The Origins of Hierarchy and the Nature of Hierarchical Structures in Prehistoric California. In *Hierarchies in Action: Cui Bono?*, edited by M. W. Diehl, pp. 221–240. Center for Archaeological Investigations, Southern Illinois University, Visiting Scholar Conference. No. 27., Carbondale.

Arnold, Jeanne E., Roger H. Colten, and Scott Pletka, 1997, Contexts of Cultural Change in Insular California. *American Antiquity* 62(2):300–318.

Bacus, E. A., A. W. Barker, J. D. Bonevich, S. L. Dunavan, J. B. Fitzhugh, D. L. Gold, N. S. Goldman-Finn, W. Griffin, and K. M. Mudar, editors, 1993, *Gendered Past: A Critical Bibliography of Gender in Archaeology*. University of Michigan, Museum of Anthropology, Technical Reports, No. 25, Ann Arbor.

Barclay, D. J., P. E. Calkin, and G. C. Wiles, 2001, Holocene History of Hubbard Glacier in Yakutat Bay and Russell Fiord, Southern Alaska. *Bulletin of the Geological Society of America* 113(3): 388–402.

Bard, E., B. Hamelin, M. Arnold, L. Montaggioni, G. Cabioch, G. Faure, and F. Rougerie, 1996, Deglacial Sea Level Record from Tahiti Corals and the Timing of Global Meltwater Discharge. *Nature* 382:241–244.

Barsch, R. L., 1985, *Karluk River Study*. Report submitted to the Kodiak Area Native Association. Kodiak, Alaska.

Bayham, Frank E, 1979, Factors Influencing the Archaic Pattern of Animal Utilization. *Kiva* 46:209–225.

Belovsky, Gary, 1988, An Optimal Foraging-Based Model of Hunter-Gatherer Population Dynamics. *Journal of Anthropological Archaeology* 7:329–372.

Bender, Barbara, 1978, From Gatherer-Hunter to Farmer: A Social Perspective. *World Archaeology* 10:204–222.

1985, Prehistoric Developments in the American Midcontinent and in Brittany, Northwest France. In *Prehistoric Hunter-Gatherers:Tthe Emergence of Cultural Complexity*, edited by T. D. Price and J. Brown, pp. 21–57. Academic Press, New York.

1990, The Dynamics of Nonhierarchical Societies. In *The Evolution of Political Systems: Sociopolitics in Small-Scale Sedentary Societies*, edited by S. Upham, pp. 247–263. Cambridge University Press, Cambridge.

Bentley, R. Alexander and Herbert D. G. Maschner, 2001, Stylistic Change as a Self-Organized Critical Phenomenon: An Archaeological Study in Complexity. *Journal of Archaeological Method and Theory* 8(1):35–66.

Berndt, Ronald M. and Catherine Berndt, 1964, *World of the First Australians*. University of Chicago Press, Chicago.

Bettinger, Robert L., 1999, From Traveler to Processer: Regional Trajectories of Hunter-Gatherer Sedentism in the Inyo-Mono Region, California. In *Settlement Pattern Studies in the Americas: Fifty Years since Viru*, edited by Brian R. Billman and Gary M. Feinman, pp. 39–55. Smithsonian Institution Press, Washington, D.C.

Biesele, Megan and Kxao Royal-/o/oo, 1999, The Ju/'hoansi of Botswana and Namibia. In *The Cambridge Ecnyclopedia of Hunters and Gatherers*, edited by R. B. Lee and R. Daly, pp. 205–209. Cambridge University Press, Camridge.

Binford, Lewis R., 1962, Archaeology as Anthropology. *American Antiquity* 28:217–225.

1968, Post Pleistocene Adaptations. In *New perspectives in archeolog*, edited by Sally R. Binford and Lewis R. Binford. Aldine, Chicago.

1980, Willow Smoke and Dog's Tails: Hunter-Gatherer Settlement Systems and Archaeological Site Formation. *American Antiquity* 45(1):4–20.

1982, The Archaeology of Place. *Journal of Anthropological Archaeology* 1:5–31.

Bird-David, Nurit, 1992, Beyond the "Original Affluent Society": A Culturalist Reformulation. *Current Anthropology* 33:25–48.

Birdsell, J., 1953, Some Environmental and Cultural Factors Influencing the Structuring of Australian Aboriginal Populations. *American Naturalist* 87:169–207.

Black, Lydia T., 1977, The Konyag (The Inhabitants of the Island of Kodiak) by Iosaf [Bolotov] (1794–1799) and by Gideon (1804–1807). *Arctic Anthropology* 14(2)79–108.

1988, The Story of Russian America. In *Crossroads of Continents: Cultures of Siberia and Alaska*, edited by William W. Fitzhugh and Aron C. Crowell, pp. 70–82. Smithsonian Institution Press, Washington, D.C.

1989, *The Round the World Voyage of Hieromonk Gideon, 1803–1809*. Translated by Lydia T. Black, edited by R. A. Pierce. The Limestone Press, Kingston, Ontario.

1992, The Russian Conquest of Kodiak. In *Contributions to the Anthropology of Southcentral and Southwestern Alaska*, edited by R. H. Jordan, F. de Laguna, and A. F. Steffian. *Anthropological Papers of the University of Alaska* 24(1–2):165–182.

Black, Robert F., 1974, Late Quaternary Sea-Level Changes, Umnak Island, Aleutians—their Effects on Ancient Aleuts and their Causes. *Quaternary Research* 4:264–281.

Blanton, Richard E., Gary M. Feinman, Stephen A. Kowalewski, and Peter N. Peregrine, 1996, Dual-Processual Theory for the Evolution of Mesoamerican Civilization. *Current Anthropology* 37(1):1–14.

Blitz, John H., 1993, Big Pots for Big Shots: Feasting and Storage in a Mississippian Community. *American Antiquity* 58(1):80–96.

Blurton Jones, N. G., 1983, A Selfish Origin for Human Food Sharing: Tolerated Theft. *Ethology and Sociobiology* 5(1):1–3.

1987b, Tolerated Theft, Suggestions about the Ecology and Evolution of Sharing, Hoarding and Scrounging. *Social Science Information* 26:31–54.

Boas, Franz, 1964, *The Central Eskimo*. University of Nebraska Press, Lincoln.

Boehm, Christopher, 1993, Egalitarian Behavior and Reverse Dominance Hierarchy. *Current Anthropology* 34:227–254.

Boone, James L., 1992, Competition, Conflict, and Development of Social Hierarchies. In *Evolutionary Ecology and Human Behavior*, edited by E. A. Smith and B. P. Winterhalder, pp. 301–338. Aldine de Gruyter, New York.

2000, Status signaling, social power, and lineage survival. In *Hierarchies in Action: Cui Bono?*, edited by M. W. Diehl, pp. 84–110. Center for Archaeological Investigations, Southern Illinois University, Visiting Scholar Conference. No. 27., Carbondale.

Boone, James L. and Eric A. Smith, 1998, Is it Evolution Yet? A Critique of Evolutionary Archaeology. *Current Anthropology* 39(Suppl.):141–173.

Boserup, Ester, 1965, *The Conditions of Agricultural Growth: The Economics of Agrarian Exchange under Population Pressure*. Aldine, Chicago.

1981, *Population and Technological Change: A Study of Long-Term Trends*. University of Chicago Press, Chicago.

Bourque, Bruce J., 1995, *Diversity and Complexity in Prehistoric Maritime Societies: a Gulf of Maine Perspective*. Plenum Press, New York.

Boyd, Robert and Peter J. Richerson, 1985, *Culture and the Evolutionary Process*. University of Chicago Press, Chicago.

Braun, David P. and Stephen Plog, 1982, Evolution of "Tribal" Social Networks: Theory and Prehistoric North American Evidence. *American Antiquity* 47:504–525.

Britiukov, Miron, 1788, Report Rendered to the Fleet Captain I. Billings on 2 November, 1788 at Yakutsk. Manuscript from the Frank A. Golder Collection of photostats from the Russian Archives, Imperial Majesty of Foreign Affairs, Box 3. Suzallo Library, University of Washington, Seattle. (cited in Black 1992)

Brody, Hugh, 2001, The Other Side of Eden: Hunter-Gatherers, Farmers and the Shaping of the World. Faber, London.

Broughton, Jack M., 1994, Declines in Mammalian Foraging Efficiency during the Late Holocene, San Francisco Bay, California. *Journal of Anthropological Archaeology* 13: 371–401.

Brubaker, Linda B., Patricia M Anderson, and Feng-Sheng Hu

2001 Vegetation ecotone dynamics in Southwest Alaska during the Late Quaternary. *Quaternary-Science-Reviews* 20(1–3):175–188.

Brumfiel, Elizabeth M., 1976, Regional Growth in the Eastern Valley of Mexico: A Test Case of the "Population Pressure" Hypothesis. *The Early Mesoamerican Village*, edited by Kent V. Flannery, pp. 234–249. Academic Press, New York.

1994, Factional competition and Political Development in the New World: An Introduction. In *Factional Competition and Political Development in the New World*, edited by E. M. Brumfiel and J. W. Fox, pp. 3–13. Cambridge University Press, Cambridge.

Burch, Ernest S., Jr., 1988a, War and Trade Oecumene. In *Crossroads of Continents: Cultures of Siberia and Alaska*, edited by W. W. Fitzhugh and A. Crowell, pp. 227–240. Smithsonian Institution, Washington, D.C.

1988b, Models of Exchange in North-West Alaska. In *Hunters and Gatherers, Vol. 2: Property, Power and Ideology*, edited by T. Ingold, D. Riches, and J. Woodburn, pp. 95–109. Berg, Oxford.

Burch, Ernest S., Jr. and Linda J. Ellana, editors, 1994, *Key Issues in Hunter-Gatherer Research*. Berg, Oxford.

Burney, David A., 1997, Tropical Islands as Paleoecological Laboratories: Gauging the Consequences of Human Arrival. *Human Ecology* 25:437–457.

Calkin, P. E., G. C. Wiles, and D. J. Barclay, 2001, Holocene Coastal Glaciation of Alaska. *Quaternary Science Reviews* 20(1–3): 449–461.

Campbell, L. J., 1992, Making a Living. In *Kodiak*, Alaska Geographic 19(3):61–74.

Cannon, Aubrey, 2002, Sacred Power and Seasonal Settlement on the Central Northwest Coast. In *Beyond Foraging and Collecting: Evolutionary Change in Hunter-Gatherer Settlement Systems*, edited by B. Fitzhugh and J. Habu, pp. 311–338. Kluwer Academic/Plenum Press, New York.

Cannon, Micheal D., 2000, Large Mammal Relative Abundance in Pithouse and Pueblo Period Archaeofaunas from Southwestern New Mexico: Resource Depression among the Mimbres-Mogollon? *Journal of Anthropological Archaeology* 19:317–347.

2002, *Large Mammal Resource Depression and Agricultural Intensification: An Empirical Test in the Mimbres Valley, New Mexico*. Ph.D. dissertation for the University of Washington. University Michrofilms International, Ann Arbor.

Carlson, Roy L., 1998, Coastal British Columbia in the Light of North Pacific Maritime Adaptations. *Arctic Anthropology* 35(1):23–35.

Cashdan, Elizabeth, 1980, Egalitarianism among Hunters and Gatherers. *American Anthropologist* 82:116–120.

1983, Territoriality among Human Foragers: Ecological Models and an Application to Four Bushman Groups. *Current Anthropology* 24:47–66.

Cavalli-Sforza, L. L. and M. W. Feldman, 1981, *Cultural Transmission and Evolution: a Quantitative Approach*. Princeton University Press, Princeton.

Chaussonnet, Valérie, 1995, *Crossroads Alaska: Native Cultures of Alaska and Siberia*. Arctic Studies Center, Smithsonian Institution, Washington, D.C.

Cherry, John F., 1984, The Initial Colonization of the West Mediterranean Islands in the Light of Island Biogeography and Paleogeography. In *Early Settlement in the Western Mediterranean Islands and the Peripheral Areas*, edited by W. H. Waldren, R. Chapman, J. Lewthwaite, and R-C Kennard, pp. 7–27. BAR International Series 229.

1992, Paleolithic Sardinians? Some Questions of Evidence and Method. In *Sardinia in the Mediterranean: a Footprint in the Sea*, edited by R. H. Tykot and T. K. Andrews, pp. 28–39. Sheffield Academic Press, Sheffield, England.

Childe, V. Gordon, 1951, *Man Makes Himself*. C. A. Watts, London.

1957, *The Dawn of European Civilization*. Routledge & Paul, London.

Clark, Donald W., 1956, Notes on a Pottery-Bearing Site at Olga Bay, Kodiak Island. In *Archaeology of the Uyak Site, Kodiak Island, Alaska*, Robert Heizer. University of California Anthropological Records 17:1 (pp. 94–97).

1960, A Decorated Stone Lamp from the Kodiak Area. *Anthropological Papers of the University of Alaska* 9(1):58–60.

1966a, Perspectives in the Prehistory of Kodiak Island, Alaska. *American Antiquity* 31(3) Part 1:358–371.

1966b, Two Late Prehistoric Pottery-Bearing Sites on Kodiak Island Alaska. *Arctic Anthropology* 3(2):157–184.

1970a, The Late Kachemak Tradition at Three Saints Bay and Crag Point, Kodiak Island, Alaska. *Arctic Anthropology* 6(2):73–111.

1970b, Petroglyphs on Afognak Island, Kodiak Group, Alaska. *Anthropological Papers of the University of Alaska* 15(1):13–17.

1974a, *Koniag Prehistory: Archaeological investigations at late prehistoric sites on Kodiak Island, Alaska.* Tubinger Monographien Zur Urgeschischte, vol. 1, Tubingen.

1974b, The Earliest Prehistoric Cultures of Kodiak Island, Alaska: 1971 Fieldwork, Preliminary Report. *Arctic Anthropology* 11(1):41–46.

1979, *Ocean Bay: An Early North Pacific Maritime Culture.* National Museum of Man, Mercury Series, Archaeological Survey of Canada Paper No. 86. Ottawa.

1980, Relationships of North Pacific and American Arctic Centres of Slate Grinding. *Canadian Journal of Archaeology* 4:27–38.

1982, An Example of Technological Change in Prehistory: the Origin of a Regional Ground Slate Industry in South-Central Coastal Alaska. *Arctic Anthropology* 19(1):103–126.

1984a, Pacific Eskimo: Historical Ethnography. In *Handbook of American Indians, vol. 5, Arctic,* edited by David Damas, pp. 185–197. Smithsonian Institution, Washington, D.C.

1984b, Prehistory of the Pacific Eskimo Region. In *Handbook of American Indians, vol. 5, Arctic,* edited by David Damas, pp. 136–148. Smithsonian Institution, Washington, D.C.

1985, Archaeological Test at the Russian Three Saints Bay Colony, Alaska. *Historical Archaeology* 19(2):114–121.

1986, Archaeological and Historical Evidence for an 18th Century "Blip" in the Distribution of the Northern Fur Seal at Kodiak Island, Alaska. *Arctic Anthropology* 39(1):39–42.

1987, On a Misty Day You Can See Back to 1805: Ethnohistory and Historical Archaeology on the Southeastern Side of Kodiak Island, Alaska. *Anthropological Papers of the University of Alaska* 21(1–2):105–132.

1988, Pacific Eskimo Encoded Prehistoric Contact. In *The Late Prehistoric Development of Alaska's Native People*, edited by Robert D. Shaw, Roger K. Harritt, and Don E. Dumond, pp. 211–224. *Aurora* vol. 4. Alaska Anthropological Association, Anchorage.

1992a, Archaeology on Kodiak: The Quest for Prehistory and its Implications for North Pacific Prehistory. In *Contributions to the Anthropology of Southcentral and Southwestern Alaska*, edited by R. H. Jordan, F. de Laguna, and A. F. Steffian, pp. 109–126. *Anthropological Papers of the University of Alaska* vol. 24(1–2).

1992b, "Only a Skin Boatload or two": The Role of Migration in Kodiak Prehistory. *Arctic Anthropology* 29(1):2–17.

1994, Still a Big Story: The Prehistory of Kodiak Island. In *Reckoning with the Dead: The Larsen Bay Repatriation and the Smithsonian Institution*, edited by T. L. Bray and T. W. Killion, pp. 137–149. Smithsonian Institution Press, Washington, D.C.

1996, The Old Kiavak Site, Kodiak Island, Alaska, and the Early Kachemak Phase. *Arctic* 49(3):211–227.

1997, *The Early Kachemak Phase on Kodiak Island at Old Kiavak.* Mercury Series, Archaeological Survey of Canada, Paper 155. Canadian Museum of Civilization, Hull, Quebec.

1998, Kodiak Island: The Later Cultures. *Arctic Anthropology* 35(1):172–186.

Clark, Gerald H., 1977, Archaeology on the Alaska Peninsula: The Coast of Shelikof Strait 1963–1965. *University of Oregon Anthropological Papers* No. 13. Eugene.

Clark, John E., 2000, Towards a better Explanation of Hereditary Inequality: A Critical Assessment of Natural and Historic Human Agents. In *Agency in Archaeology*, edited by M-A Dobres and J. Robb, pp. 92–112.

Clark, John E. and Michael Blake, 1994, The Power of Prestige: Competitive Generosity and the Emergence of Rank Societies in Lowland Mesoamerica. In *Factional Competition and Political*

Development in the New World, edited by E. M. Brumfiel and J. W. Fox, pp. 17–30. Cambridge University Press, Cambridge.

Clarke, David L., editor, 1972, *Models in Archaeology*. Methuen, London.

Cohen, Mark N., 1977, *The Food Crisis in Prehistory: Overpopulation and the Origins of Agriculture*. Yale University Press, New Haven.

1985, Prehistoric Hunter-Gatherers: The Meaning of Social Complexity. In *Prehistoric Hunter-Gatherers: The Emergence of Cultural Complexity*, edited by T. D. Price and J. Brown, pp. 99–119. Academic Press, New York.

Collier, Jane F. and Michelle Z. Rosaldo, 1981, Politics and Gender in Simple Societies. In *Sexual Meanings*, edited by S. B. Ortner and H. Whitehead, pp. 275–329. Cambridge University Press, Cambridge.

Conkey, Margaret W., 2001, Men and Women in Prehistory: An Archaeological Challenge. In *Gender in Cross-Cultural Perspective*, edited by C. B. Brettell and C. F. Sargent, pp. 61–70. Prentice Hall, Upper Saddle River.

Conkey, Margaret W. and Joan M. Gero, 1991, Tensions, Pluralities, and Engendering Archaeology: An Introduction to Women and Prehistory. In *Engendering Archaeology: Women and Prehistory*, edited by J. M. Gero and M. W. Conkey, pp. 3–30. Basil Blackwell, Oxford.

Corbett, Debra, 1998, Riverine Kachemak on the Upper Kenai River. Paper presented at the Annual Meeting of the Alaska Anthropological Association.

Coupland, Gary, 1985, Household Variability and Status Differentiation at Kitselas Canyon. *Canadian Journal of Archaeology* 9(1):39–56.

1988, Prehistoric Economic and Social Change in the Tsimshian area. In *Prehistoric Economies of the Northwest Coast, Research in Economic Anthropology, supp.* 3, edited by B. Isaac, pp. 211–243. JAI Press, Greenwich, Conn.

1996 This old house: cultural complexity and household stability on the northern Northwest Coast of North America. In *Emergent Complexity: The Evolution of Intermediate Societies*, Jeanne E. Arnold (ed.). International Monographs in Prehistory, Archaeological Series 9, Ann Arbor. pp. 74–90.

Crowell, Aron C., 1986, An Archaeological Survey of Uyak Bay, Kodiak Island, Alaska, final report. Department of Anthropology, National Museum of Natural History, Washington, D.C.

1988, Prehistory of Alaska's Pacific Coast. In *Crossroads of Continents: Cultures of Siberia and Alaska*, edited by W. W. Fitzhugh and A. C. Crowell, pp. 130–140. Smithsonian Institution Press, Washington, D.C.

1994, Koniag Eskimo Poison-Dart Whaling. In *Anthropology of the North Pacific Rim,* edited by W. W. Fitzhugh and V. Chaussonnet, pp. 217–242. Smithsonian Institution Press, Washington. D.C.

1997, *Archaeology and the Capitalist World System: A Study from Russian America.* Plenum Press, New York.

Crowell, Aron C. and Jeff Leer, 2001, *Ukgwepet*—"Our Beliefs": Alutiiq Spiritual Life and Traditions. In *Looking Both Ways: Heritage and Identity of the Alutiiq People*, edited by A. C. Crowell, A. F. Steffian, and G. L. Pullar, pp. 189–221. University of Alaska Press, Fairbanks.

Crowell, Aron C. and Sonja Lührmann, 2001, Alutiiq Culture: Views from Archaeology, Anthropology and History. In *Looking Both Ways: Heritage and Identity of the Alutiiq People*, edited by A. C. Crowell, A. F. Steffian, and G. L. Pullar, pp. 21–71. University of Alaska Press, Fairbanks.

Crowell, Aron C. and Daniel H. Mann, 1996, Sea Level Dynamics, Glaciers, and Archaeology along the Central Gulf of Alaska Coast. *Arctic Anthropology* 33(2):16–37.

Crowell, Aron C., Amy F. Steffian, and G. L. Pullar, editors, 2001, *Looking Both Ways: Heritage and Identity of the Alutiiq People*. University of Alaska Press, Fairbanks.

Crozier, Neal S., 1986, Preliminary Interpretation of Material Remains and Stratigraphy at Two Prehistoric Sites in Southwest Alaska. MS. on file, Bureau of Indian Affairs, Archaeology Division, Anchorage.

1987, Preliminary Interpretation of Material Remains at a Prehistoric Site, Kodiak, Alaska. Mitigation report on file, Bureau of Indian Affairs, Archaeology Division, Anchorage.

1989, Excavation at a Late Prehistoric Dwelling Structure on Kodiak Island, Alaska. *Arctic Anthropology* 26(2):78–95.

Davis, Nancy Yaw., 1978, Historical Indicators of Alaska Native Culture Change. Alaska OCS Socioeconomic Studies Program, Technical Report 15. Bureau of Land Management, Alaska Outer Continental Shelf Office, Anchorage.

Davis, Stanley D., 1989, Cultural Component I. In *The Hidden Falls Site*, edited by Stanley D. Davis, pp. 159–198. *Aurora*. Alaska Anthropological Association. Monograph Series, No 5.

1996, Hidden Falls. In *American Beginnings: The Prehistory and Paleoecology of Beringia*, edited by F. H. West, pp. 413–423. University of Chicago Press, Chicago.

Davydov, Gavriil I., 1976, A Selection from G. I. Davydov: An Account of Two Voyages to America. Translated by C. Bearne, edited by R. A. Pierce. *Arctic Anthropology* 8:1–30.

Dekin, A. A., Jr., M. S. Cassell, J. I. Ebert, E. Camilli, J. M. Kerley, M. R. Yarborough, P. A. Stahl, and B. L. Turcy, 1993, *Exxon Valdez Oil Spill Archaeological Damage Assessment, Final Report*. The Research Foundation of the State University of New York, Binghamton.

de Laguna, Frederica, 1934, *The Archaeology of Cook Inlet, Alaska*. Published for the University Museum by the University of Pennsylvania Press, Philadelphia. (2nd ed., Alaska Historical Society, Fairbanks, 1975).

1946, (Review) Ales Hrdlicka, "The Anthropology of Kodiak Island" (1944) and "The Aleutian and Commander Islands and Their Inhabitants" (1945). *Pacific Affairs* 55(4):478–481.

1956, (Review) "Archaeology of the Uyak Site, Kodiak Island, Alaska" by R. F. Heizer. *American Antiquity* 22(2):202.

1983, Aboriginal Tlingit Sociopolitical Organization. In *The Development of Political Organization in Native North America*, edited by Elizabeth Tooker, pp. 71–85. Proceedings of the American Ethnological Society. American Ethnological Society, Washington D.C.

1988, Potlatch Ceremonialism on the Northwest Coast. In *Crossroads of Continents: Cultures of Siberia and Alaska*, edited by W. W. Fitzhugh and A. Crowell, pp. 271–280. Smithsonian Institution Press, Washington, D.C.

1990, Tlingit. In *Handbook of North American Indians, Vol. 7, Northwest Coast*, edited by W. Suttles, pp. 203–228. Smithsonian Institution Press, Washington, D.C.

Desson, Dominique, 1995, *Masked Rituals of the Kodiak Archipelago*. Ph.D. dissertation, Department of Anthropology, University of Alaska-Fairbanks. University Microfilms International, Ann Arbor.

Dewar, Robert E., 1997, Does it Matter that Madagascar is an Island? *Human Ecology* 25:481–489.

Dixon, E. J., T. H. Heaton, T. E. Fifield, T. D. Hamilton, D. E. Putnam, and F. Grady, 1997, Late Quaternary Regional Geoarchaeology of Southeast Alaska Karst: A Progress Report. *Geoarchaeology* 12(6):689–712.

Dobres, Marcia-Anne, 2001, Meaning in the Making: Agency and Social Embodiment of Technology and Art. In *Antrhopological Perspectives on Technology*, edited by M. B. Schiffer, pp. 47–76. University of New Mexico Press, Albuquerque.

Donald, Leland, 1983, Was Nuu-chah-nulth-aht (Nootka) Society Based on Slave Labor? In *The Development of Political Organization in Native North America*, edited by E. Tooker, pp. 108–119. American Ethnological Society, Washington, D.C.

1984, The Slave Trade on the Northwest Coast of North America. *Research in Economic Anthropology* 6:121–158.

1997, *Aboriginal Slavery on the Northwest Coast of North America*. Berkeley: University of California Press.

Donta, Christopher, 1993, *Koniag Ceremonialism: An Archaeological and Ethnographic Analysis of Sociopolitical Complexity and Ritual among the Pacific Eskimo*. Ph.D. dissertation, Department of Anthropology, Bryn Mawr College. University Microfilms International, Ann Arbor.

1994, Continuity and Function in the Ceremonial Material Culture of the Koniag Eskimo. In *Reckoning With the Dead: The Larsen Bay Repatriation and the Smithsonian Institution*, edited by T. L. Bray and T. W. Killion, pp. 122–136. Smithsonian Institution Press, Washington, D.C.

Dotter, Walter R., 1988, 1988 BIA ANCSA SEC. 14(H)(1) Field Investigations in Prince William Sound and Adjacenet Areas. Paper presented at the 15th Annual Meeting of the Alaska Anthropological Association, Fairbanks.

Drucker, Philip, 1939, Rank, Wealth, and Kinship in Northwest Coast Society. *American Anthropologist* 41:55–65.

1951, The Northern and Central Nootkan Tribes, *Bureau of American Ethnology Bulletin* 144. Smithsonian Institution, Washington, D.C.

1983, Ecology and Political Organization on the Northwest Coast of America. In *The Development of Political Organization in Native North America*, edited by E. Tooker, pp. 86–96. American Ethnological Society, Washington, D.C.

Dumond, Don E., 1965, Population Growth and Cultural Change. *Southwestern Journal of Anthropology* 21(4):302–324.

1979, People and Pumice on the Alaska Peninsula. In *Volcanic Activity and Human Ecology*, edited by P. D. Sheets and D. K. Grayson, pp. 373–392. Academic Press, New York.

1981, *Archaeology on the Alaska Peninsula: the Naknek Region, 1960–1975*. Anthropological Papers of the University of Oregon 21. Eugene.

1984a, Prehistory: Summary. In *Handbook of American Indians, vol. 5, Arctic*, edited by David Damas, pp. 72–79. Smithsonian Institution, Washington, D.C.

1984b, Prehistory of the Bering Sea Region. In *Handbook of American Indians, vol. 5, Arctic*, edited by David Damas, pp. 94–105. Smithsonian Institution, Washington, D.C.

1987, *The Eskimos and Aleuts*, (second ed.). Thames and Hudson, London.

1988a, The Alaska Peninsula as Super-Highway: A Comment. In *The Late Prehistoric Development of Alaska's Native People*, edited by R. D. Shaw, R. K. Harritt and D. E. Dumond, pp. 379–388. Aurora: Alaska Anthropological Association Monograph Series 4.

1988b, Trends and Traditions in Alaskan Prehistory: A New Look at an Old View of the Neo-Eskimo. In *The Late Prehistoric Development of Alaska's Native People*, edited by R. D. Shaw, R. K. Harritt and D. E. Dumond, pp. 17–26. Aurora: Alaska Anthropological Association Monograph Series 4.

1994, The Uyak Site in Prehistory. In *Reckoning with the Dead: The Larsen Bay Repatriation and the Smithsonian Institution*, edited by T. L. Bray and T. W. Killion, pp. 43–53. Smithsonian Institution Press, Washington, D.C.

1998, Maritime Adaptations on the Northern Alaska Peninsula. *Arctic Anthropology* 35(1):187–203.

2000, A Southern Origin for Norton Culture? Anthropological Papers of the University of Alaska 25(1):87–102.

2001, *Archaeology in the Aleut Zone: Some Recent Research*, edited by D. E. Dumond, pp. 289–309. University of Oregon Anthropological Papers No. 58. Eugene.

Dumond, Don E. and Dennis G. Griffin

2002 Measurements of the Marine Reservoir Effect on Radiocarbon Ages in the Eastern Bering Sea. Arctic 55(1):77–86.

Dumond, Don E. and Richard Knecht, 2001, An Early Blade Site in the Eastern Aleutians. In *Archaeology in the Aleut Zone: Some Recent Research*, edited by Don E. Dumond, pp. 9–34. University of Oregon Anthropological Papers No. 58. Eugene.

Dunnell, Robert C., 1980, Evolutionary Theory and Archaeology. *Advances in Archaeological Method and Theory* 3:35–99.

1986, Methodological Issues in Americanist Artifact Classification. In *Advances in Archaeological Method and Theory* 9:149–207.

Dyson Hudson, R. and Eric A. Smith., 1978, Human Territoriality: An Ecological Reassessment. *American Anthropologist* 80:21–41.

Edwards, M. E., P. M. Anderson, L. B. Brubaker, T. A. Ager, A. A. Andreev, N. H. Bigelow, L. C. Cwynar, W. R. Eisner, S. P. Harrison, F.-S. Hu, D. Jolly, A. V. Lozhkin, G. M. MacDonald, C. J. Mock, J. C. Ritchie, A. V. Sher, R. W. Spear, J. W. Williams, and G. Yu, 2000, Pollen-Based Biomes for Beringia 18,000, 6000 and 0 ^{14}C yr BP. *Journal of Biogeography* 27(3): 521–554.

Ehrenreich, Robert M., Carole L. Crumley, and Janet E. Levy, editors, 1995, *Heterarchy and the Analysis of Complex Societies*. American Anthropological Association, Arlington, VA.

Ellison P., 1994, Human Reproductive Ecology. *Annual Review of Anthropology* 23:255–275.

Emmons, G. T., 1991, *The Tlingit Indians*. Edited and with additions by Frederica de Laguna. American Museum of Natural History Anthropological Papers 70, New York.

Erlandson, Jon M., 1988, The Role of Shellfish in Coastal Economies: A Protein Perspective. *American Antiquity* 53:102–109.

2001, The Archaeology of Aquatic Adaptations: Paradigms for a New Millenium. *Journal of Archaeological Research* 9(4):287–350.

Erlandson, Jon M. and Madonna L. Moss, 1996 Pleistocene-Holocene Transition along the Pacific Coast of North America. In *Humans at the End of the Ice Age: the Archaeology of the Pleistocene-Holocene Transition*, edited by L. G. Straus, Berit Valentin Eriksen, Jon M. Erlanson, and David R. Yesner, pp. 277–301. Plenum Press, New York.

2001, Shellfish Feeders, Carrion Eaters, and the Archaeology of Aquatic Adaptations. *American Antiquity* 66(3):413–432.

Erlandson, Jon M., Aron Crowell, Christopher Wooley and James Haggarty, 1992, Spatial and Temporal Patterns in Alutiiq Paleodemography. *Arctic Anthropology* 29(2):42–62.

Etnier, Michael, 2002, *The Effects of Human Hunting on Northern Fur Seals (Callorhinus ursinus) Migration and Breeding Distributions in the Holocene*. Ph.D. dissertation, University of Washington, Seattle.

Fagan, Brian M. and Herbert D. G. Maschner, 1991, Emergence of Cultural Complexity on the West Coast of North America. *Antiquity* 65(249):974–976.

Fairbanks, R. G., 1989, A 17,000 Year Glacio-eustatic Sea Level Record: Influence of Glacial Melting Rates on the Younger Dryas Event and Deep-Ocean Circulation. Nature 342:637–642.

Feinman, Gary M., 1995, The Emergence of Inequality: A Focus on Strategies and Processes. In *Foundations of Social Inequality*, edited by T. D. Price and G. M. Feinman, pp. 255–279. Plenum Press, New York.

2000, Corporate/Network: A New Perspective on Leadership in the American Southwest. In *Hierarchies in Action: Cui Bono?*, edited by M. W. Diehl, pp. 152–180. Center for Archaeological Investigations, Visiting Scholar Conference, No. 27, Southern Illinois University, Carbondale.

Feinman, Gary M. and Neitzel, Jill

1984, Too Many Types: An Overview of Sedentary Prestate Societies in the Americas. *Advances in Archaeological Method and Theory* 7:39–102.

Ferguson, Brian, 1983, Warfare and Redistributive Exchange on the Northwest Coast. In *The Development of Political Organization in Native North America*, edited by E. Tooker, pp. 133–147. American Ethnological Society, Washington, D.C.

1984, A Reexamination of the Causes of Northwest Coast Warfare. In *Warfare, Culture, and the Environment*, edited by Brian Ferguson, pp. 267–328. Academic Press, New York.

Finney, Bruce P., I. Gregory-Eaves, M. S. V. Douglas, J. P. Smol, 2002, Fisheries Productivity in the Northeastern Pacific Ocean over the Past 2,200 years. *Nature* 416(6882):729–733.

Finney, Bruce P., I. Gregory-Eaves, J. Sweetman, M. S. V. Douglas, J. P. Smol, 2000, Impacts of Climatic Change and Fishing on Pacific Salmon Abundance Over the Past 300 Years. *Science* 290(5492):795–799.

Fisher, Raymond H., 1990, Finding America. In *Russian America: the Forgotten Frontier*, edited by B. S. Smith and R. J. Barnett, pp. 17–31. Washington State Historical Society, Tacoma.

Fitch, W. Tecumseh and Marc D. Hauser, 1995, Vocal Production in Nonhuman Primates: Acoustics, Physiology, and Functional Constraints on "Honest" Advertisement. American Journal of Primatology 37(3):191–219.

Fitzhugh, Ben, 1995, Clams and the Kachemak: Seasonal Shellfish Use on Kodiak Island, Alaska (1200–800 B.P.). In *Research in Economic Anthropology*, Vol. 16, edited by B. L. Isaac, pp. 129–176. JAI Press, Greenwich, CT.

1996, The Evolution of Complex Hunter-Gatherers in the North Pacific: An Archaeological Case Study from Kodiak Island, Alaska. Ph.D. dissertation, University of Michigan. University Microfilms International, Ann Arbor.

2000, Thoughts on the Evolution of Social Inequality: A Paradigmatic Analysis. In *Alternatives to Social Evolution*, edited by N. Kradin. Archaeological Insititute of the Russian Far East: Vladivostok.

2001, Risk and Invention in Human Technological Evolution. *Journal of Anthropological Archaeology* 20:125–167.

2002, Residential and Logistical Strategies in the Evolution of Complex Hunter-Gatherers on the Kodiak Archipelago. In *Beyond Foraging and Collecting: Evolutionary Change in Hunter-Gatherer Settlement Systems*, edited by B. Fitzhugh and J. Habu, pp. 257–304. Kluwer Academic/Plenum, New York.

2003, The Evolution of Complex Hunter-Gatherers on the Kodiak Archipelago. In *Hunter-Gatherers of the North Pacific Rim*, edited by J. Habu, J. M. Savelle, S. Koyama, and H. Hongo. *SENRI Ethnological Studies* 63:13–48.

Fitzhugh, Ben and Junko Habu, editors, 2002, *Beyond Foraging and Collecting: Evolutionary Change in Hunter-Gatherer Settlement Systems*. Kluwer Academic/Plenum Publishers, New York.

Fitzhugh, William W., 1972, *Environmental Archaeologyand Cultural Systems in Hamilton Inlet, Labrador: A Survey of the Central Labrador Coast from 3000 B.C. to the Present*. Smithsonian Contributions to Anthropology, No. 16, Washington, D.C.

1974, Ground Slates in the Scandinavian Younger Stone Age with Reference to Cirumpolar Maritime Adaptations. *Proceedings of the Prehistoric Society* 40:45–58.

1975, A Comparative Approach to Northern Maritime Adaptations. In *Prehistoric Maritime Adaptations of the Circumpolar Zone*, edited by William W. Fitzhugh, pp. 339–386. Mouton, The Hague.

Fitzhugh, William W. and Susan A. Kaplan, 1988, *Inua: Spirit World of the Bering Sea Eskimo*. Smithsonian Institution Press, Washington, D.C.

Fladmark, Knut R., Kenneth M. Ames, and P. D. Sutherland, 1990, Prehistory of the North Coast of British Columbia. In *Handbook of North American Indians, vol. 7, Northwest Coast*, edited by W. Suttles, pp. 229–239. Smithsonian Institution Press, Washington, D.C.

Flanagan, James G., 1989, Hierarchy in Simple "Egalitarian" Societies. *Annual Review of Anthropology* 18:245–266.

Flannery, Kent V.,_1972, The Cultural Evolution of Civilizations. *Annual Review of Ecology and Systematics* 3:395–425.

Ford, Richard I., 1977, Evolutionary Ecology and the Evolution of Human Ecosystems: A Case Study from the Midwestern U.S.A. In *Explanation of Prehistoric Change*, edited by James N. Hill, pp. 153–184. University of New Mexico Press, Albuquerque.

Fried, Morton H., 1967, *The Evolution of Political Society: An Essay in Political Anthropology.* Random House, New York.

Friedman, Jonathan and Michael J. Rowlands, 1978, Notes Towards an Epigenetic Model of the Evolution of 'Civilization'. In *The Evolution of Social Systems*, edited by J. Friedman and M. Rowlands, pp. 201–276. University of Pittsburgh Press, Pittsburgh.

Frisch, Rose, 1978, Nutrition, Fatness, and Fertility: The Effect of Food Intake on Reproductive Ability. In *Nutrition and Reproduction*, edited by W. W. Moseley, pp. 99–122. Plenum Press, New York.

Gailey, Christine Ward, 1987, *Kinship to Kingship: Gender Hierarchy and State Formation in the Tongan Islands*. University of Texas Press, Austin.

Gibson, Jon L., 2001, *The Ancient Mounds of Poverty Point: Place of Rings*. University Press of Florida, Gainsville.

Giddens, Anthony, 1984, *The Constitution of Society: Outline of the Theory of Structuration*. Polity Press, Cambridge.

Gideon, Hiermonk, 1989, *The Round the World Voyage of Hiermonk Gideon, 1803–1809*, translated by Lydia T. Black, edited by Richard A. Pierce. Limestone Press, Kingston, Ont.

Gilman, Antonio, 1991, Trajectories towards social complexity in the later prehistory of the Mediterranean. In *Chiefdoms: Power, Economy, and Ideology*, edited by T. Earle, pp. 146–168. Cambridge University Press, Cambridge.

Gilpin, Lou M., 1995, *Holocene Paleoseismicity and Coastal Tectonics of Kodiak Islands, Alaska*. Ph.D. dissertation, University of California-Santa Cruz. University Microfilms International, Inc., Ann Arbor.

Goggin, J. M. and W. C. Sturtevant, 1964, The Calusa: A Stratified, Nonagricultural Society (with Notes on Sibling Marriage). In *Explorations in Cultural Anthropology: Essays in Honor of George Peter Murdock*, edited by W. H. Goodenough, pp. 179–219. McGraw Hill, New York.

Goland, Carole, 1991, The Ecological Context of Hunter-Gatherer Storage: Environmental Predictability and Environmental Risk. In *Foragers in Context: Long Term, Regional, and Historical Perspectives in Hunter-Gatherer Studies*, edited by P. T. Miracle, L. E. Fisher, and J. Brown, pp. 107–125. Michigan Discussions in Anthropology Vol. 10, Ann Arbor.

Gould, Richard, 1982, To Have and Have Not: The Ecology of Sharing among Hunter-Gatherers. In *Resource Managers: North American and Australian Hunter-gatherers*, edited by E. Hunn and N. Williams, pp. 69–92. Westview Press, Boulder, CO.

Grayson, Donald K., 1984, *Quantitative Zooarchaeology: Topics in the Analysis of Archaeological Faunas*. Academic Press, New York.

Grayson, Donald K. and Michael Cannon, 1999, Human Paleoecology and Foraging Theory in the Great Basin. In *Models for the Millenium: Great Basin Anthropology Today*, edited by Charlotte Beck, pp. 141–151. University of Utah Press, Salt Lake City.

Griggs, Robert F., 1918, The Recovery of Vegetation at Kodiak. In "Scientific Results of the Katmai Expeditions of the National Geographic Society." *Ohio Journal of Science* 19(1):1–57.

1934, The Edge of the Forest in Alaska and the Reasons for its Position. *Ecology* 15(2):80–96.

Habu, Junko, n.d., *Ancient Jomon of Japan*. Cambridge University Press, Cambridge (in press.)

Haggarty, James C., Christopher B. Wooley, Jon M. Erlandson, and Aron Crowell, 1991, *The 1990 Exxon Cultural Resource Program: Site Protection and Maritime Cultural Ecology in Prince William Sound and the Gulf of Alaska*. Exxon Shipping Company and Exxon Company, Anchorage.

Halpin, Marjorie M. and Margaret Seguin, 1990, Tsimshian Peoples: Southern Tsimshian, Nishga, and Gitksan. In *Handbook of North American Indians: Volume 7, Northwest Coast*, edited by W. Suttles, pp. 267–284. Smithsonian Institution Press, Washington, D.C.

Halstead, Paul and John O'Shea, 1989, Introduction: Cultural Responses to Risk and Uncertainty. In *Bad Year Economics: Cultural Responses to Risk and Uncertainty*, edited by Paul Halstead and John O'Shea, pp. 1–7. Cambridge University Press, Cambridge.

Hames R. and W. Vickers, 1982, Optimal Foraging Theory as a Model to Explain Variability in Amazonian Hunting. *American Ethnologist* 9:358–378.

Hausler-Knecht, Philomena, 1991, An Expanded View of the Ocean Bay Period: Preliminary Findings from the KOD-363 Site. Paper presented, 18th Annual Meeting of the Alaska Anthropological Association, Anchorage. (cited in Haggarty et al. 1991).

Hawkes, Kristen, 1990, Why Do Men Hunt?: Benefits for Risky Choices. In *Risk and Uncertainty in Tribal and Peasant Economies*, edited by E. Cashdan, pp. 145–166. Westview Press, Boulder.

1991, Showing Off: Tests of an Hypothesis about Men's Foraging Goals. *Ethology and Sociobiology* 12(1):29–54.

1992, Sharing and Collective Action. In*Evolutionary Ecology and Human Behavior*, edited by E. A. Smith and B. P. Winterhalder, pp. 269–300. Aldine de Gruyter, New York.

1993, Why Hunter-Gatherers Work: An Ancient Version of the Problem of Public Goods. *Current Anthropology* 34(4):341–361.

2000, Hunting and the Evolution of Egalitarian Societies: Lessons from the Hadza. In *Hierarchies in Action: Cui Bono?* edited by M. M. Diehl, pp. 59–83. Center for Archaeological Investigations, Southern Illinois University, Carbondale. Visiting Scholar Conference, No. 27.

Hayden, Brian, 1981, Research and Development in the Stone Age: Technological Transitions among Hunter-Gatherers. *Current Anthropology* 22:519–528.

1994, Competition, Labor, and Complex Hunter-Gatherers. In *Key Issues in Hunter-Gatherer Research*, edited by Ernest S. Burch and Linda J. Ellanna, pp. 223–239. Berg Press, Oxford.

1995, Pathways to Power: Principles for Creating Socioeconomic Inequalities. In *Foundations of Social Inequality*, edited by T. D. Price and G. M. Feinman, pp. 15–85. Plenum Press, New York.

Heizer, Robert F., 1956, *Archaeology of the Uyak Site, Kodiak Island, Alaska.* University of California Anthropological Record 17.

Henn, Winfield, 1978, Archaeology on the Alaska Peninsula: the Ugashik Drainage, 1973–1975. *Anthropological Papers of the University of Oregon* 14. Eugene.

Henning, R. A., M. Loken, B. Olds, and L. Morgan, editors, 1978 *Alaska Whales and Whaling. Alaska Geographic* vol. 5, no. 4.

Heritage (Office of History and Archaeology, State of Alaska), 1983, Investigating Brickmaking in Russian America. *Heritage Newsletter of the Alaska Office of History and Archaeology* 14:1–2.

Heusser, C. J., L. E. Heusser, and D. M. Peteet, 1985, Late-Quaternery Climate Change on the American North Pacific Coast. *Nature* 315:485–487.

Hildebrandt W. R. and T. L. Jones, 1992, Evolution of Marine Mammal Hunting: A View from the California and Oregon Coasts. *Journal of Anthropological Archaeology* 11:360–401.

Hildebrandt, William R. and Kelly, R. McGuire, 2002, The Ascendance of Hunting during the California Middle Archaic: An Evolutionary Perspective. *American Antiquity* 67(2):231–256.

Hill, Kim and Kristen Hawkes, 1983, Neotropical Hunting among the Ache of Eastern Paraguay. In *Adaptive Responses of Native Amazonians*, edited by R. Hames and W. Vickars, pp. 223–267. Academic Press, New York.

Hill, Kim and A. Magdalena Hurtado, 1996, *Aché Life History: The Ecology and Demography of a Foraging People*. Aldine de Gruyter, New York.

Hill, Kim and Hillard Kaplan, 1993, On Why Male Foragers Hunt and Share Food. *Current Anthropology* 34(5): 701–710.

Holmberg, Heinrich J., 1985, *Holmberg's Ethnographic Sketches*, translated by Fritz Jaensch, edited by M. W. Falk. University of Alaska Press, Fairbanks.

Holmes, C. D., R. J. Dale, and J. D. McMahan, 1989, *Archaeological Mitigation of the Thorne River Site (CRG-177)*. Office of History and Archaeology, Report Number 15. Division of Parks and Outdoor Recreation, Alaska Department of Natural Resources, Anchorage.

Hood, D. W., 1986, Physical Setting and Scientific History. In *The Gulf of Alaska: Physical Environment and Biological Resources,* edited by D. W. Hood and S. T. Zimmerman, pp. 5–27. National Oceanic and Atmospheric Administration (U.S. Dept. of Commerce) and Minerals Management Service (U.S. Dept. of Interior), Washington, D.C.

Hrdlička, Ales, 1944, *The Anthropology of Kodiak Island.* The Wistar Institute of Anatomy and Biology, Philadelphia.

Hu, Feng Sheng, Emi Ito, Linda B. Brubaker, and Patricia M. Anderson, 1998, Ostracode Geochemical Record of Holocene Climatic Change and Implications for Vegetational Response in the Northwestern Alaska Range. *Quaternary Research* 49(1):86–95.

Huggins, Ely L., 1947, Kodiak and Afognak Life, 1868–1870. *Western Folklore* 6:119–142.

Ingold, Tim, 1999, On the Social Relations of the Hunter-Gatherer Band. In *The Cambridge Encyclopedia of Hunters and Gatherers*, edited by R. B. Lee and R. Daly, pp. 399–410. Cambridge University Press, Cambridge.

Isaac, Barry L., 1992, Discussion. In *Long-term Subsistence Change in Prehistoric North America* (Supplement 6, *Research in Economic Anthropology)*, edited by D. R. Croes, R. A. Bennett, and B. L. Isaac, pp. 441–452. JAI Press, Greenwich, CT.

Izmailov, Gerasim G., 1790, Testimony Given Before Captains Billings and Sarychev, 1st July, 1790 at Kodiak, Witnessed by the Ship's Chaplain Vasilii Sivtsov. Manuscript from the Yudin Collection, Box 2, Folder 23. Manuscript Division, Library of Congress, Washington, D.C. (cited in Black 1992).

Jacob, Klaus H., 1986, Seismicity, Tectonics, and Geohazards of the Gulf of Alaska Regions. In *The Gulf of Alaska: Physical Environment and Biological Resources*, edited by D. W. Hood and S. T. Zimmerman, pp. 145–184. National Oceanic and Atmospheric Administration (U.S. Dept. of Commerce) and Minerals Management Service (U.S. Dept. of Interior), Washington, D.C.

Jewitt, John R., 1987, The*Adventures and Sufferings of John R. Jewitt: Captive of Maquinna*, illustrated and annotated by Hilary Stewart. University of Washington Press, Seattle.

Johnson, Gregory A., 1982, Organizational Structure and Scalar Stress. In *Theory and Explanation in Archaeology*, edited by C. Renfrew, M. Rowlands, and B. Segraves, pp. 389–421. Academic Press New York.

Jordan, Richard H., 1992, A Maritime Paleoarctic Assemblage from Crag Point, Kodiak Island, Alaska. In *Contributions to the Anthropology of Southcentral and Southwestern Alaska*, edited by R. H. Jordan, F. de Laguna, and A. F. Steffian, pp. 127–140. *Anthropological Papers of the University of Alaska* vol. 24, nos. 1–2.

1994, Qasqiluteng: Feasting and Ceremonialism among the Traditional Koniag of Kodiak Island, Alaska. In *Anthropology of the North Pacific Rim*, edited by W. W. Fitzhugh and V. Chaussonnet, pp. 147–173. Smithsonian Institution Press, Washington, D.C.

Jordan, Richard H. and Richard A. Knecht, 1988, Archaeological Research on Western Kodiak Island, Alaska: the Development of Koniag Culture. In *Late Prehistoric Development of Alaska's Native People*, edited by R. D. Shaw, R. K. Harritt, and D. E. Dumond, pp. 225–306. *Aurora IV*, Alaska Anthropological Association, Anchorage.

Jorganson, J. G., 1980, *Western Indians: Comparative Environments, Languages, and Cultures of 172 Western American Tribes*. W. H. Freeman, San Francisco.

Josenhans, H., D. Fedje, R. Pienitz, and J. Southon, 1997, Early Humans and Rapidly Changing Holocene Sea Levels in the Queen Charlotte Islands—Hecate Strait, British Columbia, Canada. *Science* 277:71–74.

Junker, Laura Lee, 1999, *Raiding, Trading, and Feasting: The Political Economy of Philippine Chiefdoms*. University of Hawaii Press, Honolulu.

Kaplan, Hillard and Kim Hill, 1985, Food Sharing among Ache Foragers: Tests of Explantory Hypotheses. *Current Anthropology* 26(2):223–246.

1992, The Evolutionary Ecology of Food Acquisition. In *Evolutionary Ecology and Human Behavior*, edited by E. A. Smith and B. P. Winterhalder, pp. 167–201. Aldine de Gruyter, New York.

Karlstrom, Thor N. V., 1969, Regional Setting and Geology. In *The Kodiak Island Refugium: Its Geology, Flora, Fauna and History*, edited by Thor N. V. Karlstrom and George E. Ball, pp. 20–54. The Boreal Institute, University of Alberta, Calgary.

Keeley, Lawrence H., 1988, Hunter-Gatherer Economic Complexity and "Population Pressure": A Cross-Cultural Analysis. *Journal of Anthropological Archaeology* 7:373–411.

Kelly, Robert L., 1991, Sedentism, Sociopolitical Inequality, and Resource Fluctuations. In *Between Bands and States*, edited by S. A. Gregg, pp. 135–158. Center for Archaeological Investigations, Occasional Paper No. 9. Southern Illinois University Press, Carbondale.

1995, *The Foraging Spectrum*. Smithsonian Institution Press, Washington D.C.

Kessler, Doyne W., 1985, *Alaska's Saltwater Fishes and Other Sea Life*. Alaska Northwest Publishing Co., Anchorage.

Kienle, J. and S. E. Swanson, 1983, Volcanism in the Eastern Aleutian Arc: Late Quaternary and Holocene Centers, Tectonic Setting and Petrology. *Journal of Volcanology and Geothermal Research* 17:393–432.

Kirch, Patrick V., 1984, *The Evolution of Polynesian Chiefdoms*. Cambridge University Press, Cambridge.
1990, Monumental Architecture and Power in Polynesian Chiefdoms: A Comparison of Tonga and Hawaii. *World Archaeology* 22(2):206–222.
Kirch, Patrick V. and Joanna Ellison, 1994, Palaeoenvironmental Evidence for Human Colonization of Remote Oceanic Islands. *Antiquity* 68(259):310–321.
Knecht, Richard A., 1995, *The Late Prehistory of the Alutiiq People: Culture Change on the Kodiak Archipelago from 1200–1750 AD*. Ph.D. Dissertation, Department of Anthropology, Bryn Mawr College. University Microfilms International, Inc., Ann Arbor.
Knecht, Richard A. and Richard Davis, 2001, A Prehistoric Sequence for the Eastern Aleutian Islands. In *Archaeology in the Aleut Zone: Some Recent Research*, edited by Don E. Dumond, pp. 269–288. University of Oregon Anthropological Papers No. 58, Eugene.
Knecht, Richard A., Richard S. Davis, and Gary A. Carver, 2001, The Margaret Bay Site and Eastern Aleutian Prehistory. In *Archaeology in the Aleut Zone: Some Recent Research*, edited by Don E. Dumond, pp. 35–69. University of Oregon Anthropological Papers No. 58, Eugene.
Knecht, Richard A. and Richard H. Jordan, 1985, Nunakakhnak: An Historic Period Koniag Village in Karluk, Kodiak Island, Alaska. *Arctic Anthropology* 22(2):17–35.
Kopperl, Robert, 2002, Early Alutiiq Subsistence on Kodiak Island, Alaska: An Analysis of Faunal Remains from the Rice Ridge Site (KOD 363). Paper presented at the 67th Annual Meeting of the Society for American Archaeology, Denver, March 21.
Lambert, Patricia M. and P. L. Walker, 1991, Physical Anthropological Evidence for the Evolution of Social Complexity in Coastal Southern California. *Antiquity* 65:963–973.
Lantis, Margaret, 1938, The Alaskan Whale Cult and Its Affinities. *American Anthropologist* 40:438–464.
1947, *Alaskan Eskimo Ceremonialism*. University of Washington Press, Seattle.
1984, Nunamiut Eskimo. In *Handbook of American Indians, vol. 5, Arctic,* edited by David Damas, pp. 209–223. Smithsonian Institution, Washington, D.C.
Laughlin, William S.,_1975, Aleuts: Ecosystem, Holocene History, and Siberian Origin. *Science* 189:507–515.
Laughlin, William S. and Jean S. Aigner, 1966, Preliminary Analysis of the Anangula Unifacial Core and Blade Industry. *Arctic Anthropology* 3(2):41–56.
Laughlin, William S. and W. G. Reeder, 1962, Rationale for the Collaborative Investigation of Aleut-Konyag Prehistory and Ecology. *Arctic anthropology*. 1(1):104–108.
Leacock, Eleanor B, 1978, Women's Status in Egalitarian Society: Implications for Social Evolution. *Current Anthropology* 19(2):247–275.
Leacock, Eleanor B. and Richard B. Lee, 1982, Introduction to Politics and History in Band Societies. In *Politics and History in Band Societies*, edited by E. B. Leacock and R. B. Lee, pp. 1–20. Cambridge University Press, Cambridge.
Lee, Richard B., 1972, The !Kung Bushmen of Botswana. In *Hunters and Gatherers Today*, edited by M. Bicchieri, pp. 327–368. Holt, Rhinehart and Winston, New York.
1979, *The !Kung San: Men, Women, and Work in a Foraging Society*. Cambridge University Press, Cambridge.
1982 Politics, Sexual and Non-Sexual, in an Egalitarian Society. In *Politics and History in Band Societies*, edited by E. Leacock and R. B. Lee, pp. 37–59. Cambridge University Press, Cambridge.
1990, Primitive Communism and the Origin of Social Inequality. In *Evolution of Political Systems: Sociopolitics in Small-Scale Sedentary Societies*, edited by Steadman Upham, pp. 225–246. Cambridge University Press Cambridge.
Lee, Richard B. and Richard Daly, 1999, *The Cambridge Encyclopedia of Hunter-Gatherers*. Cambridge University Press, Cambridge.
Lee, Richard B. and Irven DeVore, 1968, *Man the Hunter*. Aldine, Chicago.
Legros, Dominique, 1985, Wealth, Poverty, and Slavery among the 19th Century Tchutchone Athabascans. *Research in Economic Anthropology* 7:37–64.
Lisianskii, Yurii, 1814, *A Voyage Around the World, 1803–1806*. Booth, London.

Lucier, Charles V. and James W. VanStone, 1995, *Traditional Beluga Drives of the Iñupiat of Kotzebue Sound, Alaska*. Fieldiana, Anthropology, new series, no. 25, Publication 1468. Field Museum of Natural History, Chicago.

Lyman, R. Lee., 1995 On the Evolution of Marine Mammal Hunting on the West Coast of North America. *Journal of Anthropological Archaeology* 14(1): 45–77.

MacArthur, Robert H., 1972, *Geographical Ecology*. Harper and Row, New York.

MacArthur, Robert H. and E. O. Wilson, 1967, *The Theory of Island Biogeography*. Princeton University Press, Princeton.

Madsen, David B. and Dave N. Schmitt, 1998, Mass Collecting and the Diet Breadth Model: A Great Basin Example. *Journal of Archaeological Science* 25(5):445–455.

Mann, Daniel H. and Aron L. Crowell, 1996, A Large Earthquake Occurring 700 to 800 Years Ago in Aialik Bay, Southern Coastal Alaska. *Canadian Journal of Earth Sciences* 33:117–126.

Mann, Daniel H., Aron L. Crowell, T. D. Hamilton, Bruce P, Finney, 1998, Holocene Geologic and Climatic History around the Gulf of Alaska. *Arctic Anthropology* 35(1):112–131.

Mann, Daniel H. and Dorothy M. Peteet, 1994, Extent and Timing of the Last Glacial Maximum in Southwestern Alaska. *Quaternary Research* 42:136–148.

Manson, Joseph H., 1996, Rhesus Macaque Copulation Calls: Re-Evaluating the "Honest Signal" Hypothesis. *Primates* 37(2):145–154.

Marquardt, William H., 1986, Development of Cultural Complexity in Southwest Florida: Elements of a Critique. *Southeastern archaeology* 5(1):63–70.

1988, Politics and Production among the Calusa of South Florida. In *Hunters and Gatherers 1: History, Evolution, and Social Change*, edited by T. Ingold, D. Riches, J. Woodburn, pp. 161–188. Berg, London.

1992, *Culture and Environment in the Domain of the Calusa*, edited by W. H. Marquardt. Institute of Archaeology and Paleoenvironmental Studies, University of Florida, Gainesville.

Maschner, Herbert D. G., 1991, The Emergence of Cultural Complexity on the Northern Northwest Coast. *Antiquity* 65:924–934.

1992, The Origins of Hunter-gatherer Sedentism and Political Complexity: A Case Study from the Northern Northwest Coast. Ph.D. Dissertation, University of California-Santa Barbara. University Microfilms International, Inc., Ann Arbor.

1996, Settlement, Sedentism, and Village Organization on the Lower Alaska Peninsula and Eastern Aleutian Islands. Paper presented at the 1996 Annual Meetings of the Society for American Archaeology, New Orleans, LA.

1997, Settlement and Subsistence in the Later Prehistory of Tebenkof Bay, Kuiu Island, Southeast Alaska. *Arctic Anthropology* 34(2):74–99.

Maschner, Herbert D. G. and John Q. Patton, 1996, Kin Selection and the Origins of Hereditary Social Inequality: A Case Study from the Northern Northwest Coast. In *Darwinian Archaeologies*, edited by H. D. G. Maschner, pp. 89–107. Plenum, New York.

Maschner, Herbert D. G. and Katherine L. Reedy-Maschner, 1998, Raid, Retreat, Defend (Repeat): The Archaeology and Ethnohistory of Warfare on the North Pacific Rim. *Journal of Anthropological Archaeology* 17(1):19–51.

Mason, Owen K., 1998, The Contest Between the Ipiutak, Old Bering Sea, and Birnik Polities and the Origin of Whaling During the First Millennium A.D. along Bering Strait. *Journal of Anthropological Archaeology* 17(3):240–325.

Matson, Roy G. and Gary Coupland, 1995, *The Prehistory of the Northwest Coast*. Academic Press, New York.

Mauss, Marcel, 1969, *The Gift: Forms and Functions of Exchange in Archaic Societies*. Translated by I. Cunnison. Norton, New York.

McAllister, Carol., 1990, Women and Feasting: Ritual Exchange, Capitalism, and Islamic Revival in Negeri Sembilan, Malaysia. *Research in Economic Anthropology* 12:23–51.

McCartney, Allen P., 1974, Maritime Adaptations on the North Pacific Rim. *Arctic Anthropology* 11(supp):153–162.

1984, Prehistory of the Aleutian Region. In *Handbook of American Indians, vol. 5, Arctic,* edited by David Damas, pp. 119–135. Smithsonian Institution Press, Washington, D.C.

Meillassoux, Claude, 1978, Kinship Relations and Relations of Production. In *Relations of Production: Marxist Approaches to Economic Anthropology*, edited by D. Seddon, pp. 289–330. Frank Cass, London.

1981, *Maidens, Meal and Money: Capitalism and the Domestic Community*. Cambridge University Press, Cambridge.

Meindl, Richard S. and Katherine F. Russell, 1998, Recent Advances in Method and Theory in Paleodemography. *Annual Review of Anthropology* 27:375–399.

Merck, Carl H., 1980, *Siberia and Northwestern America 1788–1792: The Journal of Carl Heinrich Merck.* Materials for the Study of Alaska History, No. 17, traslated by Fritz Jaensch, edited by Richard A. Pierce. Limestone Press, Kingston, Ontario.

Miller, Jay, 1984, Feasting with the Southern Tsimshian. In *Tsimshian*, edited by Margaret Seguin, pp. 27–39. University of British Columbia Press, Vancouver.

Mills, Robin O., 1994, Radiocarbon Calibration of Archaeological Dates from the Central Gulf of Alaska. *Arctic Anthropology* 31(1):126–149.

Mitchell, Donald, 1983, Seasonal Settlements, Village Aggregations, and Political Autonomy on the Central Northwest Coast. In *The Development of Political Organization in Native North America*, edited by E. Tooker, pp. 97–107. American Ethnological Society, Washington, D.C.

1984, Predatory Warfare, Social Status and the North Pacific Slave Trade. *Ethnology* 23:39–48.

Mitchell Donald and Leland Donald, 1988, Archaeology and the Study of Northwest Coast Economies. In *Prehistoric Economies of the Northwest Coast, Research in Economic Anthropology, supp. 3*, edited by B. Isaac, pp. 293–351. JAI Press, Greenwich, Conn.

Moss, Madonna L., 1993, Shellfish, Gender and Status on the Northwest Coast: Reconciling Archaeological, Ethnographic, and Ethnohistorical Records of the Tlingit. *American Anthropologist* 95:631–652.

1998, Northern Northwest Coast: Regional Overview. *Arctic Anthropology* 35(1):88–111.

Moss, Madonna L. and Jon M. Erlandson, 1992, Forts, Refuge Rocks, and Defensive Sites: The Antiquity of Warfare along the North Pacific Coast of North America. *Arctic Anthropology* 29(2): 73–90.

Nakamura, O'oki. 1999. Bosei kara yomu Jomon shakai no kaisoka [Social stratification in Jomon society on the basis of burial analysis]. In Jomon-gaku no Sekai [The World of Jomon Archaeology], ed. Tatsuo Kobayashi, pp. 48–60. Tokyo: Asahi Shinbun-sha (in Japanese).

Niebauer, H. J., 1998, Variability in Bering Sea Ice Cover as Affected by a Regime Shift in the North Pacific in the Period 1947–1996. *Journal of Geophysical Research C: Oceans*. 103(C12): 27, 717–27, 737.

Neiman, Fraser, 1997, Conspicuous Consumption as Wasteful Advertising: A Darwinian Perspective on Spatial Patterns in Classic Maya Terminal Monument Dates. In *Rediscovering Darwin: Evolutionary Theory in Archaeological Explanation*, edited by C. M. Barton and G. A. Clark, pp. 267–290. Archaeological Papers of the American Anthropological Association No. 7.

Nelson, Robert E. and Richard H. Jordan, 1988, A Postglacial Pollen Record for Western Kodiak Island, Alaska. *Arctic* 41(1):59–63.

Nowak, Michael, 1978, U.S. Fish and Wildlife Service Archaeological Reconnaissance of Kodiak of Kodiak National Wildlife Refuge. Unpublished manuscript on file, U.S. Fish and Wildlife Service, Anchorage.

O'Connell, James, 1995, Ethnoarchaeology Needs a General Theory of Behavior. *Journal of Archaeological Research*. 3(3): 205–255.

Oeschger, H., T. Riesen, and J. C. Lerman, 1970, Bern Radiocarbon Dates VII. *Radiocarbon* 12(2):358–384.

Ogden, Adele, 1941, *The California Sea Otter Trade 1784–1848*. University of California Publications in History, vol. 26. University of California Press, Berkeley.

1991, Russian Sea Otter and Seal Hunting on the California Coast, 1803–1841. In *Fort Ross: California Outpost of Russian Alaska, 1812–1814*, edited by E. O. Essig, A Ogden, and C DuFour, pp. 35–59. The Limestone Press, Kingston, Ont.

Okada, Hiroaki, A. Okada, Y. Kotani, and B. Yamaguchi, 1984, The Hot Springs Village Site (4): Preliminary Report of the 1980 and 1982 Excavations at Port Moller, Alaska. In *The Qaluyaarmiut* (2)edited by H. Okada et al., pp. 1–52. Department of Behavioral Science, Hokkaido University, Sapporo.

Okada, H., A. Okada, Y. Kotani, K. Yajima, W. M. Olson, T. Nishimoto, and S. Okino, 1989, *Haceta Island, Southeastern Alaska Anthropological Survey in 1987*. Department of Behavioral Science, Faculty of Letters, Hokkaido University, Sapporo, Japan.

Oleksa, Michael J., 1992, *Orthodox Alaska: A Theology of Mission*. St. Vladimir's Seminary Press, Crestwood, NY.

Orians, G. H. and N. E. Pearson, 1979, On the Theory of Central Place Foraging. In *Analysis of Ecological Systems*, edited by D. H. Horn, R. D. Mitchell, and G. R. Stairs, pp. 154–177. Ohio State University Press, Columbus.

Ortner, Sherry B.

1994 Theory in anthropology since the sixties. In *Culture/ Power/ History a Reader in Contemporary Social Theory*, N. B. Dirks, G. Eley, and S. B. Ortner (eds.). Princeton: Princeton University Press. pp. 372–411.

O'Shea, John M., 1981, Coping with Scarcity: Exchange and Social Storage. In *Economic Archaeology: Towards an Integration of Ecological and Social Approaches*, edited by Andrew Sheridan and Geoffrey Bailey, pp. 167–186. British Archaeological Reports, International Series 96, Oxford.

O'Shea John M. And Alex W. Barker, 1996, Measuring Social Complexity and Variation: A Categorical Imperitive? In *Emergent Complexity: The Evolution of Intermediate Societies*, edited by J. E. Arnold, pp. 13–24. International Monographs in Prehistory, Ann Arbor.

Oswalt, Wendall, 1955, Prehistoric Sea Mammals Hunters of Kaflia, Alaska. *Anthropological Papers of the University of Alaska* 4:23–61.

1987, Technological Complexity: The Polar Eskimos and the Tareumiut. *Arctic Anthropology* 24(2):.82–98.

Partlow, Megan A., 2000, Salmon Intensification and Changing Household Organization in the Kodiak Archipelago. Ph.D. dissertation, University of Wisconsin-Madison. University Microfilms International, Inc., Ann Arbor.

Pavlinskaya, Larisa R., 1994, Shaman Costume: Image and Myth. In *Ancient Traditions: Shamanism in Central Asia and the Americas*, edited by G. Seaman and J. S. Day, pp. 257–264. University Press of Colorado, Niwot.

Peebles, Christopher S. and Susan M. Kus, 1978, Some Archaeological Correlates of Ranked Societies.*American Antiquity* 42:421–448.

Peterson, Roger T., 1990, *Western Birds*. Peterson Field Guides. Houghton Mifflin New York.

Philibert, Sylvie, 1994, Ocre et le Traitement des Peaux: Révision d'une Conception Traditionelle par l'Analyse Fonctionelle des Grattoirs Ocrés de la Balma Margineda (Andorre). [Ochre and the Treatment of Skins: Revision of a Traditional Concept for the Functional Analysis of Ochre Grinders in the Balma Margineda (Andorre).] *Anthropologie* 98:447–453. (In French)

Pianka, E. R., 1970, On *r* and *K* Selection or *b* and *d* Selection? *American Naturalist* 106:581–588.

Plafker, G., K. R. Lajolie, and M. Rubin, 1992, Determining the Recurrence Intervals of Great Subduction Zone Earthquakes in Southern Alaska by Radiocarbon Dating. In *Radiocarbon After Four Decades*, edited by R. E. Taylor, A. Long, and R. Kra, p. 27. University of Arizona Press, Tucson.

Potter, James M., 2000, Pots, Parties, and Politics: Communal Feasting in the American Southwest. *American Antiquity* 65(3):471–492.

Price, T. Douglas and J. Brown, editors, 1985, *Prehistoric Hunter-Gatherers: the Emergence of Cultural Complexity*. Academic Press, New York.

Price, T. Douglas and Gary M. Feinman, editors, 1995, *Foundations of Social Inequality*. New York, Plenum Press.

Pritchard, Diane S., 1990, Joint Tenants of the Frontier: Russian-Hispanic Relationships in Alta California. In *Russian America: the Forgotten Frontier*, edited by B. S. Smith and R. J. Barnett, pp. 81–93. Washington State Historical Society, Tacoma.

Pullar, Gordon L., 1992, Introduction. In *Contributions to the Anthropology of Southcentral and Southwestern Alaska*, edited by R. H. Jordan, F. de Laguna, and A. Steffian, pp. 1–2. Anthropological Papers of the University of Alaska, 24, Fairbanks.

Pullar, Gordon L. and Richard Knecht, 1995, Alutiiq. In *Crossroads Alaska: Native Cultures of Siberia and Alaska*, edited by V. Chaussonnet, pp. 14–15. Arctic Studies Center, Smithsonian Institution, Washington. D.C.

Rappaport, Roy A., 1978, Adaptation and the Structure of Ritual. In *Human Behaviour and Adaptation*, edited by N. Blurton Jones and V. Reynolds, pp. 77–102. Taylor & Francis, London.

Rausch, R. L., 1953, On the Status of Some Arctic Mammals. *Arctic* 6:91–148.

1969, Origin of the Terrestrial Mammalian Fauna of the Kodiak Archipelago. In *The Kodiak Island Refugium: Its Geology, Flora, Fauna, and History*, edited by T. N. V. Karlstrom and G. E. Ball, pp. 216–235. Boreal Institute, University of Alberta, Edmonton.

Reeburgh, William S. and George W. Kipphut, 1986, Chemical Distribution and Signals in the Gulf of Alaska, its Coastal Margins and Estuaries. In *The Gulf of Alaska: Physical Environment and Biological Resources*, edited by D. W. Hood and S. T. Zimmerman, pp. 77–91. National Oceanic and Atmospheric Administration (U.S. Dept. of Commerce) and Minerals Management Service (U.S. Dept. of Interior), Washington, D.C.

Reeves, Randall R., Stephen Leatherwood, Stephen A. Karl, and Evelyn R. Yohe, 1985, Whaling Results at Akutan (1912–39) and Port Hobron (1926–37), Alaska. *Reports of the International Whaling Commission*. 35:441–457.

Reger, Douglas R., 1977, An Eskimo Site Near Kenai, Alaska. *Anthropological Papers of the University of Alaska* 18(2):37–52.

1981, A Model for Culture History in Upper Cook Inlet, Alaska. Ph.D. dissertation in Anthropology, Washington State University, Pullman.

1987, Archaeology of a Late Prehistoric Subsistence Locality, the Clam Gulch Site (49KEN-045). *Anthropological Papers of the University of Alaska* 18(2):37–52.

1998, Archaeology of the Northern Kenai Peninsula and Upper Cook Inlet. *Arctic Anthropology* 35(1):160–171.

Reger, Douglas R. and Alan Boraas, 1996, An Overview of the Radiocarbon Chronology in Cook Inlet Prehistory. In, *Adventures Through Time: Readings in the Anthropology of Cook Inlet, Alaska*, edited by N. Davis and W. Davis, pp. 157–171. Cook Inlet Historical Society, Anchorage.

Reger, Douglas R., J. David McMahan, and Charles E. Holmes, 1991, Effects of Crude Oil Contamination on Some Archaeological Sites in the Gulf of Alaska, 1991 Investigations. MS., Office of History and Archaeology, Alaska Division of Parks and Outdoor Recreation. Anchorage.

Richardson, Tim and Dave Cline, editors, 2001, *Kodiak Bears and the Exxon Valdez*. Kodiak Brown Bear Trust, Kodiak.

Richerson, Peter J. and Robert Boyd, 1992, Cultural Inheritance And Evolutionary Ecology. In *Evolutionary Ecology and Human Behavior*, edited by E. A. Smith and B. P. Winterhalder, pp. 61–92. Aldine de Gruyter, New York.

Rogers, A. R., 1992, Resources and Population Dynamics. In *Evolutionary Ecology and Human Behavior*, edited by E. A. Smith. and B. P. Winterhalder, pp. 375–402. Aldine de Gruyter, New York.

Roppel, Patricia, 1982, *Alaska's Salmon Hatcheries*. Alaska Historical Commission Studies in History, No. 20.

1986, *Salmon from Alaska*. Alaska Historical Commission, Anchorage.

Russell, Priscilla N., 1991, *Kodiak Alutiiq Plantlore*. Manuscript on file at the Alutiiq Museum, Kodiak, Ak.

Ruyle, Eugene E., 1973, Slavery, Surplus and Stratification on the Northwest Coast: The Ethnoenergetics of an Incipient Stratification System. *Current Anthropology* 14(5): 603–631.

Sahlins, Marshall, 1968, Notes on the Original Affluent Society. In *Man the Hunter*, edited by R. B. Lee and I. DeVore, pp. 85–89. Aldine, Chicago.

1972, *Stone Age Economics*. Aldine, Chicago.

Saltonstall, Patrick G., 1995, Excavations at Settlement Point (AFG-015): Preliminary Report of the 1995 Field Season. Report prepared for Afognak Native Corporation, Kodiak, Alaska.

1996, Settlement Point (AFG 015) Preliminary Report from the1996 Field Season. Report prepared for the Afognak Native Corporation, Kodiak, Alaska.

1997, Archaeology at Settlement Point: 1997 Preliminary Report. Report prepared for the Afognak Native Corporation, Kodiak, Alaska.

Saltonstall, Patrick and Gary A. Carver, 2002, Earthquakes, Sibsidence, Prehistoric Site Attrition and the Archaeological Record: A View from the Settlement Point Site, Kodiak Archipelago, Alaska. In *Natural Disasters and Cultural Change,* edited by Robin Torrence and John Grattan. One World Archaeology series #45. Routledge, New York.

Sambrotto, Raymond N. and Carl J. Lorenzen, 1986, Phytoplankton and Primary Productivity. In *The Gulf of Alaska: physical environment and biological resources,* edited by D. W. Hood and S. T. Zimmerman, pp. 249–282. National Oceanic and Atmospheric Administration (U.S. Dept. of Commerce) and Minerals Management Service (U.S. Dept. of Interior), Washington, D.C.

Sarychev, Gavriil, 1802, *Atlas. The Travels of Capitan Sarychev in the Northeast Part of Siberia, Icy Sea, and Eastern Ocean, during Eight Years, as Part of Geographical and Astronomical Ocean Expedition, formerly under the heading of the leadership of the fleet of Capitan Billings* (Puteshestv'ie flota kapitana Sarycheva po s′i‵everovostochnoi chasti Sibiri, Ledovitomu mor′i‵u i Vostochnomu okeanu, v" prodolzhen'ie os'mi l′i‵et", pri Geograficheskoi i astronomicheskoi morskoi ·ekspedi′t‵s'ii, byvshei pod" nachal'stvom" flota kapitana Billingsa s" 1785 po 1793 god".) Shnora, St. Petersburg.

Savelle, James M., 2002, Logistical Organization, Social Complexity, and the Collapse of Prehistoric Thule Whaling Societies in the Central Canadian Arctic Archipelago. In *Beyond Foraging and Collecting: Evolutionary Change in Hunter-Gatherer Settlement Systems*, edited By B. Fitzhugh and J. Habu, pp. 73–90. Kluwer Academic/Plenum , New York.

Schaaf, Jeanne and L. Johnson, 1990, Upland Investigations at SEL-188, McArthur Pass, Kenai Fjords National Park. National Park Service (Alaska Region) and Chugach Alaska Corporation, Anchorage.

Schink, Timothy D. Katherin A. McGraw, and Kenneth K. Chew, 1983, *Pacific Coast Clam Fisheries*. Washington Sea Grant, Technical Reports.

Schiffer, Michael B., 1987, *Formation Processes and the Archaeological Record*. University of New Mexico Press, Albuquerque.

Scott, G. Richard, 1991, Continuity or Replacement at the Uyak site: A Physical Anthropological Analysis of Population Relationships. In *The Uyak Site on Kodiak Island: Its Place in Alaskan Prehistory*, edited by Don E. Dumond and G. Richard Scott, pp. 1–56. University of Oregon Anthropological Papers No. 44, Eugene.

1992, Affinities of Prehistoric and Modern Kodiak Islanders and the Question of Kachemak-Koniag Biological Continuity. *Arctic Anthropology* 29:150–166.

1994, Teeth and Prehistory on Kodiak Island. In *Reckoning with the Dead: The Larsen Bay Repatriation and the Smithsonian Institution*, edited by T. L. Bray and T. W. Killion, pp. 67–74. Smithsonian Institution Press, Washington, D.C.

Shelikhov, Gregorii I., 1787, Note by Shelikhov about his Voyage in the Eastern Sea, May–November 1787. In *Russkie otkrytiia v Tikhom okeane I Severnoi Amerike XVIII veka*. A. I. Andreev (comp. and ed.). Moscow: Ogiz. pp. 226–249. (cited in Black 1992 as 1787c).

1981, *A Voyage to America 1783–1786*. Translated by M. Ramsay, edited by R. A. Pierce. Limestone Press, Kingston, Ontario.

Shinkwin, Ann D. and Elizabeth F. Andrews, 1979, *Archaeological Excavations at the Erskine House, Kodiak, Alaska, 1978*. University of Alaska Department of Anthropology, Fairbanks.

Shostak, Marjorie, 1981 *Nisa, the Life and Words of a !Kung Woman*. Harvard University Press, Cambridge.

Shubin, Valerii O., 1994, Aleut in the Kurile Islands 1820–1870. In *Anthropology of the North Pacific Rim,* edited by W. W. Fitzhugh and V. Chaussonnet, pp. 337–345. Smithsonian Institution Press, Washington, D.C.

Simon, James J. K., 1992, Mortuary Practices of the Late Kachemak Tradition in Southcentral Alaska: A Perspective from the Crag Point Site, Kodiak Island. *Arctic Anthropology* 29(2):130–149.

Simon, James J. K. and Amy F. Steffian, 1994, Cannibalism or Complex Mortuary Behavior? An Analysis of Patterned Variability in the Treatment of Human Remains from the Kachemak Tradition of Kodiak Island. In *Reckoning with the Dead: The Larsen Bay Repatriation and the Smithsonian Institution*, edited by T. L. Bray and T. W. Killion, pp. 75–100. Smithsonian Institution Press, Washington, D.C.

Smith, Eric A., 1981, The Application of Optimal Foraging Theory to the Analysis of Hunter-Gatherer Group Size. In *Hunter-Gatherer Foraging Strategies*, edited by B. P. Winterhalder and E. A. Smith, pp. 36–65. University of Chicago Press, Chicago.

1988, Risk and Uncertainty in the "Original Affluent Society": Evolutionary Ecology of Resource Sharing and Land Tenure. In *Hunters and Gatherers 1: History, Evolution, and Social Change*, edited by T. Ingold, D. Riches, and J. Woodburn, pp. 222–252. Berg, Oxford.

1991, *Inujjuamiut Foraging Strategies: Evolutionary Ecology of an Arctic Hunting Economy*. Aldine de Gruyter, New York.

Smith, Eric A. and Rebecca Bleige-Bird, 2000, Turtle Hunting and Tombstone Opening: Public Generosity as Costly-Signaling. *Evolution and Human Behavior* 21:245–261.

Smith, Eric A. and Bruce P. Winterhalder, editors, 1992, *Evolutionary Ecology and Human Behavior*. Aldine de Gruyter, New York.

Sowles, A. L., S. A. Hatch, and C. J. Lensink., 1978, *Catalog of Alaskan Seabird Colonies*. Biological Services Program, U. S. Fish and Wildlife Service, Washington, D.C.

Spaulding, Albert, 1982, Structure in Archaeological Data: Nominal Variables. In *Essays on Archaeological Typology*, edited by R. Whallon and J. Brown, pp. 1–22. Center for American Archaeology, Evanston, Ill.

Speth, John D., 1990, Seasonality, Resource Stress, and Food Sharing in So-Called "Egalitarian" Foraging Societies. *Journal of Anthropological Archaeology* 9:148–188.

Speth, John D. and Katherine A. Spielmann, 1983, Energy Source, Protein Metabolism, and Hunter-Gatherer Subsistence Strategies. *Journal of Anthropological Archaeology* 2:1–31.

Spielmann, Katherine A., 1986, Interdependence among Egalitarian Societies. *Journal of Anthropological Archaeology* 5:279–312.

Spiess, Arthur, Deborah Wilson, and James Bradley, 1998, Paleoindian Occupation in the New England-Maritimes Region: Beyond Cultural Ecology. *Archaeology of Eastern North America* 26:201–264.

Steadman, David W., 1989, Extinction of Birds in Eastern Polynesia: A Review of the Record, and Comparisons with other Pacific Island groups. *Journal of Archaeological Science* 16(2):177–205.

Steffian, Amy F., 1992a, Archaeological Coal in the Gulf of Alaska: A View from Kodiak Island. *Arctic Anthropology* 29(2):111–129.

1992b, Fifty Years after Hrdlicka: Further Excavation of the Uyak Site, Kodiak Island, Alaska. In *Contributions to the Anthropology of Southcentral and Southwestern Alaska*, edited by Richard H. Jordan,

Frederica de Laguna, and Amy F. Steffian, pp. 141–164. Anthropological Papers of the University of Alaska vol. 24, Fairbanks.

n.d., *Economic and Social Organization Among the Kachemak Tradition Foragers of Kodiak Island, Alaska*. Ph.D. dissertation, Department of Anthropology, University of Michigan, Ann Arbor. (In Prep).

Steffian, Amy F., Jim Beget, and Patrick G. Saltonstall, 1996, Prehistoric Alutiiq Artifact from Kodiak Island Provides Oldest Documentary Record of Ancient Volcanic Eruptions in Alaska. *Alaska Volcano Observatory BiMonthly Report* 8(2):13–14.

Steffian, Amy F., Elizabeth Pontti, and Patrick G. Saltonstall, 1998, Archaeology of the Blisky Site: A Prehistoric Camp on Near Island, Kodiak Archipelago, Alaska. Manuscript of the Alutiiq Museum and Archaeological Repository. (March 1998).

Stefffian, Amy F., Elizabeth Pontti Eufemio, and Patrick G. Saltonstall, n.d., Early Sites and Microblade Technologies from the Kodiak Archipelago. *Anthropological Papers of the University of Alaska*. In Press.

Steffian, Amy F. and James J. K. Simon, 1994, Metabolic Stress among Prehistoric Foragers of the Central Alaskan Gulf. *Arctic Anthropology* 31(2):78–94.

Steffian, Amy F. and Patrick Saltonstall, 2001, Markers of Identity: Labrets and Social Evolution on Kodiak Archipelago. *Alaska Journal of Anthropology* 1(1):1–27.

Stein, Julie K., 1984a, Organic Matter and Carbonates in Archaeological Sites. *Journal of Field Archaeology* 11(2):239–246.

1984b, Interpreting the Stratigraphy of Northwest Shell Middens. *Tebiwa* 21:26–34.

2000, *Exploring Coast Salish Prehistory: The Archaeology of San Juan Island*. Burke Museum of Natural History and Culture/University of Washington Press, Seattle.

Stein, Julie K., editor, 1992, *Deciphering a Shell Midden*. Academic Press, San Diego.

Stephens, David W. and John R. Krebs, 1986, *Foraging Theory*. Princeton University Press, Princeton.

Stiner, Mary C., 2002, Carnivory, Coevolution, and the Geographic Spread of the Genus *Homo*. *Journal of Archaeological Research* 10(1):1–63.

Stuckenrath, Robert, Jr., Juergen Wittenberg, and Roscoe Wilmeth, 1966, University of Pennsylvania Radiocarbon Dates IX. *Radiocarbon* 8:348–385.

Stuiver, M., and T. F. Braziunas, 1993, Modeling Atmospheric ^{14}C Influences and ^{14}C Ages of Marine Samples Back to 10,000 BC. *Radiocarbon* 35:137–189.

Stuiver, Minze and Austin Long, editors, 1993, *Calibration*. A special issue of *Radiocarbon* 35(1).

Stuiver, Minze, Paula J Reimer, E. Bard, J. W. Beck, G. S. Burr, K. A. Hughen, B. Kromer, F. G. McCormac, J. Plicht, and M. Spurk, 1998a, INTCAL98 Radiocarbon Age Calibration 24,000–0 cal BP. *Radiocarbon* 40:1041–1083.

Stuiver, Minze, Paula J. Reimer, and T. F. Braziunas, 1998b, High-precision Radiocarbon Age Calibration for Terrestrial and Marine Samples. *Radiocarbon* 40:1127–1151.

Suttles, Wayne, 1968, Coping with Abundance: Subsistence on the Northwest Coast. In *Man the Hunter*, edited by R. B. Lee and I. Devore, pp. 56–67. Chicago, Aldine.

Talma, A. S. and J. C. Vogel, 1993, A Simplified Approach to Calibrating (super 14) C Dates. *Radiocarbon* 35(2):317–322.

Testart, Alain, 1982, The Significance of Food Storage among Hunter-Gatherers: Residence Patterns, Population Densities, and Social Inequalities. *Current Anthropology* 23:523–537.

Thomas, David Hurst, 1983, *The Archaeology of Monitor Valley*. American Museum of Natural History, New York.

Tinbergen, Niko, 1963, On Aims and Methods in Ethology. *Zeitschrift für Tierpsychologie* 20:410–433.

Townsend, Joan B., 1980, Ranked Societies of the Alaska Pacific Rim. In *Alaska Native Culture and History*, edited by Y. Kotani and W. B. Workman, pp. 123–156. SENRI Ethnological Studies 4. National Museum of Ethnology: Osaka.

1983, Pre-Contact Political Organization and Slavery in Aleut Society. In *The Development of Political Organization in Native North America*, edited by Elizabeth Tooker, pp. 120–132. Proceedings of the American Ethnological Society. American Ethnological Society, Washington D.C.

Trivers, R. L., 1971, The Evolution of Reciprocal Altruism. *Quarterly Review of Biology* 46:35–57.

Urcid, Javier, 1994, Cannibalism and Curated Skulls: Bone Ritualism on Kodiak Island. In *Reckoning with the Dead: The Larsen Bay Repatriation and the Smithsonian Institution*, edited by T. L. Bray and T. W. Killion, pp. 101–121. Smithsonian Institution Press, Washington, D.C.

Veblen, Thorstein, 1953, *The Theory of the Leisure Class; An Economic Study of Institutions.* New American Library, New York. [First published, 1912].

Vehrencamp, S. L., 1983, A Model for the Evolution of Despotic Versus Egalitarian Societies. *Animal Behavior* 31:667–682.

Veniaminov, Ivan E. P., 1984, *Notes on the Islands of the Unalaska District*, translated by L. Black and R. H. Geoghegan, and edited by R. A. Pierce. Limestone Press, Kingston, Ontario.

Viereck, Leslie A. and Elbert L. Little, 1991, *Alaska Trees and Shrubs: Agricultural Handbook 410*, (Fourth Printing). USDA, Washington, D.C.

Voland, Eckart, 1998, Evolutionary Ecology of Human Reproduction. *Annual Review of Anthropology* 27: 347–374.

Vogel, J. C., AnneMarie Fuls, Ebbie Visser, and Bernd Becker, 1993, Pretoria Calibration Curve for Short-lived Samples, 1930–3350 BC. *Radiocarbon*. 35(1):73–86.

Wenzel, George, 1995, Ningiqtuq: Resource Sharing and Generalized Reciprocity in Clyde River, Nunavut. *Arctic Anthropology* 32(2):43–60.

Widmer, Randolph J., 1988, *The Evolution of the Calusa: A Non-Agricultural Chiefdom on the Southwest Florida Coast*. University of Alabama Press Tuscaloosa.

Widner, Leslie A., 2002, A Question of Colonization: Lithic Debitage at Tanginak Springs (KOD 481), Sitkalidak Island, Kodiak, Alaska. Senior Honors Thesis, Department of Anthropology, University of Washington, Seattle.

Wiessner, Polly, 1977, *Hxaro: A Regional System of Reciprocity for Reducing Risk Among the !Kung San*. Ph.D. dissertation, University of Michigan. Ann Arbor: University Microfilms International, Inc. Ann Arbor.

1982, Risk, Reciprocity and Social Influences on !Kung San Economics. In *Politics and History in Band Societies*, edited by E. Leacock and R. B. Lee, pp. 61–84. Cambridge University Press, Cambridge.

Wiles G. C., 1992, Holocene Glacial Fluctuations in the Southern Kenai Mountains, Alaska. Ph.D. dissertation, State University of New York. University Microfilms International, Ann Arbor.

Wiles, G. C., A. Post, E. H. Muller, and B. F. Molnia, 1999, Dendrochronology and Late Holocene History of Bering Piedmont Glacier, Alaska. *Quaternary Research* 52(2):185–195.

Wilmsen, E. N., 1982, Studies in Diet, Nutrition, and Fertility Performance. *Social Science Information* 21:95–125.

Wilson, Judith G. and James E. Overland, 1986, Meteorology. In *The Gulf of Alaska: Physical Environment and Biological Resources*, edited by D. W. Hood and S. T. Zimmerman, pp. 31–54. National Oceanic and Atmospheric Administration (U.S. Dept. of Commerce) and Minerals Management Service (U.S. Dept. of Interior), Washington, D.C.

Winslow, Margaret A. and L. Lewis Johnson, 1989, Prehistoric Human Settlement Patterns in a Tectonially Unstable Environment: Outer Shumagin Islands, Southwestern Alaska. *Geoarchaeology* 4(4):297–318.

Winterhalder, Bruce, 1981, Optimal Foraging Strategies and Hunter-Gatherer Research in Anthropology: Theory and Models. In *Hunter-Gatherer Foraging Strategies: Ethnography and Archeological Analyses.* Edited by B. P. Winterhalder and E. A. Smith, pp. 13–35. Prehistoric Archaeology and Ecology Series, University of Chicago Press, Chicago.

1986, Diet Choice, Risk, and Food Sharing in a Stochastic Environment. *Journal of Anthropological Archaeology* 5:369–392.

1993, Work, Resources, and Population in Foraging Societies. *Man* 28:321–340.

1996, Social Foraging and the Behavioral Ecology of Intragroup Resource Transfers. *Evolutionary Anthropology* 5(2):46–47.

2002, Models. In *Darwin and Archaeology: A Handbook of Key Concepts,* edited by J. P. Hart and J. E. Terrell, pp. 201–223. Bergin & Garvey, Westport, Connecticut.

Winterhalder, Bruce, W. Baillargeon, F. Cappelletto, J. I. R. Daniel, and C. Prescott, 1988, The Population Ecology of Hunter-Gatherers and their Prey. *Journal of Anthropological Archaeology* 7: 289–328.

Winterhalder, Bruce P. and Paul Leslie, 2002, Risk-Sensitive Fertility. The Variance Compensation Hypothesis. Evolution and Human Behavior 23(1):59–82.

Winterhalder, Bruce P., Flora Lu, and Bram Tucker, 1999, Risk-Sensitive Adaptive Tactics: Models and Evidence from Subsistence Studies in Biology and Anthropology. *Journal of Archaeological Research* 7(4):301–348.

Winterhalder, Bruce and Eric A. Smith,_ 1992, Evolutionary Ecology and the Social Sciences. In *Evolutionary Ecology and Human Behavior*, edited by E. A. Smith and B. P. Winterhalder, pp. 3–23. Aldine de Gruyter, New York.

Wood, James W., 1998, A Theory of Preindustrial Population Dynamics: Demography, Economy, and Well-Being in Malthusian Systems. *Current Anthropology* 39(1):99–135.

Woodburn, James C., 1968, Stability and Flexibility in Hadza Residential Groupings. In, *Man theHunter*, edited by R. B. Lee and I. DeVore, pp. 103–110. Aldine, Chicago.

1970, *Hunters and Gatherers: The Material Culture of the Nomadic Hadza*. British Museum, London.

1979, Minimal Politics: The Political Organization of the Hadza of North Tanzania. In *Politics in Leadership*, edited by W. A. Shack and P. S. Cohen, pp. 244–266. Clarendon Press, Oxford.

1980, Hunters and Gatherers Today and Reconstruction of the Past. In *Soviet and Western Anthropology*, edited by E. Gellner, pp. 95–117. Duckworth, London.

1982. Egalitarian Societies. *Man* 17:431–451.

Woodhouse-Beyer, Katherine E., 2001, Gender Relations and Socio-Economic Change in Russian America: an Archaeological Study of the Kodiak Archipelago, Alaska, 1741–1867 A.D. Ph.D. dissertation, Brown University. University Microfilms International, Ann Arbor.Workman, Karen W. and William B. Workman, 1988, The Last 1300 Years of Prehistory in Kachemak Bay: Where Later is Less. In, *The Late Prehistoric Development of Alaska's Native People*, edited by R. D. Shaw, R. K. Harrit, and D. E. Dumond, pp. 339–354. *Aurora* 4, Alaska Anthropological Association Monograph Series, Anchorage.

1997, 1996 Test Excavations at a Late Prehistoric Site at Port Graham, Kenai Peninsula, Alaska. Paper presented at the 24th Annual Meeting of the Alaska Anthropological Association, Whitehorse.

Workman, William B., 1966, Archaeological Reconnaissance of Chirikof Island, Kodiak Group: A Preliminary Report. *Arctic Anthropology* 3(2):185–192.

1977, New Data on the Radiocarbon Chronology of the Kachemak Bay Sequence. *Anthropological Papers of the University of Alaska* 18(2):31–36.

1992, Life and Death in a First Millennium AD Gulf of Alaska Culture: The Kachemak Tradition Ceremonial Complex. In *Ancient Images, Ancient Thought: The Archaeology of Ideology*, edited by A. S. Goldsmith, S. Garvie, D. Selin, and J. Smith, pp. 19–25. Proceedings, Chacmool Annual Conference, University of Calgary Archaeological Association, No. 23. Calgary.

1993, Human Colonization of the Cook Inlet Basin Before 3000 Years Ago. Paper presented at the 20th Annual Meeting of the Alaska Anthropological Association, Anchorage.

Workman, William B., J. Lobdell, and Karen W. Workman, 1980, Recent Archaeological Work in Kachemak Bay, Gulf of Alaska. *Arctic* 33(3):385–399.

Workman, William B., Janet Klein, Marlene Testaguzza and Peter Zollars, 1993, "1992 Test Excavations at the Sylva Site (SEL 245): A Stratified Late Ocean Bay Occupation in Upper Kachemak Bay, Kenai Peninsula, Alaska." Paper presented at the 20th Annual Meeting of the Alaska Anthropological Association, Anchorage.

Workman, William B. Allen P. McCartney, 1998, Coast to Coast: Prehistoric Maritime Cultures in the North Pacific. *Arctic Anthropology* 35(1):361–370.

Yesner, David R., 1980, Maritime Hunter-Gatherers: Ecology and Prehistory. *Current Anthropology* 21(6): 727–750.

1981, Archaeological Applications of Optimal Foraging Theory: Harvest Strategies of Aleut Hunter-Gatherers. In *Hunter-Gatherer Foraging Strategies: Ethnography and Archeological Analyses*, edited by B. P. Winterhalder and E. A. Smith, pp. 148–170. Prehistoric Archaeology and Ecology Series. University of Chicago Press, Chicago.

1989, Osteological Remains from Larsen Bay, Kodiak Island, Alaska. *Arctic Anthropology* 26(2):96–106.

1992, Evolution of Subsistence in the Kachemak Tradition: Evaluating the North Pacific Maritime Stability Model. *Arctic Anthropology* 29(2):167–181.

1998, Origins and Development of Maritime Adaptations in the Northwest Pacific Region on North America: A Zooarchaeological Perspective. *Arctic Anthropology* 35(1):204–222.

Zahavi, A., 1987, The Theory of Signal Selection and Some of its Implications. In *Proceedings of the International Symposium of Biological Evolution*, edited by P. V. Delfino, pp. 305–325. Adriatica Editricia, Bari.

Zollars, Peter and Janet R. Klein, 2002, The Faulkner Site: Overview of an Early Holocene Lithic Assemblage. Paper presented at the 28th Annual Meeting of the Alaska Anthropological Association, Anchorage.

Index